Weekend Pilots

Weekend Pilots

TECHNOLOGY, MASCULINITY, AND PRIVATE AVIATION IN POSTWAR AMERICA

ALAN MEYER

Johns Hopkins University Press
Baltimore

Printed in the United States of America on acid-free paper

2 4 6 8 9 7 5 3 1

Johns Hopkins University Press
2715 North Charles Street
Baltimore, Maryland 21218-4363
www.press.jhu.edu

Library of Congress Cataloging-in-Publication Data
Meyer, Alan, 1965– author.
Weekend pilots : technology, masculinity, and private aviation in postwar America / Alan Meyer.
pages cm
Includes bibliographical references and index.
ISBN 978-1-4214-1858-2 (hardcover : alk. paper) — ISBN 978-1-4214-1859-9 (electronic) — ISBN 1-4214-1858-4 (hardcover : alk. paper) — ISBN 1-4214-1859-2 (electronic) 1. Private flying—United States—History—20th century. 2. Air pilots—United States—Psychology. 3. Air pilots—Sex differences—United States. 4. World War, 1939–1945—Influence. I. Title.
TL721.4.M47 2015
797.5—dc23 2015010636

A catalog record for this book is available from the British Library.

Special discounts are available for bulk purchases of this book. For more information, please contact Special Sales at 410-516-6936 or specialsales@press.jhu.edu.

Johns Hopkins University Press uses environmentally friendly book materials, including recycled text paper that is composed of at least 30 percent post-consumer waste, whenever possible.

To my mom and dad, who made the mistake of teaching me that I could accomplish anything if I tried hard enough . . .

And to my wife and children, who patiently endured the consequences of that lesson as I completed this book.

CONTENTS

ACKNOWLEDGMENTS

From my first research and writing seminar in graduate school, when I stumbled across Jane and Ralph Nelson and the Aircraft Owners and Pilots Association's Pinch-Hitter Course, to the process of copyediting and indexing the finished manuscript for publication, *Weekend Pilots* was more than fifteen years in the making. Along the way I incurred more debts to both individuals and institutions than I can possibly recognize in this limited space.

A four-year graduate fellowship from the University of Delaware's Hagley Program got me started on this project, while a Guggenheim Fellowship from the Smithsonian Institution's National Air and Space Museum, a Smithsonian Predoctoral Fellowship at the National Museum of American History, and the Melvin Kranzberg Dissertation Fellowship from the Society for the History of Technology made it possible for me to finish my dissertation. Arwen Palmer Mohun was an exceptional dissertation advisor and continues to be my mentor, intellectual role model, and friend. Every member of my dissertation committee—Joseph Corn, Deborah Douglas, David Shearer, and Susan Strasser—took an early, active, and enduring interest in my project as well as my education as a historian. Pete Daniel and Dominick Pisano proved excellent advisors during my tenure as a Smithsonian fellow and beyond. During my stint as a U.S. Air Force historian, Dr. Charles "Joe" Gross, my boss at the Air National Guard History Office, was supportive of my efforts to complete my Ph.D. even as I worked a full-time job, as were my fellow Air Force historians, David Anderson and Dr. Janet Valentine.

At Auburn University, a summer research and writing fellowship from the College of Liberal Arts greatly assisted my efforts to turn a sprawling dissertation into a more focused book. In addition, every one of my colleagues in the history department deserves at least a full sentence of thanks, but that would push my acknowledgments section (and most readers' patience) beyond allowable limits. Therefore, I have singled out the contributions of a few to represent the supportive environment I found when I arrived six years ago. Department chair Charles Israel protected me and other new faculty members from the burdens of departmental and university service during those crucial, crowded years leading to tenure. Morris Bian, who succeeded Charles as chair, repeatedly took time from his busy schedule to check on my progress, proffered guidance on the ins and outs of academic publishing, and worked tirelessly to ensure that my tenure and promotion file was in order. I gained much from David Lucsko's expertise on post-World War II car culture (and the history of technology in general) as he served as a sounding board for arguments and case studies found in every chapter; Dave also enthusiastically read the page proofs and caught several errors that had slipped through the rigorous copyediting process. As a graduate student I had admired William Trimble and James Hansen from afar for their significant scholarly contributions to the field of aerospace history. When I joined the faculty at Auburn, suddenly their offices were literally right down the hall from mine, and thanks to their generous open-door philosophy (which applied even when they were hard at work behind closed doors) in short order I counted both Bill and Jim as friends and mentors. At the other end of the spectrum in terms of both seniority and academic interests, Matt Malczycki, an accomplished scholar of early Arabic papyrology and my loyal running partner, helped me maintain some semblance of mental and physical health as I struggled to balance the competing demands of teaching, family, and finishing this book. At one time or another Melissa Blair, Cari Casteel, Kate Craig, Adam Jortner, Sarah Hamilton, Kelly Kennington, Monique Laney, Eden McLean, Mark Sheftall, and Tiffany Sippial had an office near mine in our little close-knit cul-de-sac on the third floor of Thach Hall; their friendship, camaraderie, and conversations on many topics buoyed my spirits and constantly reminded me why I so love this profession. And to everyone else in my department, a heartfelt thank you.

If Auburn University is my intellectual home base, my second home will always be the Smithsonian's National Air and Space Museum. Even before I arrived as a predoctoral fellow, Phil Edwards of the museum's library took a close personal interest in my topic and tracked down sources I would have never known to ask for. Roger Launius, first as a curator and later as an associate director of the museum, never missed a chance to ask pointed questions about my progress, first on my dissertation, then on this book (and, now that it's finally finished, I can hear him demanding, "So, what's next?"). Curator Jeremy Kinney has become over the years both a collaborator on several aviation history projects and a close personal friend. I owe much to other curators and staff members as well, including John Anderson, Paul Ceruzzi, Dorothy Cochrane, Martin Collins, Roger Connor, Tom Crouch, Peter Jakab, Christopher Moore, Michael Neufeld, Dominick Pisano, Alex Spencer, Bob van der Linden, and Margaret Weitekamp.

I incurred many other debts along the way. The Experimental Aircraft Association's Adam Smith not only facilitated my access to important archival material, but also arranged a one-on-one oral history interview with that organization's founder and longtime president Paul Poberezny. Former Hagley Program classmate Katherine Leonard Turner provided expert help in tracking down potential images for this book, thus saving me an expensive and time-consuming research trip. Historians Aaron Alcorn, John Davies, Becky Martin, Katina Manko, Christine Sears, Mary Tinti, and the aforementioned Katie Turner not only helped me refine my arguments through their careful readings of draft chapters, but their friendship, moral support, and occasional long-distance kick in the butt proved invaluable.

Assistance takes many forms, some more easily categorized than others. My parents, Sarah and David Meyer, long ago made the "mistake" of teaching me that I could accomplish anything if I tried hard enough. For nearly a half century they lived with the consequences, urging me along uncritically, propping me up whenever I fell short, and proudly basking in my successes. More than three decades ago, high school aviation teacher and private pilot Les Greene and flight instructor Rex E. Copenhaver taught me lessons that have served me well both in the cockpit and in the world beyond the airport fence. Since then, I have learned much from other pilots, but two excellent instructors stand out: Corey Ransom, who introduced me to the old adage "there are two kinds of taildragger pilots . . ." and always helped me bridge the gap between scholarly history and real-world flying, and Marianne Buckley, who not only helped me to "tame the taildragger" but also taught me how to fly upside down and enter spins without fear.

There's an old saying that you choose your spouse, not your in-laws, but in my case I could not have done better if I had tried. Over the years J. P. and Mary Causey provided support in more ways than I can hope to describe here. My sister Anne and sister-in-law Beth, along with their partners, offered unending encouragement through all the ups and downs that life has sent my way. Paula Causey and Bruce Gregory always supplied a welcoming home away from home on my visits to Washington, D.C., while Carol and Carter Fox, Deborah Douglas, and Loretta Kuhn all gave me the priceless gift of time, space, and solitude to write. In my fifth year of graduate school, when I was recalled to active duty for the invasion of Iraq, Chad Storlie, comrade in arms and lifelong friend, went above and beyond to make sure I made it home from our tour together in Baghdad. Many others have provided support and encouragement of all kinds over the years: Phoebe Acheson, Andrew Bozanic, Eric Brittingham, Chris and Sarah Ferguson, John and Annie (and Neil and Andrew) Jones, Jim Hoogerwerf, Roger Horowitz, Richard Jereski, Erinn McComb, Del Meyer, Lt. Col. Craig Morris, Ken and Nancy Noe, Pat Orendorf, Rick Ostericher, Erik Rau, Amit Ray, Jim Robyn, Mike Tryby, and Brian Turner.

I am forever grateful to Bob Brugger of Johns Hopkins University Press, who expressed interest after I delivered my first conference paper on gender and aviation as a fledgling graduate student in 2001, then unflaggingly kept tabs on both me and my project for years until I at last had a finished manuscript to submit for consideration. Courtney Bond, also with Johns Hopkins, managed the details of converting that manuscript into this book with equal measures of patience and skill. Brian MacDonald did an excellent job copyediting, and Beatrice Burton of Human Enterprises helped create a far better index than I ever could produce on my own. As this book neared completion, several individuals helped me navigate the complexities of finding images and securing permissions, including Chris Rose at *AOPA Pilot*, Kate Igoe at the Smithsonian's National Air and Space Museum, Erin Rushing of Smithsonian Institution Libraries, Marguerite Roby and the rest of the staff at the Smithsonian Institution Archives, and Linda Gross, Max Moeller, and Vicki Wasserman at the Hagley Museum and Library in Wilmington, Delaware. I also want to express my gratitude to the family of K. O. Eckland for generously granting permission to use their late father's wonderful caricatures of pilots in this book.

Several years ago, one of the first photos my son took with his new Fisher Price camera was a toddler's-eye view of daddy sitting at the kitchen table hunched over the keyboard and staring into a computer screen. This image, taken innocently enough by a child who was simply recording an everyday event, is burned into my mind for what it truly represents: no one has sacrificed more for this book than my wife and soulmate Evelyn, our son Harrison, and our daughter Virginia. From start to finish, this project took many thousands of hours, much of this on nights and weekends when I might have been doing something else with my family. Lyn, a talented historian who completed her Ph.D. three full years ahead of mine, started her own successful consulting business, managed our growing household mostly on her own, and somehow arranged things so that I could divide my scant free time between working on this book and being a dad. In addition, she edited multiple drafts of every chapter, and her substantive and stylistic suggestions have made it a far better work. Any errors that remain are mine and mine alone.

My children grew up with a father who was too often there and yet not there, lost in concentration trying to make sense of some obscure historical fact, wrangle a half-formed argument into shape, or streamline yet another balky paragraph. To Lyn, and to my children, I owe a debt that I can never possibly repay, though I vow to make the effort for the rest of my days.

ABBREVIATIONS

ABS	American Bonanza Society
AOPA	Aircraft Owners and Pilots Association
BPPP	Bonanza Pilot Proficiency Program
CAA	Civil Aeronautics Authority (1938–40), Civil Aeronautics Administration (1940–58)
CAP	Civil Air Patrol
CG	center of gravity
CPTP	Civilian Pilot Training Program
EAA	Experimental Aircraft Association
ERCO	Engineering and Research Corporation
FAA	Federal Aviation Agency (1958–67), Federal Aviation Administration (1967–present)
FIP	Flight Indoctrination Program
hp	horsepower
mph	miles per hour
NACA	National Advisory Committee for Aeronautics
NASA	National Aeronautics and Space Administration
NTSB	National Transportation Safety Board
QB	Quiet Birdmen
WASP	Women Airforce Service Pilots

Weekend Pilots

Introduction

One pleasant October evening in 1963, pilot Ralph Nelson threw his wife, Jane, fully dressed in an evening gown, into the swimming pool at the posh Riviera Hotel in Palm Springs, California. Some 2,000 pilots and their wives attending the Aircraft Owners and Pilots Association (AOPA) annual convention witnessed the dunking from their poolside tables, and nearly 100,000 more AOPA members had an opportunity to read about the event two months later in the December issue of the organization's full-length magazine, *AOPA Pilot*. According to the caption accompanying photos of this dinnertime drama, "Janie's crime was that she learned to fly without letting her husband know."[1]

Jane Nelson had just graduated from the first ever "Pinch-Hitter Course," which had debuted with much ado at that year's convention. Unlike the AOPA's numerous other flight instruction clinics, which were designed to help licensed pilots maintain or upgrade their existing flying skills, the Pinch-Hitter Course was specifically geared toward the nonflying wives of private pilots. This new program was not intended to turn graduates into full-fledged aviators or even bring them to the point where they could fly solo without an instructor sitting beside them in the cockpit. Instead, it imparted just enough knowledge and hands-on experience to allow a front-seat passenger of a small personal plane to take over the controls, call for help on the radio, and then land should the pilot have a heart attack or become otherwise incapacitated. In addition to a half day of "ground school" in the classroom, the "Pinch-Hitters" (so named for a baseball player who is called upon to bat in place of another) spent four hours in the air with an instructor, less than one-half the flight time that most student pilots needed before they were considered ready to make their first solo flight.

According to the *AOPA Pilot*'s version of events, when Jane's husband, Ralph, stopped by the Palm Desert Airpark just down the road from the convention hotel, he expected to see his wife embark upon her first Pinch-Hitter flight.

Instead, he was reportedly shocked to see her take off alone in a plane without an instructor aboard, a feat no graduate of the Pinch-Hitter Course was remotely qualified to perform. It was only then that Ralph supposedly discovered that Jane had started learning to fly before they left home for Palm Springs. In truth, he could not have been so completely ignorant of his wife's new skills as the article claimed. The author made no secret of the fact that Ralph Nelson was the AOPA official who had developed the Pinch-Hitter Course during the previous year in conjunction with the Ohio State University's School of Aviation; however, he neglected to mention that in August 1963, just a few months before the course debuted in Palm Springs, Jane had published her own article in the same magazine in which she explained how she started taking lessons because she was tired of being an "airport widow." In this piece directed at the wives of pilots, Jane described how learning the basics had cured her of her fear of flying and thus allowed her to become a willing traveling partner for her pilot husband, and she even hinted (without directly admitting) that Ralph had served as her instructor.[2] Yet Jane apparently crossed an invisible line when she publicly surprised Ralph by making her first solo flight at Palm Springs without his knowledge.

That night, as several thousand AOPA members and their wives looked on from their dinner tables, the master of ceremonies announced that Mr. and Mrs. Nelson would promenade around the hotel's swimming pool as a "post-solo stunt." At a prearranged spot, one of Ralph's co-workers, Victor "Vic" Kayne, joined the couple, and together the two men attempted to toss an unsuspecting Jane into the pool. According to the article, "Kayne and Nelson had met their match—Mrs. Nelson was determined not to go into the water at that time of night." But, the author assured readers, the matter wasn't settled yet. "Kayne's quick thinking preserved male superiority for all time. He grabbed Janie around the waist and jumped into the water with her while Husband Ralph pushed from the rear." A helpful member of the audience reportedly gave Ralph a shove, and he ended up in the pool as well. From the essay's gleeful tone, it would seem that the pilots in the crowd, as well as their wives, loved every minute of this impromptu dinnertime show.[3]

Throughout the postwar era, wives of private pilots were expected to ride along as willing passengers. Ideally, they were also supposed to help out in the cockpit in a role that combined the duties of housewife with those of an unofficial, and definitely unequal, copilot. These women served another key function: flying was an act of performance, and for this performance to have any meaning, pilots needed an admiring audience on hand. The AOPA specifically created the

Pinch-Hitter Course in 1963 to convert what one observer described as "private aviation's No. 1 enemy—the anti-flying wife," into a willing, even enthusiastic cockpit companion.[4] Completing the course did not, however, grant her license to compete with her husband for the title of "pilot-in-command" of the family plane. While the woman who did so might not end up literally getting thrown into a pool like Jane Nelson, at the very least she would find herself swimming upstream against the current of accepted norms within the predominantly male community of pilots.

Jane Nelson's experience brings up a truism in all forms of flying, be it private, professional, or military aviation: most pilots are men. This was true in the earliest days of powered flight, and it remains true today.[5] In 1959, when the federal government began gathering this statistic on a regular basis, women represented fewer than 3 percent of civilian aviators in America. By 1977 this figure had doubled to 6 percent, but then it stopped rising. Over the next three and a half decades, the total number of active fliers, men and women combined, would fluctuate by more than 30 percent. During that same period, the proportion of women within the overall pilot population remained stuck at 6 percent, varying by a minuscule 0.5 percent.[6] (See fig. I.1.)

Several factors help explain this stark statistical reality during the first decades following World War II. Despite highly publicized contributions by the Women Airforce Service Pilots (WASP) program during the war, both the armed forces and the airline industry continued to bar women fliers from their ranks for more than a quarter century. When the country entered a "space race" with the Soviet Union in the late 1950s, NASA deliberately excluded women—even those with considerable aviation experience—from becoming astronauts.[7] Thanks to these formal barriers, an entire generation of young women born after World War II had few high-status, well-paying aviation careers to aspire to. In addition, whereas boys and young men grew up surrounded by examples of male pilots in popular culture and real life, women and girls lacked high-profile role models such as female fighter pilots, airline captains, and astronauts to inspire them to take up flying. Even if a young woman happened to know a family member, neighbor, friend, or co-worker who flew for the military, for a living, or for fun, chances are that this pilot was a man.[8]

Most of these factors, however, start to lose their powers of explanation beyond the mid-1970s. In the wake of the feminist movement of the 1960s and the introduction of new laws prohibiting gender discrimination in the workplace, barriers that had long kept women out of the cockpit finally began to fall. In 1973 major U.S. airlines ended their long-standing policy against hiring

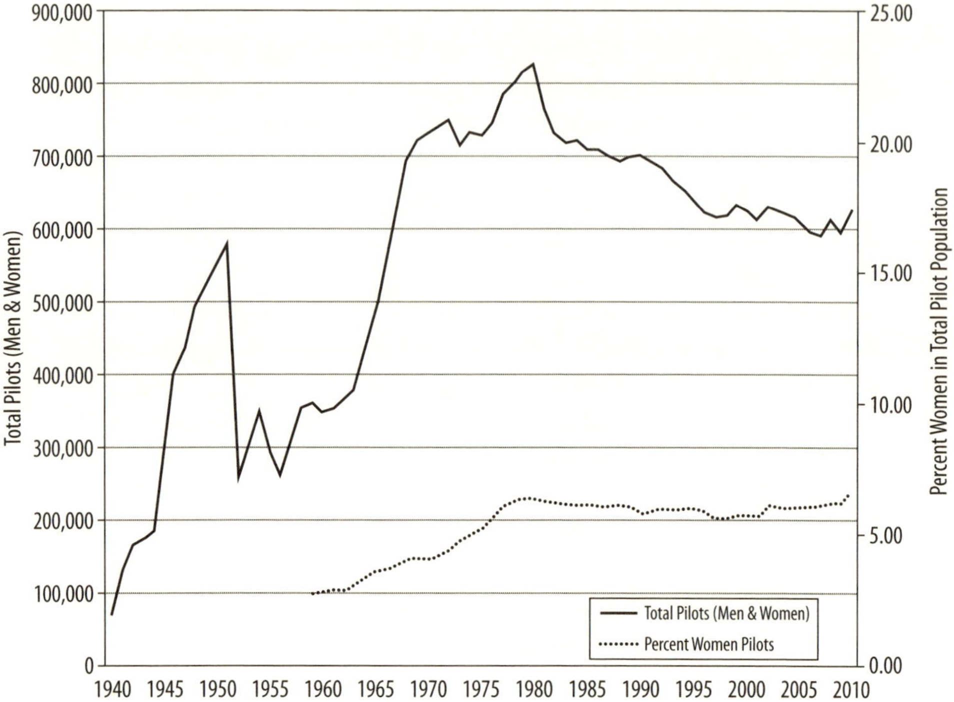

Fig.I.1. Total U.S. civilian pilots versus women pilots, 1940–2010. Solid line (keyed to left-hand vertical axis) depicts total civilian pilots (men and women) in the United States; dotted line (keyed to right-hand vertical axis) indicates the percentage of these pilots who were women. Data from CAA, *Statistical Handbook of Civil Aviation*; FAA, *Statistical Handbook of Aviation*; FAA, *U.S. Airmen Statistics*. Graph by author.

women pilots.[9] The military opened its doors to women fliers as well: the first female aviators in the U.S. Army and Navy graduated from flight school in 1974; two years later, the U.S. Air Force announced its first two classes of women pilot trainees. And in 1978, NASA selected its first group of women astronaut candidates, including Sally Ride, who in 1983 became the first American woman to fly into space.[10] With increasing job opportunities for women in aviation and, perhaps more important, a growing pool of modern-day role models to emulate, it made sense that more girls and women might be enticed to take up flying. But, as the statistics show, this was not the case. Instead, aviation as both pastime and profession remained dominated by men.

With most of the old obstacles out of the way, there must be other reasons for the continued scarcity of women in the cockpit. Nearly five decades after Jane Nelson's unexpected swim, longtime pilot, aviation writer, and editor emeritus of *Flying* magazine Richard Collins asked readers in a January 2012 column for *Air Facts Journal*: "Why hasn't aviation attracted more females?" After noting, "I have always thought that females can be better pilots than males," he mused, "Could it be that females are not attracted in number because they take one look at male pilots and think, 'Yuk, I don't want to be like *that*.'" He went on to conclude, "It is my opinion that we males have created a fraternal bond in flying that largely excludes females."[11] After more than half a century of firsthand experience and observation, Collins was convinced that a long-standing, pervasive culture of masculinity had made aviation fundamentally unattractive to most members of roughly half the nation's population—namely, women.[12]

As Richard Collins suggests, the culture of private flying plays a significant role in determining who becomes a pilot and who does not. Throughout the postwar era, private fliers deliberately constructed, embraced, and perpetuated a masculine pilot identity, creating a gendered culture that defined the norms for how pilots were supposed to act and served to deter individuals who did not fit this mold. Past issues of aviation magazines abound with not-so-subtle hints that flying was—or at least was supposed to be—a manly endeavor. A 1961 advertisement for a steam pressure washer in *AOPA Pilot* features a shapely, scantily clad woman perched atop the apparatus under the provocative title "Hot to go in 70 seconds!" The ad's subtitle, "The sweetheart of the airport is designed especially for the airport operator," allowed readers to speculate whether the "sweetheart" who was "hot to go" referred to machine or woman. A 1969 cartoon in the same magazine depicts a man climbing out of the cockpit of a small plane that has just been doused with firefighting foam by the airport "crash truck." Taken at face value, his sardonic comment to the firefighter—"There wasn't any emergency . . . my wife always lands that way"—meant that his wife was to blame, but it seems just as likely this was a cover story for his own botched landing. Either way, the cartoonist assumed that readers would both understand and find amusing this aviation version of widespread popular assumptions (and jokes) of the era regarding the technical skills, or lack thereof, demonstrated by women drivers.[13]

Many male pilots treated any woman who *was* a good flier, or even one who liked to fly, as an anomaly. In an ironic twist on the punch line "my wife always lands that way," after former WASP Margaret Ringenberg landed at Eagle Creek Airport in Indianapolis on a particularly gusty day in the mid-1960s, the line boys (aviation's equivalent of service station attendants) completely ignored her as they rushed to congratulate her nonpilot husband for his skillful handling of the airplane.[14] Whether intended or not, through these and countless other actions, men involved in private aviation constantly reinforced the clichéd phrase that Ringenberg chose as the title for her autobiography: *Girls Can't Be Pilots.*

More than a few people believed that women, by their nature, were fundamentally unfit to be good pilots. For instance, in a 1969 *Flying* magazine article about why most women supposedly hated private aviation, Robert Peterson wrote: "Men who fly are hot for speed and willingly forego comfort. Men love speed—for good, sound, masculine reasons. The lines of a hot plane evoke instant masculine response." Peterson was clearly convinced that biological differences explained why men loved to fly and women did not. However, he inadvertently acknowledged that nurture played a role as well when he attempted to explain why male pilots were reluctant to express apprehension or fear in the midst of stressful situations aloft, noting that men's facade of "stoicism . . . marks years of training going back to Daddy, who said boys are brave, only girls cry." Although Peterson's article reflects one individual's opinion, a flurry of positive reader responses, in the form of letters to the editor, suggests that he was far from alone.[15]

Postwar academic research also reported that masculinity and private flying went hand in hand. For instance, a 1974 University of Michigan study of civilian fliers indicated that men who flew personal planes had more in common with U.S. Navy jet pilots—a group viewed both by the public and by naval aviators themselves as the epitome of manliness, skill, and courage—than they did with the average adult America male. The psychiatrist and psychologist who ran the study concluded that the typical pilot, regardless of whether he flew military aircraft into combat or personal planes for pleasure, was the "active-masculine" type "who is oriented toward demonstrating his strength and competency, who thrives on adventure, who finds pleasure in mastering complex tasks, and whose manifest sexual orientation is decidedly heterosexual." In addition, "this aviation profile fits well with the popular description of pilots in song, movie and verse, as [a] courageous, romantic, 'he man.' "[16] According to the same study, female pilots possessed a personality profile similar to male civilian pilots, characterized by quantifiable traits including achievement, exhibition, dominance, and change (and radically different from the typical American woman's profile, defined by high degrees of deference, order, affiliation, succorance [dependence], abasement, nurturance, and endurance). The researchers, however, were quick to add that despite displaying so many masculine traits, "these female pilots are 'feminine' in that their own sexual and gender identification is feminine and in that libidinal drive activity is directed toward heterosexual (male) objects."[17]

Having concluded that all fliers—regardless of whether they were civilian or military, amateur or professional, male or female—possessed plenty of masculine personality traits, the University of Michigan team made no attempt to explore the broader implications of its findings. How did this high concentration of individuals with such a pervasive masculine identity shape the behavior of pilots and the culture of private aviation? How might private fliers' strong identification with "active-masculine" values affect practical aspects of aviation, such as the design of airplanes or the appearance and ambience of airports? More recent studies have implicated masculine culture in a range of aviation-related issues, from complaints that widespread sexism among male airline pilots helps account for persistently low numbers of women in the profession to warnings that a "macho" cockpit environment can lead to a communication breakdown between members of airline flight crews that in turn contributes to accidents. Written by nonhistorians concerned with present-day problems, these works treat masculinity as an important factor but offer little insight into how this culture operated and evolved over time within the aviation community.[18]

This book addresses these questions as it explores the role of masculinity in private flying during the first four decades after World War II. I argue that throughout this era, private pilots created and maintained a community and culture that demonstrated and celebrated ideals and behaviors traditionally associated with masculinity in America. This masculine culture in turn influenced the evolution of private aviation throughout the postwar era in ways both obvious and obscure and with results that were sometimes subtle and in other cases substantive. As fliers performed—and, in the process, perpetuated—this masculine pilot identity, they also shaped all aspects of the training and acculturation of aspiring aviators, the social and cultural fabric of the community of pilots with whom they associated, and even the very technology and complexity of the aircraft that were available for private pilots to fly.

No one familiar with popular Hollywood movies like *Top Gun* (1986) or *The Right Stuff* (1983, based on Tom Wolfe's 1979 best-selling book by the same name) should be surprised to learn there is a long-standing connection between aviation and masculinity.[19] Perceptions that aviation was an activity best left to men date to the first years of powered flight. For instance, in 1911, world renowned British flier Claude Grahame-White left no doubt where he stood on the matter: "I have taught many women to fly and I regret it. My experience has taught me that the air is no place for a woman."[20] A few years later as World War I raged across Europe, government propagandists, members of the press, and war-weary populations on both sides of the conflict collectively spawned the widely popular myth of the chivalrous "ace" fighter pilot as a heroic alternative to the anonymous, inglorious blood and mud realities of trench warfare. During the two decades that followed, movies, radio shows, dime novels, and comics simultaneously profited from and reinforced this manly pilot image in popular culture. Meanwhile, real-life explorers, air racers, and airmail pilots dominated newsreel footage and made international headlines.[21] Although few everyday people had ever flown in a plane, the public became obsessed with aviation during the interwar era: for instance, 30 million Americans—roughly a quarter of the nation's population—turned out to catch a glimpse of Charles Lindbergh on his coast-to-coast tour following his 1927 transatlantic solo flight.[22] Seeking to cash in on this widespread fascination with flying, itinerant barnstormers performed impromptu air shows in rickety, war surplus biplanes over small towns across the country to drum up interest before landing at a farmer's field to sell airplane rides to the locals.

Displaying behavior reminiscent of young men who felt freed from traditional social strictures by the tough, egalitarian environment of America's Western frontier in the latter half of the nineteenth century, pioneering aviators of the early twentieth century were infamous for their drinking, carousing, and womanizing on the ground, as well as for their daring, often foolhardy feats in the air. Men pushed the limits of both their machines and themselves as they sought to show off to each other, to set new records, or in many cases to merely avoid being labeled "chicken" by their peers.[23] Although the money was good, with airmail pilots earning $3,600 a year in the early 1920s (more than twice the national average income), there were intangible rewards as well. As early airmail pilot Dean Smith recalled in his autobiography, "People often asked why I liked being a pilot, why I flew the mail and took such chances of getting killed. I would try to explain, but never could find the words to explain it all. I knew that I could fly and fly well, and this skill set me apart from the run of the mill." Although he conceded that "one of the most rewarding things about a mail pilot's job was the high pay and the high percentage of leisure time," Smith also freely admitted that there were rewards that money could not buy: "Sometimes I was called a hero, and I liked that."[24] Echoing the same sentiment, barnstormer Walt Ballard declared, "Well, yeah, I felt like a king because that's the way people treated you. Treated you like you were a king. People around the country had never seen an airplane before and that's what made the barnstorming popular."[25] More than a few early aviators translated this kind of status into success with women. For instance, in his final memoir, Charles Lindbergh described how "many a barnstorming pilot slept with one woman one night, another the next, and never saw them again as he followed the wind and season." He drove this point home with an example: "A pilot I met in Nebraska was famed for his directness of approach on arriving at a new location. 'Do you or don't you? That's all I want to know,' he said as he canvassed a town's feminine possibilities."[26]

As Grahame-White's 1911 rueful admission that he had "taught many women to fly" reveals, almost from the start a few daring women managed to take to the sky despite staunch resistance from male pilots. The press quickly coined special terms—*aviatrice* in 1910 and *aviatress* in 1911, which were replaced by the mid-1920s with the more familiar *aviatrix*—to set female fliers apart from male *aviators*.[27] If aviators were considered exceptional because they possessed both the skill and courage to conquer the dangers of the air, then aviatrixes were all the more extraordinary, if not abnormal. After all, these women were trespassing in a male domain that was fueled by a volatile mix of technology, competition, and risk, hardly considered proper behavior for members of the "gentler sex." Precisely because they were viewed as exceptions to the rule, the presence of a few women in the cockpit during the interwar years, including Ruth Law, Phoebe Omlie, and Bessie Coleman (who as an African American was all the more uncommon) tended to reinforce rather than challenge cultural associations between aviation and masculinity.[28] Even women whose fame and aerial feats arguably matched those of their male counterparts could not break free from the long shadow of masculinity. Amelia Earhart, for instance, was quickly dubbed "Lady Lindy" by a press and public eager to compare her to the most famous male pilot of the day, Charles Lindbergh, rather than celebrate her accomplishments as an aviator who also happened to be a woman.[29]

Joseph J. Corn's pathbreaking 1983 book, *The Winged Gospel: America's Romance with Aviation, 1900–1950*, describes how paradoxically even as the American public worshiped pilots as the personification of American manliness, millions longed for a day in the not-so-distant future when the rough and tumble frontier of the sky would be tamed. This, they hoped, would lead to a "democratization" of the air, where everyday men and women could soar aloft at the controls of personal planes that were as safe, economical, and easy to operate as automobiles.[30] Optimistically proclaiming, "Get ready today, groundlings, for tomorrow we fly," a December 1940 *Saturday Evening Post* article predicted that, thanks to recent innovations in aviation technology, "the eagle is changing into the sparrow and uncounted thousands of us will be flying soon."[31] This dream of "an airplane in every garage" continued unabated into World War II. According to one wartime survey, 39 percent of men and women reported having "a desire to learn to fly some day [*sic*]," with men only somewhat more inclined than women

(46 vs. 33 percent). Enthusiasm ran highest among younger respondents, with 69 percent of those between 18 and 24 years old aspiring to become pilots.[32] Even people who did not personally want to fly believed that the nation was poised on the brink of a new aerial age. A nationwide survey of 2,900 Americans published in 1945 concluded: "Nine out of ten think that a lot of people will want to own and fly their own planes after the war, and 35% even go so far as to say that the day will come when as many people will own airplanes as now own automobiles."[33] These and similar polls fueled government and industry forecasts that nearly a half a million personal planes, and perhaps as many as 4.5 million private pilots to fly them, would fill America's skies by 1955.[34]

As war raged across the globe, Cessna Aircraft Company ran a nationwide marketing campaign in *Time, Flying,* and other popular magazines that promised readers they would soon be able to own a "Cessna Family Car of the Air . . . the postwar airplane that you'll be able to *buy and fly* if you can *buy and drive* an automobile."[35] Full-page color advertisements showed men and women of all ages cruising over the countryside behind the wheel of sleek, carlike planes. The accompanying captions promised that carefree, long-distance personal air travel for everyone was just around the corner. One ad described how golfers would use their family plane to "leave home after breakfast to tee off *500 miles away* at noon."[36] Another declared: "*Dreams Do Come True.* And this is one of them. It's the age-old dream of magic trips to everywhere. It's a dream of vacations every week end [*sic*] instead of once a year, of trips as far as from Louisville to Florida in an easy afternoon's cruise."[37] Operating on the assumption that, like automobiles a generation earlier, postwar airplanes would become commonplace only when both men and woman incorporated this form of transportation into their everyday lives, roughly a third of Cessna's wartime advertisements featured smartly dressed, well-coifed women behind the controls. This approach also reinforced the message that the future Family Car of the Air would be so safe and simple to operate that "almost anyone" could learn to fly.[38] In short, even as hundreds of thousands of young men were earning their military wings to engage in what many people at the time considered the ultimate test of manhood—one-on-one aerial combat against the nation's enemies—Cessna sought to re-gender aviation so that *anyone*—man or woman, young or old—would feel at home in sky.

Ultimately, the postwar skies would indeed be democratized, but not by the Cessna (or any other manufacturer's) Family Car of the Air. Instead, ever-increasing numbers of Americans took to the air as passive, paying passengers aboard large, complex commercial aircraft owned by a handful of major corporations and operated by highly trained professional pilots. By 1980, nearly two-thirds of all adult Americans had taken at least one trip by airline in their lifetimes, and approximately one-quarter of American adults—nearly 47 million people—had made at least one such flight within the previous year. By comparison, that same year there were 827,000 active civilian pilots in the United States, which translates to only 2 percent of recent airline travelers and 0.5 percent of the nation's total adult population. What's more, after 1980 the total number of pilots (including private pilots) began to decline while the percentage of adults who had flown by airline continued to grow, so that by 1997 the ratio of licensed pilots (including private pilots) to recent airline travelers had dropped below 1 percent.[39] Americans were flying more than ever before, but they weren't flying themselves.

Still, prewar and wartime enthusiasm for private aviation had not completely disappeared. Hundreds of thousands of Americans became pilots after World War II, swelling the ranks of licensed fliers from fewer than 34,000 in 1939 to more than 580,000 by 1951. In those 12 years alone, the proportion of American adults who knew how to fly skyrocketed from just 1 in 2,700 to around 1 in 180.[40] And most active postwar fliers in any given year did not fly for a living. For instance, just 2 percent of all licensed civilian pilots in 1950 listed flying as their primary occupation, slowly rising to 9 percent by 1980, 16 percent by 1990, and 20 percent by the year 2000.[41] This indicates that throughout the half century following World War II, at least 80 percent of pilots in any given year flew exclusively for pleasure or personal transport, not for hire. For this reason alone, private flying provides the widest possible window into connections between aviation and gender identity.

At first glance, the behavior of these postwar private pilots, both in and out of the cockpit, bore only passing resemblance to the swashbuckling, macho barnstormers and airmail pilots who roamed the skies, bars, and brothels during an earlier era of open cockpit biplanes, leather helmets, and silk scarves. Even so, this new generation of fliers—who were mostly middle-class (and, by the late 1950s, increasingly middle-aged) white men—in many ways still acted more like the prewar "eagle" than the timorous "sparrow" whom the *Saturday Evening Post* had once predicted would soon dominate the skies. With women barred from military flight schools well into the mid-1970s, and only a few thousand new military flight trainees, male or female, accepted annually in the decades that followed, most women who wanted to learn to fly throughout the postwar era had little choice but to start by working toward a private pilot's license with a civilian instructor. Thus, for reasons of numbers alone, it was in the world of private aviation that most women likely came face-to-face with the "fraternal bond" that Richard Collins theorized led many to decide "Yuk, I don't want to be like *that*." And it is here, among the majority of pilots who do not fly for a living, that we are most likely to find clues to answer this question: Why, more than a century after the first powered flight, more than a half century since Betty Friedan's book *The Feminine Mystique* kickstarted the feminist movement of the 1960s, and after decades of concerted efforts by the aviation industry to attract more women, is it still true that most pilots are men?

As the title suggests, this is a book about pilots, not about airplanes. That said, I do focus on a few particular aircraft in two of the chapters. Fans of the Ercoupe, Beechcraft V-tail Bonanza, Cessna Skymaster, or tailwheel-equipped aircraft may feel that I have singled out these models for undue criticism. This was certainly not my intent. Because these examples generated genuine controversies and debates—deserved or not—that reverberated throughout the entire private aviation community, these aircraft reveal much about attitudes held by many members of that community. The same is true regarding flight training, especially when it comes to the question of spin training. I consciously avoid taking sides or passing judgment on any of these case studies; instead, I interpret the words of those actively involved in these debates as evidence for how masculine ideals pervaded even practical and technical arguments about postwar private aviation.

The end of World War II serves as the starting point for this study, since private aviation barely existed before the war, and the war laid the foundations for this activity as we know it today. Although hundreds of thousands of people continue to participate in this pastime, and more than a few vestiges of masculine culture described in this book remain a part of the experience, the mid-1980s represent a turning point. The number of active pilots, generally on the rise since the mid-1950s, reached an all-time high of 827,000 in 1980. But then it plummeted, dropping by more than 117,000 pilots (nearly 15 percent) over the next five years. At the same time, manufacturers ceased or severely curtailed production of several aircraft used mainly by private pilots. These decisions were some time in the making. Starting in the mid-1970s, both the number of lawsuits and the monetary damages awarded as a result of accidents began to increase dramatically, even though fatal accident rates for general aviation had declined slowly but steadily over the past decade.[42] Some pilots responded by dropping insurance coverage, while others quit flying altogether. Meanwhile, complaining that skyrocketing liability insurance now accounted for nearly a third of the price of new personal planes, one by one the major general aviation manufacturers either closed their doors or stopped building small, single-engine aircraft to concentrate on the more lucrative (and statistically less risky) market for multimillion-dollar business jets.[43]

By 1980, the outlook seemed so grim that one industry observer used stark terms to describe the prevailing mood in the city where both Cessna and Beechcraft were headquartered: "If they had tall buildings in Wichita, [Kansas,] people would be jumping out of them."[44] Cessna Aircraft, the same company that four decades earlier had promised war-weary Americans their very own Family Car of the Air, was the last holdout among the "big three" aircraft manufacturers after Piper and Beechcraft stopped selling entry-level personal planes in the early 1980s. But in 1985 Cessna halted production of its two-seat model 152, and one year later the company stopped making the best-selling personal plane of all time, its four-seat model 172.[45] Cessna's decision to withdraw from the field in 1986, combined with the continued decline in active pilots throughout the decade, seemed at the time like a death knell for private flying to many members of the aviation community. Combined, these reasons explain my decision to focus this study on the four decades spanning 1945 to 1985.

It is far more difficult to define *masculinity* than it is to determine the dates to begin and end this study. In 1969 sociologist Patricia Cayo Sexton observed that in contemporary American society, "male norms stress values such as courage, inner direction, certain forms of aggression, autonomy, mastery, technological skill, group solidarity, adventure, and a considerable amount of toughness in mind and body."[46] Although the book in which this statement appeared was problematic at the time and is outdated today, scholars continue to quote Sexton's description of masculine traits as a starting point for further analysis. I too found Sexton's brief list useful, if only because many postwar private pilots openly celebrated and cultivated the same traits that she described. However, Sexton's list was descriptive rather than analytical and, at best, represents only a snapshot of mainstream public perceptions about what was (and was not) considered masculine behavior at the time.

Several generations of psychologists, sociologists, anthropologists, and historians have since struggled to come up with more precise and inclusive definitions of gender identity. The title of one scholarly essay on the subject, "Masculinity and Femininity: Defining the Undefinable," neatly sums up the complexity of the issue.[47] Part of the problem is that masculine identity is not a monolithic, unchanging concept. Starting in the mid-1980s, sociologists formulated the concept of *hegemonic masculinity* to describe "the pattern of practice (i.e., things done, not just a set of role expectations or an identity) that allowed men's dominance over women to continue." These scholars argue that because public and individual perceptions of masculinity involve constant renegotiation and reinforcement, popular notions regarding masculinity evolve over time. Furthermore, different versions of masculinity not only coexist in the same time and place, but an individual or group may also adopt and exhibit multiple versions despite apparent contradictions. Finally, proponents of hegemonic masculinity argue that only a few men need to act out "a strong version of masculine dominance" in order to influence everyone in society. Other men who did not actually act that way still "received the benefits of patriarchy . . . [and] could be regarded as showing a complicit masculinity."[48] When applied to postwar private flying, this concept suggests that only a minority of participants had to act like the "courageous, romantic, 'he man'" pilots depicted in popular culture to perpetuate a pervasive ethos of masculinity in aviation.[49]

Gail Bederman, one of the pioneering historians of masculinity, reminds scholars that another reason this term defies precise definition is that masculinity is not a "thing." As such, it cannot be measured. Instead, masculinity is best viewed as a concept we can use to help unpack and understand the complicated relationships that members of society have constructed throughout history in order to sort and categorize people and behaviors.[50] Perhaps this multiplicity of meanings helps explain why relatively few historians who study the relationship between gender and technology seem willing to define masculinity in print.

Beyond the obvious daring feats and "he man" boasting like those depicted in *The Right Stuff* and *Top Gun*, many aspects of masculine culture in aviation are so pervasive that at first glance they are effectively invisible. Bederman notes that until the late twentieth century, "existing histories of politics, war, and almost everything else were chock-full of men, but illuminated nothing at all about the sex-gender system."[51] The same is true of the history of aviation. However, the fluid and constantly renegotiated nature of masculinity created points of friction in the past, which in turn can provide revealing insights into the role of masculine identity and ideals in private flying. For instance, virtually all pilots agreed that flight training should be comprehensive enough to prepare aspiring aviators for the many potentially dangerous situations they might encounter in flight. Yet stark differences of opinion arose regarding how best to accomplish this goal, and members of the aviation community argued endlessly about whether certain standards and instructional techniques were too rigorous or not tough enough, with some venturing so far as to warn against allowing "sissies" to take to the sky. Similar disagreements emerged about how best to reduce accidents, with some favoring pilot skill and others promoting advances in technology. In both cases, one side championed ideals traditionally associated with masculinity (technical competence, mental and physical toughness, and courage in the face of physical danger), whereas the opposing viewpoint argued that overreliance on these same traits could be counterproductive to the growth of private flying, or even dangerous to the point of contributing to accidents. These and other examples of conflict and controversy provide opportunities to explore both the role of masculinity in private aviation and the contested nature of gender identity in the postwar United States.

Unlike the 1974 University of Michigan study described earlier, I make no attempt to discover the relationship between sexual orientation and private flying. Decades ago, psychiatrists and scholars routinely linked sexual orientation to gender identity, associating, for instance, the presence of so-called masculine personality traits with heterosexuality in males, these same traits with lesbianism in females (conversely, a dearth of masculine traits supposedly indicated heterosexuality in properly "feminine" women and homosexuality in men). This probably explains why the authors of the 1974 study seemed almost surprised with their findings that although female pilots had personality profiles that resembled men more than they did women, the same battery of tests also indicated that these women fliers were decidedly heterosexual.[52] More recent scholarship suggests that the relationship between sexual orientation and gender identity is far more complex than previously believed. Not surprisingly, almost all the sources that I used for this project—including archival records, popular magazines, aviation journals, oral history interviews, and artifacts—implicitly described a heterosexual ideal for the postwar community of pilots. Likewise, I found almost no evidence of pilots publicly challenging this norm during the first four decades following World War II. Proponents of the concept of hegemonic masculinity would argue that this lack of evidence is a direct result of "the policing of heterosexuality" by the dominant masculine culture.[53] Although this dearth of evidence renders gay and lesbian fliers invisible, it seems likely that they accepted the same masculine pilot identity as their heterosexual counterparts. Whether they did so wholeheartedly, with some degree of ambivalence, or as a necessary price to pay for admittance into the flying fraternity is another story, one that cannot be discerned from the sources that I used.

Although private aviation certainly changed over time, while working on this book I was repeatedly struck by the impression that a pilot from 1950 would feel right at home at the local airport if he (or she) was somehow magically transported forward in time to 1980, despite the massive changes that had rocked American society outside the airport gates. For this reason, I organized the book topically instead of chronologically to highlight the striking cultural continuities within the community of pilots throughout the postwar era. Because I focus on the role of masculinity, readers who desire a comprehensive "soup to nuts" history of private flying may wish to consult one of several broad surveys of the subject.[54] These works more completely describe the various makes and models of aircraft that were available in the marketplace, the legislative and regulatory actions that affected public access to the sky, and the constantly changing fortunes of the industry as manufacturers flourished, merged, went bankrupt, and rose again from the ashes. Not surprisingly, broader surveys cover these details at the expense of the kind of in-depth analysis that I hope to provide here regarding the social and cultural factors that so profoundly shaped private flying.

The first two chapters describe the pilots themselves. Chapter 1 examines why, from a purely practical perspective, most new pilots of the immediate postwar era were men, while chapter 2 examines how World War II continued to influence flight training long after the guns fell silent. The next pair of chapters explores the relationship between pilot identity, individual skill, and technology. Chapter 3 describes how postwar pilots relied on skill and technology to differentiate themselves from the nonflying public and, in the process, rejected innovations that threatened to make flying too easy. Chapter 4 examines how some pilots extended this celebration of skill to create a hierarchy *within* the aviation community. The final two chapters examine the social aspects of private aviation. There was far more to being a pilot than simply going aloft, and chapter 5 explores how individuals defined themselves as pilots through their actions in and around the geographic center of aviation activities: the airport. The sixth and final chapter describes how pilots used formal organizations to define postwar aviation in terms of gender and, as a result, reinforced rather than revolutionized the traditional exclusion of women from mainstream private flying.

In addition to considering the relationship between masculinity and aviation, this book engages historical arguments about the role of technological enthusiasm in American society. In the standard historical narrative, people embraced a particular technology (the means to an end) mainly because they expected it to deliver something practical and tangible (the desired end). But recent work on the history of drag racing and hot rodding reveals that enthusiasts did not simply tinker with their cars to make them go faster; rather, they tinkered with cars because they loved to tinker with cars. In other words, tinkering itself became both the means *and* the ends to their hobby.[55] This kind of technical enthusiasm is hardly limited to the automotive world; for instance, the same fundamental motivations helped spur another male-dominated technical pastime in America, ham radio.[56]

Flying differs from souped-up cars or ham radios in that, throughout the postwar era, manufacturers and pilots alike argued that private flying was a practical means to achieve a rational end: fast and efficient long-distance personal transportation. And yet, in many cases, pilot behavior does not support this argument. Acknowledging technical enthusiasm as a rational end—and one that was deeply intertwined with masculine identity—helps explain much about how private aviation evolved in the postwar era, from the kinds of airplanes that pilots chose to buy or rent to which side they took in "technique versus technology" debates about how to make flying safer. This book thus bolsters arguments that technological enthusiasm plays a major role in shaping both technology and human behavior.[57]

Weekend Pilots cannot by itself tell the full, rich story of private aviation. But by focusing on one specific aspect of the community of pilots—the role of masculine culture—it explains much of what transpired in private flying between 1945 and 1985, as well as why this activity looks the way it does today. Any attempt to analyze the history of private aviation—or most technology-related activities, for that matter—is likely to come up short unless it recognizes the far-reaching influence of gender. This book provides an example of just how deeply embedded gender is in technology-related activities, and how inconspicuous and yet profound its long-term effects can be.

In the process of exploring the role of masculinity in postwar private aviation, I also hope to provide readers with insight into how gender identity and technology relate to issues of accessibility, exclusion, and power in the postwar United States. Gender does not just happen. Instead, it is a part of human identity that shapes how people view themselves and each other as women or men. At the same time, gender also serves a larger role in society, defining boundaries and limiting access to certain activities and institutions. As such, gender itself becomes symbolic as objects and activities take on masculine or feminine connotations.[58] Working from these premises, the editors of *Technology and Culture*'s special 1997 issue on gender argue, "It is as impossible to understand gender without technology as to understand technology without gender."[59] Because society is made up of people, and it is people who create and use both gender and technology for multiple reasons, closely examining the relationship between gender and technology throughout history provides us with a window for understanding not only the past but also the present day in which we live. Thus, the fact that a husband would conspire to throw his wife into a hotel swimming pool in front of several thousand onlookers, just because she learned to fly behind his back, can tell us as much about postwar American society as it does about the private pilots who were sitting in the audience that October evening in 1963.

CHAPTER ONE

Who Is "Mr. General Aviation"?

The Origins and Demographics of Postwar Private Flying

In the spring of 1969, the Aircraft Owners and Pilots Association published the results of a survey mailed earlier that year to each of its roughly 150,000 members. This "Pilot Poll" allowed the organization to create a "composite picture" of the typical private flier in the United States, whom it labeled "Mr. General Aviation." To illustrate the AOPA's growing influence, the article included a bar graph to chart the organization's growth from its humble beginnings in 1939. But in place of ordinary rectangular bars, the graph featured the image of a private pilot to depict the number of members during selected years. For 1939, this figure stood less than a quarter of an inch tall; by 1969, he towered nearly three inches high (fig. 1.1).[1]

This image of Mr. General Aviation bore little resemblance to the daring, devil-may-care barnstormer who had roamed the skies and dominated the imagination of Americans a generation earlier. Nor did he look anything like the heroic young fighter and bomber pilots popularized in film and print during World War II. Mr. General Aviation was no astronaut or test pilot, and he certainly had nothing in common with that other boundary-testing male of the era: the long-haired, tie-dyed, counterculture youth of the late 1960s. Instead, AOPA's typical private flier looked like a smiling, slightly rotund, middle-aged businessman who, at least in terms of appearance, was remarkable for being unremarkable. In place of a leather jacket and silk scarf, he wore a suit with a handkerchief tucked neatly into his breast pocket. And rather than a helmet and goggles, he carried a briefcase and sported a fedora perched atop his almost certainly balding head. In short, Mr. General Aviation could have been "the guy next door," especially if you happened to live in a typical white, middle- to upper-middle class suburb of the era.

So who, according to this poll, was the typical private pilot of 1969? About three-quarters were between 25 and 49 years old, married, had three or fewer

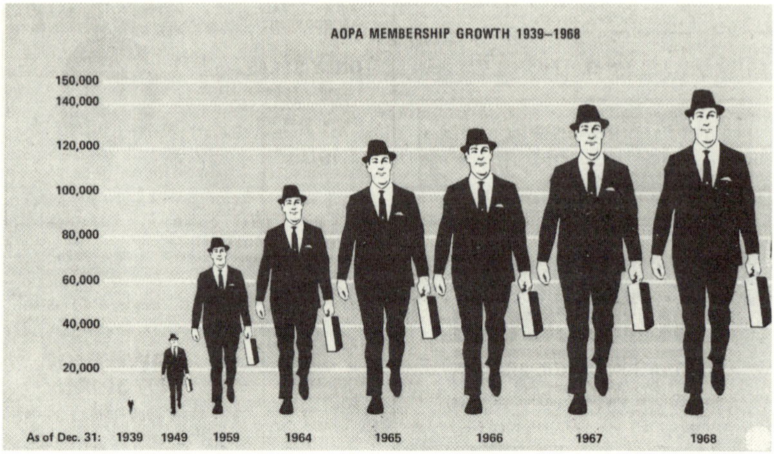

Fig. 1.1. "Mr. General Aviation" circa 1969. Artist's rendering of the typical private pilot, nicknamed "Mr. General Aviation," as used in 1969 to depict membership growth since the Aircraft Owners and Pilots Association's founding three decades earlier. "AOPA Membership Growth, 1939–1968," in Lew Townsend, "AOPA Members Cast Large Shadow," *AOPA Pilot* 12, no. 5 (May 1969): 94 (graph). Courtesy of AOPA and the Hagley Museum & Library.

children, and owned rather than rented their homes. Mr. General Aviation earned anywhere from $9,000 to $40,000 per year and claimed a net worth ranging from $20,000 to more than $100,000. While income placed him comfortably within the top two-fifths of U.S. households, his net worth is even more telling in an era when fewer than 10 percent of U.S. households had $10,000 or more in total assets. Regarding occupation, Mr. General Aviation was "situated in almost every area of activity," but four broad categories stood out: manufacturing; architecture and engineering; civil aviation (which presumably included professional pilots, flight instructors, airport managers, and mechanics); and a professional group that encompassed medicine, dentistry, law, and education. Within these mostly white-collar career fields, more than one-third of respondents served in top management positions, another 14 percent were in middle management, and some 30 percent identified themselves as holding professional or technical positions. Only about 3 percent of those surveyed worked as professional airline or air-taxi pilots. Thus, most respondents flew mainly for pleasure and personal business, not for a living.[2]

In addition to flying, Mr. General Aviation preferred to spend his leisure time fishing, playing golf, hunting, and boating—all activities commonly associated with middle-class America at the time. Just over half of the pilots who responded to the survey owned two cars in an era when single-car households were still the norm. One-quarter owned a pleasure boat, and nearly a fifth maintained a summer or vacation residence. The survey also revealed that 1 out of 15 held some type of public office, and 1 in 5 served on the board of directors "of some business concern or association."[3]

AOPA's visual and statistical depiction of Mr. General Aviation provides a telling glimpse into the world of private flying after World War II. It should come as no surprise that the organization did not bother to describe a Mrs. (or Miss, or Ms.) General Aviation. After all, in 1969, only 4 percent of civilian pilots in the United States were women. It is also not surprising that the AOPA did not assess the racial or ethnic background of its members. There are no reliable statistics regarding how many nonwhite minorities participated in private flying, but anecdotal and contextual evidence overwhelmingly indicates that most pilots were white. Thus, this "composite picture" of Mr. General Aviation is a fairly accurate portrayal of who was flying personal planes in postwar America.[4]

The ranks of active fliers would rise and fall over the ensuing years as a result of social, political, and economic developments outside the aviation community. However, even as the rest of the nation experienced dramatic cultural changes related to the baby boom, women's liberation, and the civil rights movement, the community of pilots remained a remarkably homogeneous group. Examining how most people become pilots reveals that the pathways aspiring aviators followed into private flying go far toward explaining this remarkable degree of demographic continuity.

The Making of Mr. General Aviation

Before World War II, private flying barely existed in the United States. Aviation was still widely perceived as the domain of wealthy playboys, rough-and-tumble barnstormers, and pioneering airmail pilots. Americans young and old worshiped famous fliers such as Eddie Rickenbacker, Charles Lindbergh, and Amelia Earhart, while newspaper reports and newsreel footage followed the exploits of popular air racers like Jimmy Doolittle and Roscoe Turner.[5] During the Great Depression, families flocked to local airfields on weekends when the weather was fine, not only to attend scheduled air shows but also simply to watch airplanes take off and land. According to a 1930 study by Harvard University, airports at several cities—including Indianapolis, Louisville, Milwaukee, and Minneapolis—hosted 10,000 or more visitors on a typical summer weekend. Love Field, in Dallas, Texas, experienced even higher numbers, ranging from

15,000 to 25,000 people per weekend, even though the airport averaged only six airline flights per day. From this and other evidence, one aviation historian concludes that before World War II "most people experienced flying as spectators, not participants."[6]

The Second World War changed this dramatically. In 1939, as Nazi Germany invaded Poland, there were still fewer than 34,000 licensed pilots in the United States, or about 1 for every 2,700 adults. By 1946, some 400,000 Americans held a pilot's license, and up to 250,000 more were taking flight instruction. These numbers continued to climb over the next five years, topping 580,000 by 1951, not including students, which meant that 1 out of 183 adults held a license to fly.[7] It seemed as if the nation was well on its way to realizing the longtime dream of private aviation's most ardent supporters: millions of ordinary citizens taking to the skies and "an airplane in every garage."[8] But how had so many people found their way into the cockpit in such a short time? Directly or indirectly, most of these new fliers earned their wings as a result of World War II. And this, in turn, had a profound effect on the long-term demographics of postwar private flying and what "Mr. General Aviation" looked like throughout the postwar era.

Despite strict restrictions on pleasure and "nonessential" civilian flying, the bulk of wartime flight training took place outside of the military. Even before the Japanese attacked Pearl Harbor, the Civilian Pilot Training Program (CPTP) offered young Americans an unprecedented opportunity to take to the skies at government expense. Launched in 1939 as part of President Franklin D. Roosevelt's New Deal, the CPTP originally had two distinct purposes: provide a much-needed boost to the civil aviation industry, which had been crippled by the Great Depression; and create a pool of trained aviators for the armed forces in case of war. This second goal dictated who could enroll in the program, which was initially limited to individuals between 18 and 25 years old. As American involvement in World War II seemed increasingly imminent, Congress changed the rules to require all CPTP participants to agree to join the military after they graduated from college. Thus, although this program was technically defined as civilian flight training, it became a de facto extension of the armed forces. By mid-1944, some 435,000 college-age Americans, nearly 99 percent of them young white men, had learned to fly through the CPTP.[9]

Meanwhile, nearly 200,000 American servicemen completed all phases of U.S. Army Air Forces flight training during the war to earn their aviator's wings, and the U.S. Navy produced another 66,000 naval and Marine Corps aviators.[10] Because the army accepted relatively few graduates of the CPTP into its own flight schools, only a small proportion of military aviators learned to fly in this civilian program.[11] As a result, between the CPTP and military flight schools, well over 500,000 (and by some estimates more than 750,000) Americans earned their wings at government expense between 1939 and 1945.[12]

Whereas CPTP cadets received a private pilot's license upon graduating from the program, civilian flying credentials were not automatically granted to military aviators. However, as the war drew to a close, the federal government streamlined the transition from military to civilian aviation in a deliberate effort to give postwar private flying a boost. If a former military pilot received an honorable discharge, passed a civilian flight physical exam, and applied within 12 months of leaving the service, he could obtain a civilian license commensurate with his military flying experience without any further training or flight checks.[13]

Despite the relative ease with which former military aviators could transition to private flying, not all chose to do so. Some simply lacked a lasting interest in aviation. Others, according to Dr. Douglas D. Bond, a psychiatrist who served with the Eighth Air Force in Europe and later published a scholarly book about his wartime observations and experiences, could not disassociate going aloft from the stress and fear of combat. Longtime pilot and aircraft salesman Frank Coviello agreed. Looking back on the previous two decades, Coviello wrote in 1966 that, after returning home from World War II, "the 'tigers' of the Air Corps wanted tranquility. Apparently, flying memories were associated with military 'must go' missions." Other ex-military pilots reportedly found postwar private flying *too* tranquil. One insurance company complained in 1946 about high accident rates because former military aviators demanded too much performance from less robust private aircraft, sometimes with tragic consequences. For these fliers, the relatively slow and docile personal planes that were available for rent or purchase represented a letdown after flying some of the fastest and most powerful warplanes in the world. Finally, like other returning veterans, many ex-military pilots simply had other pressing postwar priorities. Coviello summed this up in 1966 as well as anyone: "Former pilots therefore returned to school under the G.I. Bill, went into business with a G.I. Loan, or both. Each married his sweetheart, bought a cottage and settled down to raise tiger cubs."[14]

Such was the case with Donn Munson. Writing in the early 1960s, the former army liaison pilot recalled, "In 1945, I collected a handful of campaign medals, three fat log books [official records of his flight experience] and an honorable discharge. I promptly quit flying. In the 16 years that passed, I acquired a wife, a daughter, a series of sports-racing cars but absolutely no desire to fly." Jack Wilson's military flying career "came to a halt with a well-placed blast of German flak" in an engine of his B-17 bomber in 1944, and he spent the remainder of the war as a prisoner of war in Europe. As the former wartime pilot explained in 1967, upon returning home at war's end he had "just melted into society" without giving aviation a second thought for years. Instead, he graduated from the Carnegie Institute of Technology's College of Fine Art in Pittsburgh, moved to the greater Chicago area where he found work as a photographer and art director, and fathered six children. But other former military aviators did not "promptly quit flying" or "fade away" as Munson and Wilson did. And they, along with numerous CPTP graduates, helped to swell the ranks of civilian pilots more than 10-fold between 1939 and 1946.[15]

World War II continued to influence the demographics of private flying in America even after the guns fell silent. On June 22, 1944, President Franklin D. Roosevelt signed into law the Servicemen's Readjustment Act. Better known as the GI Bill of Rights, this unprecedented program was designed to help 16 million returning servicemen transition back to civilian life. Benefits included assistance in finding a job; loan guaranties for a home, farm, or business; and money to pay for higher education or vocational training.[16] Although the GI Bill is best remembered for putting a generation of World War II veterans through college and helping them purchase their first homes, it also created an opportunity for many to learn to fly.

Aspiring pilots came from the ranks of the millions of soldiers, sailors, airmen, and marines who had not become aviators as a direct result of America's wartime effort. Some had worked on or around airplanes during the war. All had undoubtedly seen sleek, powerful planes roar through the sky overhead. And every last one had witnessed, through the newspapers and newsreel footage at the very least, the glory and honor lavished upon America's wartime pilots. Upon returning home, government subsidized vocational training meant that they could suddenly stop being mere spectators and become pilots themselves, and many did exactly that. The Veterans Administration estimated that it spent $125 million on flight training for former GIs in 1946 alone, and a study published in 1949 reported, "With the government paying the bills, approximately 350,000 veterans had enrolled in flight schools by November 1, 1947."[17] Several thousand new flight schools opened at airports around the country to take advantage of this business opportunity, up from just 403 at the start of 1946, helping spark a postwar boom in civil aviation.[18]

At least a few aspiring aviators of the immediate postwar era paid for lessons out of their own pockets. Such was the case for middle-aged businessman Harry Burnett and flying grandma Anna Hartman, both of whom were too old to qualify for any form of war-related flight training. Bill Baxter represented the other end of the spectrum. Too young to have served in the war, the teenager followed the classic prewar "airport kid" tradition and worked odd jobs at a nearby airfield during the late 1940s to earn his flight time. But Baxter, Burnett, and Hartman were exceptions to the general rule, and most new pilots of the immediate postwar era found their way into private flying through government programs associated with World War II.[19]

The GI Bill, wartime military flight training programs, and the CPTP would shape the demographics of private flying for decades to come. An estimated two-thirds of civilian fliers were 30 years old or younger in 1946, the same age group that had borne the brunt of military service during the recent hostilities.[20] Although the federal government did not track how many of these newly minted pilots were actually veterans, this strong demographic correlation, combined with the widespread availability of pilot training through the CPTP, military flight schools, and the GI Bill for men in this age group, suggests that most who took up civilian flying in the early postwar era were former servicemen. As they aged, many remained involved in private aviation. In the late 1950s, more than a decade after World War II ended, nearly two-thirds of all active civilian fliers were members of this first cohort of postwar pilots, and they maintained a sizable presence at the airport for years to come (as late as 1987, some 9 percent of all active pilots belonged to this generation, by now in their 60s or older).[21] Several articles published in the mid- to late 1960s commented on this phenomenon, including one that described how many a young wartime pilot who had "turned in his B-4 [military flight] bag for schoolbooks and a baby carriage is [now] back on the airport ramp." The author went on to describe this old-new pilot as a boon for private aviation: "He's 43 . . . and means business."[22] Former army liaison pilot Donn Munson and bomber pilot Jack Wilson fit this profile perfectly: both had quit flying when they left the military at war's end, then reentered aviation as private pilots some two decades later during the mid-1960s.[23]

The same cultural norms and laws that prohibited American women from serving in combat also limited their access to military-related flight training programs during and after World War II. As a result, most members of the first generation of postwar private fliers were men. For instance, reflecting its origins in the New Deal and its stated goal of providing a much-needed boost to the civil aviation market, the CPTP had initially accepted women applicants from colleges and universities at a rate of 1 woman for every 10 men. Despite these limits, nearly 2,500 women earned their pilot's license through this venue between 1939 and 1941. As it became increasingly apparent that America would be drawn into World War II, however, Congress changed the rules to require all CPTP participants to join the military immediately after they graduated from college. Because women were exempt from the draft and could not be otherwise compelled to serve in the U.S. military for any reason (including participating in the CPTP), they were henceforth automatically excluded from the program.[24]

The Women Airforce Service Pilots program, created in 1943 by combining two competing organizations that began the previous year, became without doubt the single most celebrated group of women fliers in U.S. history. Before the program was abruptly canceled in December 1944, slightly more than 1,000 women completed the rigorous flight training that qualified them to fly military aircraft on stateside missions, which in turn freed at least that many male pilots to perform other assignments. Members of the WASP ferried fighters, bombers, and transport aircraft from factories to military bases throughout North America, flight-tested newly delivered airplanes to confirm their airworthiness, and towed the aerial targets that male service members used for gunnery practice. Contrary to a widespread popular belief—inspired no doubt by the fact that these female fliers wore military-style uniforms and flew military aircraft—women serving in the WASP were civilian contractors, not members of the U.S. military (it took an act of Congress in 1977 to finally grant official veteran status, including benefits, to these women). But for a fortunate few—some 25,000 applied, 1,830 were accepted, and only 1,074 completed the training—this program provided young women with an unprecedented opportunity to fly top-of-the-line military aircraft in wartime service for their country. It also demonstrated, at least to some skeptics, that women could fly as well as men. While he never came close to endorsing the idea of sending women pilots into combat, General Henry H. "Hap" Arnold, commanding general of the U.S. Army Air Forces during World War II, declared: "Their very successful record of accomplishment has proved that in any future total effort the nation can count on thousands of its young women to fly any of its aircraft."[25]

Although there is no record of WASP alumnae flying for major U.S. airlines, some of these women went on to become civilian flight instructors or commercial charter pilots after the war. They also served as role models for at least some of those few women who learned to fly in the postwar era. In 1946 *Skylady* magazine announced: "Watch for an announcement about the formation of WISPS. They are the women who have been taught to fly by ex-WASPS [*sic*]."[26] While no such group was formed (or, if it was, it did not last long), more than a half century after World War II ended, young women still reported being inspired to take up flying by the tenacity and successes of their pioneering wartime predecessors.[27] However, with only 1,074 graduates, the WASP program had little direct impact on the demographics of civilian flying. Furthermore, unlike military flight schools, which took men with no aviation experience and turned them into experienced aviators, the WASP program accepted only applicants who already knew how to fly. Thus, even though this all-woman program significantly upgraded the skills of a few existing pilots, it did not, unlike wartime military flight schools or the CPTP, create large numbers of new ones.[28]

While the women who flew for the WASP were not officially members of the U.S. military, more than 350,000 other women did serve in uniform during World War II in a variety of nonflying, noncombatant jobs. As veterans, these women were eligible for the same GI Bill as their male counterparts, and at least a few applied their benefits toward earning a private pilot's license. For instance, *Flying* magazine's 1946 article titled "Flight Training and the G.I. Bill" included a brief photo essay that followed former sailor John Wick and ex-WAVE (Women Accepted for Voluntary Emergency Service) Bernice Conley as they filled out the necessary paperwork at a Veterans Administration office, signed up for lessons at a local airport, and took their first flying lesson. This equal representation of men and women (albeit only one of each) seemed to echo prewar and wartime dreams that the postwar "Air Age" would be open to everyone. However, Bernice Conley was a rare exception in postwar literature produced by, for, and about pilots. Furthermore, since women made up roughly 2 percent of the 16 million returning veterans, even if they took flying lessons at the same rate as their male counterparts (and there is no evidence to indicate that they did), female veterans would have represented a tiny minority among the numerous male student pilots who flocked to local airports during the boom years immediately after 1945.[29]

Despite the many barriers that limited women's access to war-related flight training, there were far more female fliers in the United States at the end of the conflict than before it began. According to one source, the total rose from fewer than 1,000 in 1939 to just over 6,100 by the end of 1945. However, they remained such a small proportion of the aviation community that a visitor to any airport across the country in late 1945 was likely to encounter only 1 female flier for every 50 men.[30]

As with women, World War II provided a few nonwhite minorities with an opportunity to enter aviation. Before the war, African Americans were flatly denied entrance to U.S. military flight training. Under pressure from leaders in the African American community and civil rights activists (including First Lady Eleanor Roosevelt), the CPTP allowed a handful of all-black colleges and universities to participate in the prewar program. Then, in early 1941 the U.S. Army Air Corps established a segregated training program for black pilots, airmen, and ground support personnel in Tuskegee, Alabama, laying the foundations for the group that would later become known as the Tuskegee Airmen. These men overcame significant obstacles, including extreme racism from within the military, hostile treatment by white civilians when they ventured beyond the gates of stateside bases, and even a brazen attempt to sabotage the so-called experiment by a white commander who refused to assign the unit to combat missions, then criticized them in an official report for failing to shoot down as many enemy aircraft as white combat pilots (it took an investigation by the War Department to resolve this issue). The fact that these black aviators not only persevered but excelled under such extreme circumstances—eventually racking up an enviable combat record in the skies over Europe—represented a significant symbolic victory in the long, slow fight by African Americans for equal treatment.[31] Considering that only 269 of the nation's roughly 70,000 licensed civilian pilots in 1940 were black, the total number of African Americans who completed CPTP and/or army flight training during World War II boosted the number of black fliers by approximately 1,000 percent in just five short years.[32]

Yet African Americans remained underrepresented among military and CPTP trainees. According to the 1940 U.S. Census, roughly 10 percent of the American population was black.[33] Had wartime flying opportunities been equal, 1 in 10 CPTP and military flight school candidates should also have been black. The numbers tell a very different story. Only about 2,000 of the CPTP's 435,000 graduates were black, less than half of 1 percent of the total.[34] In addition, just under 1,000 men earned their wings at the segregated U.S. Army training facility in Tuskegee, Alabama, again representing less than half of 1 percent of all military aviators trained throughout the war.[35] Thus, while there were indeed far more black pilots at war's end than in 1940, thanks to wartime policies that severely limited African American access to the cockpit, they made up only a tiny proportion of the pool of young men who swelled the ranks of postwar private aviation.

Unlike black service members, who were relegated to racially segregated units, most of the estimated 250,000 to 500,000 Hispanic Americans who served during World War II were assigned to regular units alongside whites.[36] Although the federal government did not keep records on the number of Hispanic American pilots, anecdotal evidence reveals that at least some Hispanic Americans learned to fly as a result of the war. Don Lopez earned his private license in the CPTP and then went on to complete military flight training and fly fighters with the famed Flying Tigers in China.[37] Eugene A. Valencia Jr., Richard G. Candelaria, and Michael Brezas all became "aces" by shooting down five or more enemy aircraft.[38] Henry "Hank" Cervantes completed pilot training and flew B-17s during the war; Federico Gonzales also became a B-17 bomber pilot and flew combat missions until he was shot down over Europe and taken prisoner. Although Gonzales never flew again after he returned home, both Lopez and Cervantes pursued postwar careers as officers in the U.S. Air Force. But examples like these are rare in the historical record, and Cervantes later recalled that he never met another "Mexican pilot" during a 22-year military career that lasted into the mid-1960s.[39]

Of course, military flight school and the CPTP were not the only pathways into private aviation, and at least some nonwhite veterans used their GI Bill benefits to learn to fly after the war. Two enterprising white pilots, Jim Vercellino and Buell Maddux, set up a successful flight school in Arizona (where Vercellino would later become director of the state's Department of Aeronautics) for black veterans who were turned away elsewhere because of their race. According to another pilot who knew the pair, "It wasn't long before they and the instructors they hired were rolling in dough," so apparently the business thrived, at least during the brief heyday of GI Bill sponsored private pilot training.[40] Edward A. Gibbs, who learned to fly through the CPTP and then served as a civilian instructor at Tuskegee during the war, was later hired by the white owners of a flight school in Wilmington, North Carolina, who hoped that having a black instructor on staff would attract more black veterans—and their green GI Bill dollars—to their airport.[41] Roy Comeaux, a black soldier who took his first unofficial flying lessons at a jungle airstrip while serving in the Philippines (most likely in a small army liaison aircraft such as a militarized Piper J-3 Cub), used his GI Bill benefits in 1947 to earn his private license at Harlem Airfield, a traditionally black airport on Chicago's south side.[42] But with few civilian flight schools open to African Americans and far fewer career opportunities for black pilots than for their white counterparts, the possibility of learning to fly through the GI Bill did not lead to a large influx of black pilots.

Unfortunately, neither the Civil Aeronautics Administration (CAA) nor its successor, the Federal Aviation Administration (FAA), compiled information about the race or ethnicity of postwar civilian pilots for the voluminous *Statistical Handbook of Aviation*, published annually starting in 1944. Instead, the U.S. Census Bureau provides the only reliable data on this topic. According to reports on race and occupation compiled every 10 years as part of the nationwide census, black pilots made up just a tiny percentage of those who flew for a living throughout the postwar period. For instance, in 1950, out of a total of 16,650 individuals employed as professional pilots or navigators, only 60, or 0.4 percent, were black.

Because the airlines barred African Americans from the cockpit at the time, these 60 individuals, all of them men, must have flown for the military or else in lower-status, lower-paying civilian jobs such as flight instructing or crop dusting.[43]

Starting in the early 1970s, a few years after the airlines began hiring black pilots at the height of the civil rights movement, aviation magazines occasionally showed a black pilot to celebrate or encourage diversity. For instance, one 1970 advertisement claimed "Cessna's new program of flight training starts with a new kind of instructor." The accompanying color photograph included four men clad in khaki slacks and blue blazers and two women dressed in matching miniskirt-length blue dresses, all of them with a Cessna Pilot Center patch prominently displayed on their chests. One of the men is black, another appears to be of Hispanic or South Asian descent. Cessna's implied message is that its "new kind of instructor" was a departure from a norm in which *all* of the instructors would have been white men. Ironically, the impact of this progressive message is somewhat undermined by the fact that the photo focuses on a middle-aged white male instructor kneeling in the foreground, with all of his more diverse—and junior, in terms of age—colleagues leaning, slightly out of focus, against a Cessna airplane in the background.[44]

Other articles and comments by readers suggest a less rosy reality regarding race and aviation. For example, after *Flying* magazine described the persistence of racial prejudice in aviation in a 1969 article titled "Can a Black Man Fly?" several readers immediately wrote angry letters to the editor to cancel their subscriptions, indicating that not all white pilots were ready to welcome minorities at "their" airports.[45] Even in 1990, by which time the major airlines had been hiring black pilots for more than a quarter century, only 1,886 (1.8 percent) out of 109,826 individuals employed as pilots were black, and other nonwhite minorities were also seriously underrepresented. Unfortunately, figures from the U.S. Census describe only the number of pilots who flew for a living. Still, even if the proportion of African Americans in the ranks of private aviation was several times higher than that of pilots who flew professionally, it never remotely approached proportional representation compared to the nation's racial demographics.[46]

Although difficult to prove conclusively, the continued absence of nonwhite minorities in commercial or private aviation can likely be traced in part to the scarcity of such pilots during and immediately after World War II. The term *microaggressions*, coined by Harvard-trained psychiatrist Chester Middlebrook Pierce in 1970 and still widely used in the field of social psychology, describes "brief and commonplace daily verbal, behavioral, or environmental indignities, whether intentional or unintentional, that communicate hostile, derogatory, or negative racial slights and insults toward people of color."[47] Expanded definitions of this term indicate that simply showing up at an airport populated almost exclusively by white, middle-class, male pilots conveys a powerful message to anyone who does not fit this demographic profile: "You do not belong." Put another way, a lack of same-race role models creates a self-reinforcing feedback loop: because members of the nonwhite public did not see many nonwhite pilots, it was more difficult for them to imagine themselves becoming a pilot than, say, for a white boy or young man who could reasonably expect one day to grow up to join the ranks of the white, middle-class, middle-aged men who by 1969 were the face of Mr. General Aviation.

Anecdotal evidence suggests not only that nonwhite fliers were rare throughout the postwar era but also that they faced constant reminders that they represented individual exceptions to a general rule about who was supposed to fly. Former air force pilot David Harris broke the color barrier to the airline cockpit in 1964 when he became the first black pilot hired by American Airlines. When an interviewer asked Harris in 1969 whether his military flight instructors had ever treated him differently in the late 1950s because he was the only African American in his class, he responded, "No, I don't think so. But this was on my back all the time. I was constantly thinking: now am I going to have to perform better than my white counterparts?" Furthermore, the specter of racism haunted Harris even when he wasn't actually experiencing it in the cockpit. For instance, as an airline pilot in the 1960s and early 1970s he often had difficulty securing overnight lodging near airports because of his race.[48]

And Harris was by no means alone. In the late 1960s, Civil Air Patrol (CAP) member Roosevelt Lewis took a Beech T-34 military trainer owned by his CAP unit out for a routine flight. After landing, he noticed a commotion on the far side of the airport fence; schoolchildren on a field trip to the airport had rushed over to watch the plane taxi to its parking place. As Lewis, wearing aviator's sunglasses and a flight suit, climbed out of the cockpit and ambled over to the fence to answer questions, he heard several of kids, apparently unable to believe that a black man could actually fly, call out in genuine curiosity, "Hey Mister, where's the pilot?"[49]

Jesse Lee Brown, who decided to learn to fly after he returned home from military service in Vietnam, recalled similar experiences in Alabama in the early 1970s. Years later he remembered, "I went out one day to get in the airplane. I was a student . . . I was going to take a cross country [flight]." As he inspected the plane, he noticed some white spectators—obviously nonpilots—approach the airport fence to watch. "This man and his family, they came out to the fence. I had laid my books, charts and things inside [the airplane, out of sight] and I was getting ready to do a preflight." As Brown untied the ropes securing the plane, he overheard the man tell his family, "That boy's probably fixin' to clean that one up, I guess." But then, as Brown continued to ready the plane for flight, he recalled with a chuckle, "One of the kids said, 'I think he's gonna fly it!' " Brown was known as a "rare bird" among white pilots at his home airport in Muscle Shoals because of his race, but he described how they were curious and helpful, not "hateful." This wasn't always the case when he flew to distant locations. "You'd get some funny looks" after landing, he remembered. And it wasn't always funny. At one stop in Tennessee, he recalled, "I had one guy that told me I wasn't supposed to be inside the [airport] fence. It was just his assumption [on the basis of Brown's race]. . . . It was just a way of life back then. So at that point I got in the airplane and cranked it up, and I waved at him as I taxied out [for takeoff]."[50] For Harris, Lewis, Brown, and other nonwhite pilots, interactions like these—where fliers and nonfliers alike simply assumed that a black man wasn't a pilot because he was black—represent the very kinds of microaggressions that continued to reinforce a long-standing tradition that white men, and *only* white men, belonged in the cockpit.[51]

While World War II did relatively little to open the ranks of private aviation to women or nonwhite minorities, it did somewhat erode traditional class barriers. Before the war, private flying was widely viewed as a rich man's pastime, but the CPTP opened the airport gates to college students, who, in the era before the GI Bill and other forms of government financial aid, came predominantly from middle-class families. The armed forces drew from an even broader spectrum of society, and more than a few military-trained pilots were from poor, working-class, or rural farming families, including Charles Elwood "Chuck" Yeager (who would go on to postwar fame as the first pilot to break the sound barrier).[52] Like military flight training, the GI Bill was also essentially blind to socioeconomic background. In fact, this particular avenue into flying may have actually favored individuals with blue-collar aspirations. Not only did the GI Bill make learning to fly an option for veterans who lacked the money to foot the bill themselves, but since its educational benefits had definite limits, many who chose flight training probably had no intention of also attending college. As a result, not everyone hanging out at the local airport immediately after the war aspired to become a white-collar professional. Thus, in 1950 most civilian fliers were young white men, but this crowd was still more diverse, at least in terms of class, than was the comfortably well-off middle- to upper-middle class community of private pilots described by AOPA's Mr. General Aviation of 1969.

Boom, Bust, and Recovery

Between 1939 and 1951, the number of active U.S. civilian fliers, not including students, rose from about 34,000 to more than 580,000. But not everyone who took to America's skies immediately after World War II continued to fly, and private aviation's postwar boom went bust. After leveling off briefly in 1952, the population of active pilots plummeted dramatically over the next four years. According to government statistics, more than half of the pilots on the books in 1951 had stopped flying by 1956, which also meant that there were fewer active aviators that year than there had been a decade earlier in 1946.[53] Two factors account for this dramatic downturn: postwar private aviation failed to deliver as promised to consumers; and the federal government severely curtailed its flight training programs, forcing most aspiring pilots to pay their own way.

Wartime marketing campaigns by aircraft manufactures had promised Americans inexpensive, easy-to-fly aircraft like Cessna's "Family Car of the Air" once the war was over. But the reality of postwar private flying fell far short of these optimistic predictions. A government report published in 1951 concluded that impracticality, high operating costs, and "poor safety record" inhibited the "widespread acceptance of the light plane by the general public."[54] Although this specific report focused on why so many people never bothered to learn to fly, the same conclusions also applied to the ranks of disillusioned fliers who earned their pilot's license with the intention of using an airplane like an automobile, then gave up on private aviation soon after they discovered a much different reality. Instead of an affordable, easy-to-use, and practical means of transportation, many former fliers complained that personal planes were cramped, slow, and expensive, not to mention complicated, dangerous, and easily grounded by weather conditions that would not even slow highway traffic. In short, postwar personal planes were nothing at all like the long anticipated Family Car of the Air.[55]

While some pilots gave up on private flying out of disappointment, others simply ran out of money. By 1951 the so-called baby boom was well underway. That year, more than three-quarters of all World War II veterans were married, and over half of them had one or more children. Young pilots who took to the skies immediately after the war began starting families and taking out mortgages and car loans. Thus, many who were genuinely enthusiastic about private flying stopped going to the airport because they lacked the disposable income to support this expensive pastime.[56]

Other new pilots were never all that committed to private aviation in the first place. Commenting in the early 1950s on the precipitous decline in active fliers, *AOPA Pilot*'s editor declared: "Pilot licenses . . . have shaken down to the actual users. The hangers-on, the disinterested GI's, have been dropping off."[57] The editor was clearly trying to make sense of why the postwar private aviation boom had suddenly gone bust, and the GI Bill seemed an easy target. More than a decade later, the AOPA still seemed surprised when it reported that "316,000 enthusiastic men . . . trained for and got their pilot licenses under the GI Bill—and, shortly afterward, gave up flying!"[58] Jim Chanault, a former bomber pilot who worked as an instructor at a civilian flight school in York, Nebraska, immediately after the war, recalled years later in an interview, "We got a lot of students from the GI Bill of Rights. Guys that had gone through the military and always wanted to fly. They could go and receive flight training [at government expense] through places like our base there." When asked what these new fliers did after their benefits ran out, Chanault declared, "Nothing. After they were done, why, for the most part, why, that's about all they ever flew."[59]

As time ran out on GI Bill benefits for World War II veterans, so too did the steady stream of new students who had sustained flight schools through the end of the 1940s. Subsequent versions of the GI Bill for veterans who served during the postwar era were written so that these benefits no longer provided easy (or at least cheap) access to flight training. In 1948, in response to criticisms that the GI Bill flight training program was "of questionable benefit to the trainees and of more than questionable benefit to the nation as a whole," the U.S. Department of Veterans Affairs redefined private pilot training as a "recreational" instead of a "vocational" endeavor. Under this new definition, advanced flight training that could lead directly to a career in aviation, such as obtaining an instructor's rating or a commercial pilot's license, remained eligible for government subsidies. But there was a catch: before veterans could qualify for advanced flight training, they first had to earn their private pilot's license at their own expense.[60] As a result, after its heyday in the mid- to late 1940s, the GI Bill never again provided an easy pathway to allow legions of veterans to take to the sky.

The federal government also cut other types of flight training soon after World War II and created no equivalent programs to replace them. For instance, despite impassioned arguments in favor of continuing the CPTP, the program was completely dismantled by mid-1946.[61] Military flight training was curtailed as well. In the 1950s, an era that included the Korean War, the U.S. Air Force produced a grand total of 41,000 new pilots. During the next decade, even with the demands of the Vietnam War, only about half as many men earned their silver air force wings. And in the relatively lean years that followed, this branch of the military seldom trained more than 2,000 new pilots per year. All told, between 1946 and 1985 the U.S. Air Force graduated fewer than 106,000 military aviators, barely more than half the number who earned their wings in the U.S. Army Air Forces between 1939 and 1945.[62] The U.S. Navy, which was responsible for training all aviators for both the U.S. Navy and Marine Corps, produced about 68,000 new fliers in the postwar era, roughly the same number it had trained during World War II.[63] The U.S. Air Force Flight Indoctrination Program (FIP) served as another potential avenue into postwar private flying, but it too had limited overall effect. Introduced in 1956 and loosely modeled on the CPTP, the FIP provided Air Force Reserve Officers' Training Corps (ROTC) cadets with free civilian flight training leading to a private license while they were still in college. However, the FIP was slow to get going, and in its first 10 years it produced only 14,000 new private pilots.[64] Even though some military aviators of the postwar era took up private flying for off-duty recreation or after they left the service, their statistical impact on the civilian pilot population had to be relatively small. Not including helicopter pilots, the military produced in all fewer than 200,000 new fliers between 1946 and 1985; meanwhile, during that same 40-year period there were almost always at least 100,000 (and sometimes more than 200,000) active civilian student pilots learning to fly each and every year.[65] Thus, compared to either the World War II era or self-funded civilian flight training in the decades that followed, the postwar military cannot be considered a major source of new private pilots.

Short of joining the military, the Civil Air Patrol provided another potential venue for learning to fly. Formed during World War II to help patrol the U.S. coastline for enemy submarines, the CAP continued into the postwar era as a civilian auxiliary of the U.S. Air Force. In this capacity, CAP volunteers conducted search and rescue operations when private aircraft were reported overdue or missing. At the height of the Cold War, they also trained to provide civil defense services in the event of natural disasters or national emergency, including a nuclear war. Members who already held a pilot's license flew CAP aircraft at reduced rates or even for free during search missions (both training and real), which often totaled more than 20,000 hours per year, making private aviation more affordable for participants. The CAP also exposed thousands of teenage cadets to aviation each year with the explicit goal of inspiring some to take the next step to become pilots themselves, so this organization played an indirect role in shaping the future demographics of private flying.[66]

In 1965 the CAP also began offering flight instruction as an incentive to its top cadets. But unlike the prewar and wartime CPTP, the postwar CAP never subsidized large numbers of new pilots, and free training was hard to come by. In the first year, the CAP selected 12 cadets from across the nation to receive their private pilot's license in airplanes at the first "cadet flying encampment" in Elmira, New York. Another 16 pursued their private license in gliders, and 64 more completed a far shorter flight orientation course in gliders. A few years later in 1971, when asked about learning to fly at government expense, Brent R. Frodsham, a teenaged CAP cadet squadron commander in California, replied, "The chances are really kind of slim." That year, out of nearly 33,000 teenage cadets nationwide, only 94 young men and women earned their wings using free flight training through the CAP. Another 250 were scheduled to solo in a matching funds program, with the National Headquarters paying one-third of the bill, the cadet's local squadron providing another third of the funding, and the cadet paying the remaining third out of pocket. Although this subsidy helped these teenagers to start flying, it ended when they soloed, which represents only about a quarter of the cost of earning a private license. Over the next several decades, the CAP funded just a handful of flight training scholarships each year—generally around 100—for its tens of thousands of teenage cadets nationwide.[67]

The government's decision to discontinue training private fliers through programs like the CPTP and GI Bill, combined with a significant reduction in military flight training, meant that after 1950, nearly all new pilots had to pay for their own flight instruction. This change directly contributed to the sharp downturn in the number of active fliers during the postwar bust. But private aviation was far from finished. After bottoming out in 1956, the pilot population rebounded in the late 1950s, plateaued for a few years, and then took off again in the mid-1960s. For the next two decades, the number of active pilots fluctuated in a manner that followed broader social and economic developments in America. For instance, when the first members of the baby boom generation reached adulthood in the mid-1960s, they swelled the ranks of student pilots considerably. When "stagflation" eroded disposable income and the oil crisis of the early 1970s drove up the cost of flying, the number of active pilots declined. As the economy recovered and fuel prices stabilized, the number of pilots rose again. Still, after the sudden boom in the late 1940s, even in the best of times, growth in private aviation barely kept pace with the rising American population.[68]

After reaching an all-time high in 1980, pilot numbers began to decline as product liability lawsuits drove up insurance rates and the price of new planes. Concerned about the future of both private and commercial aviation, the AOPA began advocating for a government-sponsored National Pilot Training program. Modeled loosely after the wartime CPTP, AOPA intended that the NPT program would provide numerous scholarships for civilian student pilots. Ultimately the initiative failed to gain congressional approval, and the burden of paying for learning to fly remained squarely on the shoulders (or pocketbook) of the aspiring aviator.[69] Throughout these ups and downs, one thing remained constant: cost acted as a significant gatekeeper to private flying, influencing the number of people flying in any given year, as well as who those fliers were.

Only for the Rich?

In 1966 M. J. Walker published a "partly factual, almost true but tongue-in-cheek" account describing the process of learning to fly. It began with a hopeful young man approaching the manager or chief flight instructor at the local airport to ask, "Do you give flying lessons here? Would it cost much?" "Forget it," the instructor replied as he pushed this potential customer out the door and locked it behind him. Walker went on to describe the reason for this brusque response. "After coming back for three weeks in a row I finally got him to talk to me. It seems he was offended by my asking about the cost. 'It's like a boat,' he said. 'If you gotta ask about the price forget it.'"[70]

Throughout the postwar era, seasoned fliers routinely insisted that rising costs were driving them out of private aviation and keeping a younger generation of Americans away from the airport. Longtime pilot Arnold Winter complained bitterly of "sky-high" prices and ruefully declared that soon he would have to "dismantle the plane, put it in the garage and have it as a keepsake, so that I may remember the good old days when a private individual could afford to own an airplane."[71] Winter was writing in 1958. Decades later, similar complaints echoed in the pages of aviation magazines and presumably off the walls of the hangars wherever private pilots gathered to commiserate.[72] When *Flying* magazine's editor Richard L. Collins dared to declare in early 1981 that "the cost of airplanes is no higher now, in relative terms, than when [Cessna 172] Skyhawks and [Piper] Warriors were selling like hotcakes" in the late 1960s, an incensed Danny S. Wright wrote in a letter to the editor that he was "absolutely shocked" by this assertion and sternly suggested, "Maybe Collins needs to take a much closer look at the grass roots of flying."[73]

At first glance, these complaints about the skyrocketing costs of private flying seem indisputable. For instance, when a young navy officer named Les Greene earned his private license in 1950, it set him back about $500. By the time Arnold Winter voiced his complaint about "sky-high" prices eight years later, the price had risen by more than 25 percent to just over $630. And when Danny Wright penned his letter to Collins in 1981, earning a private pilot's license cost around $2,000, considerably more than what Greene had paid in 1950. Taken at face value, this $1,500 difference represents a 300 percent increase over the course of three decades. However, adjusting these figures for inflation, as Collins had done for the price of new planes in his 1981 editorial, suggests that out-of-pocket costs to student pilots had in fact risen by only 5 percent in 1958, and a mere 8 percent by 1981, when compared to the "good old days" of 1950. Thus, Collins was right, and Wright was wrong: the cost of flying had actually remained remarkably stable throughout most of the postwar era. Still, whether he realized it or not, Wright was something of a prophet; by the early 1980s, the cost of flight instruction was beginning to rise faster than inflation. For instance, in 1984, more than three decades after former navy officer Les Greene learned to fly, a recent graduate of Greene's high school aviation science class paid $2,500 to become a private pilot. Other figures suggest that most student pilots probably paid even more, indicating that the inflation-adjusted cost of learning to fly was 22 percent higher in 1984 than it had been in 1950, a significant jump in just a few years following decades of stable prices (fig. 1.2).[74]

This is not to argue that flying was easily affordable. Despite nostalgic claims regarding the "good old days" of cheap (or at least more affordable) private aviation, this hobby was expensive right from the start. Regardless of whether the year was 1950, 1985, or anytime in between, the cost of learning to fly was equivalent to roughly 10 percent of the average American family's annual pretax income. Put another way, an aspiring pilot had to come up with about one-seventh to one-eighth the price of a new four-door Cadillac sedan to earn a private license.[75] If this individual hoped to use a personal plane for long-distance travel, earning an instrument rating in addition to the private license was absolutely necessary to avoid getting stuck on the ground whenever visibility dropped and the clouds closed in. Since an instrument rating cost at least as much as obtaining a private pilot's license throughout the postwar era, this effectively doubled the price of admission into the community of pilots.[76] With or without an instrument rating, learning to fly represented just the tip of the financial iceberg.

Once a student completed his or her initial flight training and received a pilot's license, it required even more money to continue flying. During and immediately after World War II, advocates of an "airplane in every garage" predicted a not-too-distant day when new designs would make flying as easy as driving and the miracles of mass production would bring the cost of airplanes in line with automobiles. In the decades that followed, manufacturers managed to build air-

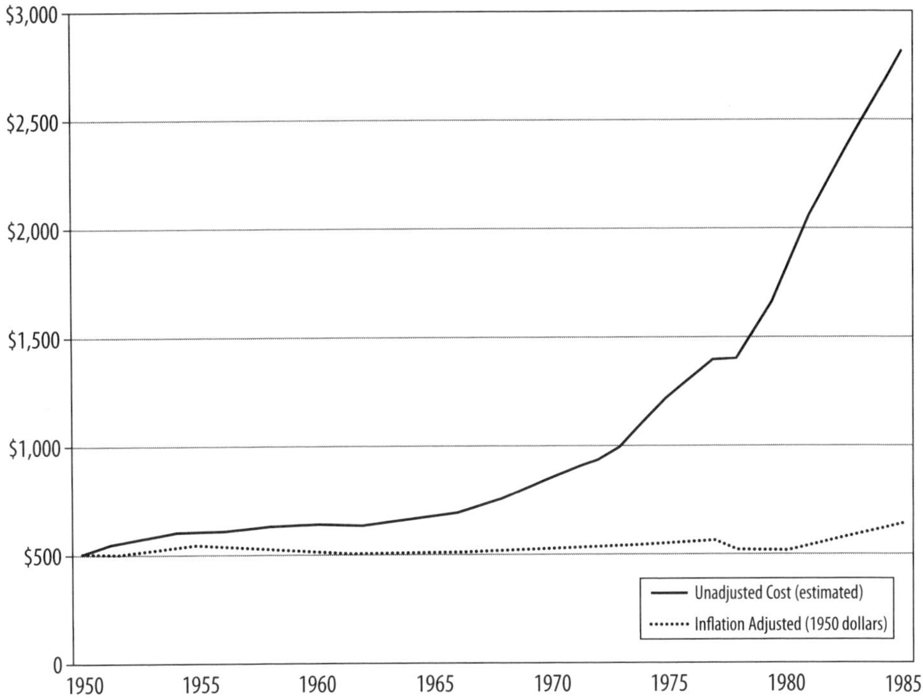

Fig. 1.2. Cost of learning to fly, 1950–1985. Although pilots routinely complained that rising costs were scaring away prospective students, adjusting for inflation reveals that the cost of earning a private pilot's license remained relatively stable for decades throughout the postwar era. Data from multiple sources (see main text and endnotes); prices adjusted for inflation to 1950 dollars using U.S. Department of Labor, Bureau of Labor Statistics, "CPI Inflation Calculator." Graph by author.

craft that delivered better performance, were somewhat more comfortable, and were at least marginally safer and easier to operate than the first postwar planes. However, not only did these newer designs fail to make flying "as easy as driving," but they were also prohibitively expensive to own, operate, and maintain for most Americans. Briefly analyzing the prices of three popular models of personal planes, each targeting a different sector of the private aviation market, illustrates this point.

While it never managed to produce its much-vaunted Family Car of the Air, Cessna consistently dominated the postwar market for two-seat, entry-level training aircraft as well as four-seat "easy fliers" that could carry a small family on a cross-country trip. These planes were extremely popular, too. During the first decade after World War II, the company produced some 7,600 of its model 120/140 two-seat aircraft and 5,100 of its larger model 170 four-seat planes. After weathering the postwar boom and bust for private aviation, in the mid-1950s Cessna upgraded these designs by replacing the original conventional (tailwheel) landing gear with more forgiving tricycle landing gear. The resulting four-seat model 172 still holds the record as "the most widely produced light aircraft in history," with some 35,000 built during a three-decade production run that ended in 1986. The smaller two-seat Cessna 150 (introduced in 1959) and its replacement, the 152 (introduced in 1978), came in close behind with combined sales of nearly 30,000. If one adds the production numbers for the predecessor models 120/140 and 170, Cessna sold roughly 40,000 entry-level four-seat planes and 37,000 two-seaters between 1946 and 1986.[77]

Among high-performance, single-engine personal planes, the Beechcraft model 35 Bonanza stands out for both its longevity and popularity within the private aviation market. First flown in late 1945, the Bonanza—with its impressive payload and speed, as well as its sleek design and distinctive V-shaped "butterfly" tail—quickly emerged as the standard against which consumers compared and other manufacturers competed. Writing in the mid-1950s, Frank Kingston Smith summed up the Bonanza's reputation by comparing it to a familiar postwar cultural icon: "What Cadillac has stood for, since World War II, in the automobile industry, the Beechcraft Bonanza has represented as a personal plane; it has 'class.'"[78] Beechcraft sold more than 10,000 model 35 V-tail Bonanzas, a remarkable number compared to most of its competitors in this high-end segment of the private aviation market, with the last one delivered to a retail customer in 1984.[79]

Because of their combined popularity and longevity, the two-seat Cessna models 120/140 and 150/152, the four-seat Cessna models 170 and 172, and the Beechcraft model 35 Bonanza provide useful benchmarks for charting the cost of buying a new personal plane throughout the postwar era. Like the cost of learning to fly, the price of a new plane remained relatively stable from the mid-1950s through 1980, just as Collins had claimed in his March 1981 editorial. In real terms, the base price for a four-seat Cessna rose by nearly 230 percent between 1955 and 1980, from $8,295 to $27,250. Once adjusted for inflation, however, this represents a far more reasonable 6.6 percent increase. But things were starting to change even as Collins published his 1981 editorial. In 1985 the AOPA's Technical Policy and Plans staff reported that new plane prices had increased at twice the rate of inflation in recent years, a trend confirmed by comparing actual aircraft prices of the era. That year, the inflation-adjusted price for a four-seat Cessna was nearly 50 percent higher than in 1955, a huge jump from just five years earlier, when the difference between 1955 and 1980 had been a mere 6.6 percent. And potential buyers seeking a new Beechcraft Bonanza faced an even great shock: even after adjusting for inflation, the plane cost more than twice as much in 1985 than it had 30 years earlier.[80] Manufacturers blamed higher costs on record lawsuit payouts to victims of personal plane accidents, and industry analysts and pilots alike began to wonder if the future of private flying might be at risk.[81]

But just how expensive was a new plane? Extending Frank Kingston Smith's analogy between airplanes and luxury cars one step further provides a useful perspective on what these figures meant to American consumers. Up through 1980, a two-seat Cessna with limited baggage capacity cost 30 to 50 percent more than a brand-new four-door Cadillac sedan. During the same period, a four-seat easy flier like the Cessna 172 cost about twice as much as the same car. Not surprisingly, planes featuring more seats, retractable landing gear, variable-pitch propellers, and more powerful engines commanded even higher prices. For instance, through 1980, a new model 35 Bonanza cost as much as five or six new Cadillacs. By 1985, the year that Cessna announced it would stop producing entry-level planes, a basic two-seater cost as much as two new Cadillacs and a four-seat easy flier cost nearly as much as three such cars. Price hikes for Bonanzas were even more dramatic, so that by 1985 one could fill a driveway with no less than 10 brand-new Cadillacs for the same price as just one of these high-performance personal planes. In fact, based on the median value of a U.S. single-family home, a new 1985 Bonanza cost nearly three times more than the typical American house.[82]

With new aircraft so expensive, most private pilots chose (or were forced) to find other ways to continue flying after they earned their license. Some purchased used planes, and it became common practice in the postwar era to continue flying carefully maintained aircraft that were 10, 20, even 30 or more years old, not, like automobiles, as "classics" or "clunkers," but as perfectly respectable alternatives to owning a newer model.[83] Prices varied widely depending on age, condition, type of radios and instruments installed, and—most importantly—how many hours of flight time since the engine was new or had been overhauled. For example, in 1965, when a new Cessna 150 two-seat trainer had a base price of $7,825 (and an average delivery price of $9,525), the first April issue of *Trade-a-Plane* listed 37 used Cessna 150s for sale nationwide ranging from $4,150 for an older model with a midlife rebuilt engine, up to $8,850 for a nicely equipped, nearly new airplane with relatively low flight time. If buyers were willing to forgo the handling ease of tricycle landing gear, they could instead purchase one of 33 1940s-era tailwheel-equipped Cessna model 120 and 140 aircraft on the market that month. With sellers asking between $2,000 and $4,850 (the latter with a newly rebuilt engine), these forerunners of the Cessna 150 cost roughly one-fifth to one-half the delivery price of a brand new plane. Listings for used Cessna 172s and Beechcraft Bonanzas in the same issue reveal a similar variation in prices.[84]

For owners of new and used planes alike, fuel, oil, and routine maintenance accounted for only a fraction of the cost of flying. Fixed expenses—including insurance, hangar rent or outdoor parking fees, and government-mandated annual inspections—were substantial, and they cost the same whether the aircraft sat idle or flew several hundred hours each year. This meant that in order to keep hourly costs down, pilots had to fly a lot. Throughout the postwar era, aviation publications consistently estimated that a pilot had to fly between 100 and 200 hours annually (sometimes more) in order to reach the "break-even point" where it made more economic sense to own rather than rent.[85] Since it took most individuals about 50 hours of flight time to earn their private license, airplane owners could easily spend far more money every year to remain active in private aviation than the already significant amount they had paid up-front to learn how to fly.

In 1961 pilot and professional aircraft mechanic J. L. Howard complained, "Flying has always been considered a rich man's plaything." He went on to lament, "The sad, sad fact is that it takes an income of about $10,000 a year to keep even the smallest airplane in the style to which it should be accustomed. The hangar rent alone in this area is equal to or exceeds what it costs to keep a roof over your own head."[86] At a time when the median annual household income in America was $5,620, fewer than 20 percent of households would have met Howard's minimum threshold for airplane ownership. By 1970, the median annual income had risen to $9,867, but the cost of flying had gone up as well.[87] That year, Archie Trammell boasted how his Cessna 172 cost him only $0.14 per mile (roughly $17.50 per hour) to operate, "*including* taxes, storage and interest on the invested money." Considering that a pilot could expect to pay $22.00 per hour (25 percent more) to rent the same airplane, Trammell's experience represents a considerable owner-versus-renter savings. Then again, he was not the typical private pilot. "In order to get it down to 14 cents . . . I fly the airplane almost 5,000 miles (40 hours) each month. There aren't many people who either need or can afford that much transportation." For most owners, he estimated, "In average recreation use (figure 180 hours per year utility), the cost would rocket to probably $35 per hour."[88]

To put Trammell's figures in perspective, even if the typical pilot earned far more than the minimum wage of $1.45 per hour in 1970, hourly rates of $22.00 for the typical renter (or $35.00 for an owner who didn't fly enough to make ownership cost effective) represented a considerable financial burden. In order to achieve his enviously low hourly rate, Trammell flew his plane nearly 500 hours annually at a total cost of nearly $8,400. Few Americans could afford to spend the $2,200 required to fly a rental aircraft for 100 hours per year (about a quarter of the medium family income), much less the considerably higher amount required to keep "even the smallest airplane in the style to which it should be accustomed." Trammell himself admitted that flying could be prohibitively expensive. Regarding his estimate of $35 an hour for owners who didn't fly as often as he did, he joked, "A man could barely spend that much for recreation in a red-light district."[89]

Most pilots rented rather than owned the planes that they used, but even then private aviation was costly. If pilots averaged 1 hour of flight time per week, or about 50 hours per year, they had to pay just as much to continue flying as they had spent to obtain a private license in the first place. And to some experts, this was barely enough flight time to maintain proficiency. In the early 1970s, for instance, Sky Roamers, a large Los Angeles–based flying club with 300 members and 36 airplanes, required all members to fly at least 40 hours per year and pass periodic check rides with club instructors.[90] Paul Poberezny, founder and president of the Experimental Aircraft Association (EAA), lamented in a 1984 editorial, "We know that the average private pilot, if flying at all today, averages little more than 30–40 hours per year. This means that the low proficiency of the individual who many times operates in adverse weather conditions beyond his or her experience, can lead to various types of accidents."[91] Just a few years earlier, Walter J. Boyne had advised prospective pilots, "Newly certified pilots, flying at the rate of 40 to 50 hours per year, lose proficiency faster than they gain it; worse, they may not be aware of it and may venture into situations that they can't handle." An enthusiastic booster of private aviation, Boyne nonetheless argued, "After you receive your private pilot certificate, you should decide to either pursue flying seriously . . . or you should give it up." As he explained, "Pursuing flying seriously means flying 100 to 200 hours a year, at a minimum."[92] Thus, unlike the one-time, up-front expense of learning to fly, renting (or, for an owner, keeping an airplane "in the style to which it should be accustomed") was a recurring expense that continued year after year for as long as a pilot continued to fly.

In a 1981 editorial titled "The Price Is Flight," Berl Brechner, executive editor of *Flying* magazine, tried to put the high cost of private aviation in perspective.

"There's an unfortunate reality all of us who love aviation must confront one day. Flying's not cheap and it never has been. But somehow, most people who have a real desire to fly find a way to manage. We might cut back on other interests, or buy a car that lacks flash, or put off some home remodeling projects."[93] Brechner's 1981 contention that even relatively well-off Americans faced tough financial choices if they wanted to keep flying are supported by statements made by pilots throughout the postwar era. In 1950 Howard F. Haines of Columbus, Ohio, notified the AOPA that he had "gone back to sailing instead of flying, much as I loved it," largely for reasons of cost.[94] One year later, Leonard J. Redding, another ex-flier, reported: "I am no longer an aircraft owner. I have sold out and am now a sports car owner."[95] Not all such stories were concessions of defeat. In the mid-1960s, one prospective pilot enthused, "I'm going to give up a cottage at the beach this summer in order to get my private certificate."[96] In the early 1970s, George H. Moloney Jr., a successful middle-aged businessman, recommended that aviation-minded individuals should consider buying a year-old car instead of a new one next time they went to the dealership, then use the savings to pay for flight instruction.[97] And few years later, George E. Burns, who identified himself as "a weekend flier who supports his habit by working for a bank," addressed the issue of cost in an article featured in the Sunday edition of the *New York Times*: "Private flying is not a nickel-and-dime hobby, but a lot of people routinely spend enough each year on golf, tennis, the theater—not to mention a European tour—to do a lot of flying."[98]

The potential sacrifices that Brechner, Moloney, and Burns suggested might help fund an aviation "habit" serve as a reminder of the relatively affluent lifestyle that many pilots enjoyed. Individuals who did not have the option of giving up their European tour, sailboat, summer beach cottage, or even a new car in the driveway every few years had to dig even deeper to pay for this expensive pastime. Vincent Gallagher, a Massachusetts police officer, wrote in 1964 that the only reason he could afford to fly was because his wife worked as a secretary.[99] Charlie and Audrey Thomas of Southern California, both private pilots, found themselves in a similar situation. Although Charlie presumably made a decent living from his longtime job as a cost analysis engineer for Union Carbide Corporation, Audrey described in a 1968 interview how she had started working at the Post Office during the Christmas rush 17 years earlier and now "can't afford to quit." She went on to explain, "If I were to quit working at the Post Office, that would chop off half of our flying."[100]

Audrey Thomas (and perhaps Mrs. Gallagher as well) was willing to give up being a housewife to help fund the family flying. Not every pilot was so lucky, and more than a few wives bristled at the sacrifices that their mates seemed willing to make in the name of this expensive hobby. Such was the attitude displayed by a pilot who, in a 1960 article titled "No Nylons for Mama," cavalierly described his penchant for expensive airplanes at the price of anything else his wife might desire. Appended to the end of the piece, the editor reported that the author had requested: "If you decide to use this article, please keep it anonymous. The last article I wrote for a magazine nearly got me a divorce." The editor concurred, commenting: "Far be it from us to get a fellow in more trouble with his wife when he is trying to explain to her why she has to give up nylons in the interest of horsepower." Obviously intended to be humorous, "No Nylons for Mama" poked fun at what many pilots and industry observers increasingly viewed as a very real problem in private aviation—the nonpilot wife.[101]

By the early 1960s, two widely recognized stereotypes of the pilot's spouse had emerged: the pro-aviation wife who supported her husband's aeronautical activities; and her polar opposite, whom one author described in 1963 as "private aviation's No. 1 problem—the anti-flying wife."[102] According to numerous articles and letters to the editor published in aviation magazines throughout the decade, the antiflying wife was the bane of many male pilots and a significant impediment to the private aviation industry as a whole. Not all articles were as blatant as one published in 1960 titled "Beware the Skirt Barrier" or the 1969 piece subtitled "How to Handle a Woman When She Won't Fly."[103] Authors also managed to comment on the antiflying-wife phenomenon when discussing topics that at first glance seem completely unrelated, such as a 1965 feature on homebuilt aircraft that mentioned the "domestic drag from wives who do not see the wonders and wisdoms of flying homebuilts."[104] All of these accounts acknowledged, sometimes implicitly but more often explicitly, that middle-class wives had a vote over how their husbands spent both their leisure time and the family's discretionary budget.

While there was nothing inherently gendered about an unenthusiastic spouse objecting to a partner's expensive hobby, the demographics of—and gendered assumptions about—private aviation are evident on the rare occasions that a woman pilot complained about her nonflying husband. In response to "How to Handle a Woman When She Won't Fly," Mrs. R. K. Fleming of Cypress, California, wrote the editor of *Flying* in 1969: "I regard non-flying wives with little or no sympathy. When may we expect your in-depth study of a real mind-boggler: the non-flying husband?" Either a pilot or at least an aviation enthusiast herself, she continued, "I'll trade in a two-year subscription to *Flying* for one 'pussycat' trick I haven't already tried." The fact that her husband's aversion to aviation was "a real mind-boggler" indicates that she accepted without question the idea that private flying was universally considered a masculine endeavor. The editor's response also acknowledged that this case ran completely contrary to gender norms regarding both private aviation and the American household, and he mocked Mr. Fleming—who was apparently not man enough to fly with his wife (much less become a pilot himself)—by advising Mrs. Fleming to "try cutting off his allowance" to force him to let her pursue her expensive hobby.[105]

In 1975 *AOPA Pilot* published an editorial titled "Only for the Rich?" that attempted to turn this rhetorical question around and argue that private flying was still dominated by the middle-class "little guy with the little plane, which has minimal equipment." In response, several despairing "little guy" pilots penned letters to the editor to lament that, actually, flying really was "only for the rich."[106] Many Americans simply could not afford the initial price of admission into the community of pilots, much less the ongoing costs of maintaining their membership by continuing to fly regularly. This goes far in explaining why AOPA's Mr. General Aviation came from a solidly middle- to upper-middle-class background. It also helps explain why after the dust settled from private aviation's boom-and-bust episode immediately after World War II, the majority of active private pilots could be described as middle-aged, since this group was the most likely to have the income needed to support an aviation "habit." However, young men were not entirely absent from the flying field. Quite the contrary: by the mid-1950s, another demographic trend had emerged that would remain true throughout the rest of the postwar era. Although most active pilots with at least a private license were age 35 or older (and, by 1963, over 40), most student pilots were younger than 30.[107]

Paul Poberezny, who founded the EAA in 1953 to represent private pilots who wished to build and fly their own planes, reflected more than five decades later in 2004 on how this demographic disparity developed in postwar private aviation. After observing that most pilots learned to fly while they were young and had few personal or fiscal responsibilities, Poberezny hypothesized that the financial burdens associated with pursuing a higher education, buying a car and house, and raising a family forced many to give up private aviation. As a result, he observed, "a great many people walk around with a pilot's certificate in their wallet and don't fly anymore. Why do you think that is? Cost." Many might never fly again, but, he concluded, at least some would return to the airport later in life when they were better established, swelling the ranks of middle-aged pilots.[108] Poberezny was hardly the first to reach this conclusion. In 1947 T. P. Wright, chief administrator of the CAA, had warned of a significant age gap between younger Americans who wanted to fly but could not afford to, and older generations who could actually afford to fly but were not particularly interested.[109] Thus, even after the government stopped footing the bill for flight instruction, learning to fly remained a young man's game throughout the postwar era, while the community of pilots actively engaged in private aviation became dominated by middle-aged men.

There is another potential explanation for the continued high proportion of youthful student pilots. Not everyone who learned to fly in the postwar era intended to use a personal plane for pleasure or business travel. Except for the relatively few aviators who earned their wings in the military after World War II, the first step on the long, expensive road to finding a high-paying, high-status flying job was to earn a private license. However, many aspiring airline pilots never achieved their final goal. Lee L. Schiek wrote in 1968 to complain that he, like so many others at the time, had been taken in by "the pilot shortage myth" allegedly propagated by flight schools seeking to fleece students of their money. "Unfortunately I realized this after spending my life savings for those hard-won ratings." As a result, Schiek successfully learned to fly an airplane, but he couldn't land a job with the airlines.[110] W. R. Caswell bemoaned spending a fortune on 500 hours of flight time for a son who was still considered underqualified by airlines that demanded more than 1,000 hours minimum experience.[111] Yet another pilot lamented in a letter to the editor, "I also was taken in by the pilot shortage myth put out by most flying schools." For this individual, obtaining a private and then commercial license to fly both airplanes and helicopters proved insufficient to secure employment, leading him to conclude: "Several thousand dollars and no job later, I am very disillusioned about the whole business."[112] One lucky fellow warned would-be airline pilots that "life was a series of miseries compounded with grief, dipped in heartbreak" until he finally succeeded in landing an airline job after 114 letters, 36 applications, and three years of persistent effort.[113] These and similar letters published in aviation magazines during the 1960s and 1970s suggest that many young student pilots who originally hoped to fly for a living had their dreams dashed by the harsh realities of the job market. But even if they gave up flying to pursue another career, some of these young pilots almost certainly followed the trajectory described by Paul Poberezny and returned to the airport later in life to fly for fun, helping to explain the disproportionate majority of youthful student pilots in a community that was otherwise dominated by middle-aged men.

Women faced even greater financial obstacles to becoming pilots than men did. Aside from Mrs. R. K. Fleming and her "non-flying husband," the historical record is mostly mute about whether, and how often, nonflying husbands reined in their wives' flying activities. For single women, the problem was more clearly defined. Working women typically earned less than their male counterparts throughout the postwar era, making it more difficult to afford something as costly as private flying. And since the airlines refused to hire women pilots until the mid-1970s, flight instruction was a waste of money for most career-minded young women for the first three decades after World War II.

It is thus tempting to conclude that most pilots were (and still are) men simply because wealthy males enjoyed the easiest access to flight training. Certainly this was a contributing factor. Immediately after World War II, the three most common pathways into the cockpit were tied to who had access to war-related flight training. Because America's social and cultural norms had barred women from combat and severely restricted access to flight training for African American males, young white men were the primary beneficiaries of the CPTP, military flight schools, and postwar GI Bill. These young men, just returned from the battlefields of Europe and the Pacific, set the cultural ground rules for the burgeoning postwar community of pilots.

As the burden of paying for flight lessons shifted from the federal government to the student pilot's pocketbook, white men from middle-class backgrounds continued to enjoy an advantage in gaining and maintaining access to what remained a prohibitively expensive pastime for much of the population. Meanwhile, a legacy of formalized discrimination against women in the workplace meant that most students who took up private flying in hopes of someday landing a career in aviation were men. After 1950, the number of new faces entering the community of pilots failed to keep pace with the expanding national population. As a result, the initial group of pilots who earned their wings thanks to the war remained a significant statistical presence at the airport for decades to come. Combined, these practical factors explain why Mr. General Aviation looked the way he did throughout the entire postwar era, be it the 1940s, 1960s, or even the 1980s.

But the fact that most pilots were men did not automatically guarantee that postwar private flying would be imbued with a culture of masculinity. This initial cohort of young, white, male war veterans constructed and perpetuated a community that valued private flying as much more than a mode of transportation, a means to get a bird's-eye view of the world, or a way to pass an otherwise idle Sunday afternoon. These fliers celebrated the mastery of a complex and potentially dangerous technology that their untrained neighbors and co-workers could not hope to operate without killing themselves. Hardly reckless daredevils in the classic sense, many pilots nonetheless believed that they could overcome risk and adversity with a combination of steady nerves and carefully honed skills. And they enforced an informal yet surprisingly strict standard of conduct among members of their community. Evidence of these ideals and behaviors shows up in nearly every aspect of private flying, from the planes that pilots chose to use, to the ways that they used these planes, to how they spent their time on the ground. And induction into this masculine culture began with the process of learning to fly.

Shouting, Shirttails, and Spins

Flight Instruction and the Acculturation of New Pilots

At first glance, Frank Kingston Smith and Diane Ackerman seem more different than alike. Smith, a World War II veteran, up-and-coming Philadelphia attorney, devoted husband and father of three, could have posed as the model for the AOPA's "Mr. General Aviation." Ackerman, a baby boomer born shortly after the war (which made her young enough to be one of Smith's children), had earned a masters of fine arts in creative writing, a Ph.D. in English from Cornell University, and was in the early stages of what would become a lifelong literary and academic career. But the lawyer and the poet did have a few things in common. Both became private pilots during the postwar era, albeit some 25 years apart. And unlike most fliers, who tended to earn their licenses in their 20s, Smith and Ackerman took up aviation in their mid-30s. Finally, both published books about their experiences soon after they completed their training, providing an unusual degree of detail about the process of learning to fly from the student's perspective.[1]

Smith, who started flying in 1955, described himself as initially reluctant to leave the ground and turned to aviation only after golf and gardening failed to calm his overtaxed nerves. After his first few lessons, Smith not only fell in love with flying but also (at least by his own account) managed to master the necessary skills without undue difficulty. In sharp contrast, Ackerman confessed that she had dreamed of flying since childhood. But when she finally began taking lessons in 1980, she was constantly frustrated by aviation's single most difficult challenge: learning to land. Their writing styles are as different as their backgrounds, ages, and experiences in the cockpit. Smith's book is lively, filled with self-effacing humor, and consciously crafted to "sell" private flying (and his book) to the average middle-class American male. And his message is unambiguous: *If I can do it, so can you.* Ackerman's work, on the other hand, is an introspective journey of self-discovery, a study in contrasts between the freedom and beauty she found while soaring above the earth and the months of exhausting effort she spent battling the plane, her instructor, and especially her own demons and self-doubts. Unlike Smith's decidedly lighthearted book, Ackerman's autobiographical poetry-in-prose explores not just flight, but the very meaning of life itself.[2]

Although they learned to fly a quarter of a century apart, and although their personal journeys from fledgling to flier were quite different, the process by which Smith and Ackerman became pilots was essentially the same. This striking similarity hints at a broader historical fact: basic flight training remained fundamentally unchanged throughout the entire postwar era. Different flight schools and independent instructors organized their syllabi into any number of stages or lessons, but all ultimately followed the same four-phase approach that was formalized before World War II: pre-solo lessons with an instructor, the first solo flight without an instructor, post-solo instruction and practice flights to hone skills, and the final practical flight test with a government-appointed examiner.[3] Most of the fundamental skills required for handling an airplane remained the same as well, as evidenced by the continued popularity (and periodic reprinting) of Wolfgang Langewiesche's highly readable 1944 manual, *Stick and Rudder: An Explanation of the Art of Flying.*[4]

Smith's and Ackerman's accounts reveal something else about learning to fly that is absent from practical-minded manuals like Langewiesche's *Stick and Rudder.* There was far more to "the art of flying" than simply signing up for flying lessons, learning to take off and land, and mastering the other skills necessary to get an airplane safely from point A to point B. Instead, becoming a pilot also involved demonstrating a checklist of character traits that could have come straight from the introduction of sociologist Patricia Cayo Sexton's 1969 book about men in America: "courage, inner direction, certain forms of aggression, autonomy, mastery, technological skill, group solidarity, adventure, and a considerable amount of toughness in mind and body."[5] The process of proving oneself worthy of entrance into the community of pilots took place both in the cockpit and on the ground, and served as a gatekeeper that helped shape the culture and demographics of private flying long after the guns of World War II fell silent.

Rites of Passage

Regardless of whether a student enrolled in a formal civilian flight school or hired a freelance instructor who gave lessons out of a hangar, the technical content of flight instruction was remarkably similar. This standardization was necessary, in part, to ensure that pilots could safely operate aircraft that were considerably more complicated than the average family car. According to a 1966 government study, the owner's manual for a typical single-engine personal plane listed "20 different pre-flight categories of items to check out, 5 specific starting maneuvers, 10 ground check operations, 10 pre-take-off checks, 7 climb operations, 10 categories of en route operations, 7 pre-landing operations, 3 landing operations, and 8 shut-down categories of items." Furthermore, it observed, "This bare minimum of 80 essential items, all of which must be individually ascertained to ensure a successful flight accomplished with a reasonable level of safety, makes an interesting comparison with the 10 items listed in current automobile owner's manuals."[6]

Aspiring aviators had to learn not only how to operate a complex machine but also how to use it safely and legally within a larger technological system that included commercial airlines, military aircraft, and other private pilots, not to mention airports, air traffic controllers, and an invisible but highly structured three-dimensional network of airspace and airways. Compared to the ground, the sky was an inherently hostile environment. If the driver of an automobile suffered a mechanical breakdown, got lost, or simply became overwhelmed by the demands of driving, he or she could simply pull over to the side of the road. Not so in the air, where even getting lost could lead to running out of gas and a deadly crash. Thus, student pilots had to learn procedures to avoid these situations and practice how to handle them should they occur. Most postwar aircraft accident reports listed "pilot error" as a contributing factor, an ambiguous term that one experienced aviation psychiatrist succinctly defined in 1974 as "failure of the pilot to cope with the demands of flight."[7] In order to simultaneously convey the necessary aeronautical skills and teach students to successfully "cope with the demands of flight," flight instruction evolved as an orderly and methodical process based on lessons learned while training hundreds of thousands of military and civilian pilots during World War II, then systematized by strict government regulations and standardized lesson plans throughout the postwar era.

But there were also other, less formal aspects of flight instruction that shaped the student's entry into private flying. After his first easygoing introductory lessons, Frank Kingston Smith discovered another side to his instructor's personality. "Now, when I made a mistake, he let me have it with both barrels," the lawyer-turned-student-pilot recalled, then added, "Most of the time it seemed he yelled at me *before* I made a mistake, which is even faster than my wife operates, after years of practice." For Smith, profanity-laced tongue lashings became a regular part of learning to land, and Bob, his instructor, was merciless. " 'Watch your airspeed, goddamnit!' he would snarl, and I would move the wheel gently to keep the airspeed at 70. . . . 'You're too high!' he would scold halfway to the runway. The engine would roar and around I would go [after aborting the landing] to make a correct approach. . . . 'Come *on*, Frank,' he would say, 'Fly it *right*!' And he would mutter to himself, fold his arms across his chest and glower out the window. It made me feel terrible."[8]

As his training continued, Smith described an emerging pattern wherein Bob at times acted disgusted or angry at his hopelessly inept student, but then would relax and even offer casual words of praise for a job well done. Whenever Bob felt that his student was getting a little too cocky, he introduced an exercise or maneuver that put the novice squarely back in his place. Smith not only accepted this treatment as a normal part of learning to fly; he appreciated it as well, recalling later: "To this day I can still hear his yell, 'Watch your airspeed,' when I cut the gun [idle the engine] to make a landing."[9] Bob's approach to flight instruction, and the relationship that developed between the two men, may have made Smith "feel terrible" at times, but it motivated the student to work harder to please someone he respected, and it made the instructor's occasional praise seem all the more meaningful.

Not surprisingly, this approach to flight instruction did not work so well on everyone. Diane Ackerman described learning to fly as a series of frustrating battles with her first instructor, a young man named Brad. According to Ackerman, Brad believed that all pilots should be "efficient, cool, stoic, strategic, regular guys, no namby-pambies, who imposed their will on lesser mortals, and knew that, even though the meek might inherit the earth, the strong would inherit the meek." No matter how hard she tried, Ackerman could never conform to this rigid definition of who a pilot should be, and her autobiography serves as a detailed case study of how a poor student-instructor relationship could potentially drive an aspiring flier away from the airport forever.[10]

Ackerman's poignant account of one in-flight confrontation with Brad is particularly revealing. "'You take all the fun out of flying,' I had said to him recently, after a particularly savage outburst at me, an outburst that ridiculed my spontaneous delight with the way you could make the wing tip hold a spot on the ground like a turning knife point. 'Fun?' he had sneered. 'FUN? Flying isn't supposed to be fun. It's *dangerous*.' Then he went on to blister me up and down about how badly I was flying, how stupid I was, what a lousy memory I had, how daft, how thick I was." Given Ackerman's description of the exchange, the results should come as no surprise. "The more he screamed at me, the worse I flew, and the louder and more brutally he screamed, until finally I crumbled up against the window, crying, my spirit like a small, thin bone, broken; and I let him fly the plane back to the airport, deftly, as I knew I never could."[11]

A quarter century after Frank Kinston Smith learned to fly, the basic message—"Come *on*. . . . Fly it *right*!"—remained the same, as did the language and manner in which it was delivered. After one lesson, Ackerman's instructor told her: "I can't have you flying the plane joking around, thinking how poetic the sky is. . . . Your attitude has to be more serious, more rigorous." In the air, Brad's haranguing never let up as Ackerman fumbled through one maneuver after another. "Why can't you do it? You know what you're supposed to do! You know you're not doing it! Why do you keep making the same mistake? Why don't you think for once! You're not thinking! Do it right this time, damn it!" Embittered, worn down by constant criticism, and starting to believe that she simply lacked whatever it took to become a pilot, Ackerman seriously considered giving up altogether on her lifelong dream of learning to fly.[12] She was hardly alone. One aviation psychiatrist noted in 1972 that "approximately 50% of the people who apply for student pilot certificates give it up before completing the program" and attributed this high attrition rate in no small part to poor student-instructor relationships.[13] Only after Ackerman moved away for a new job, forcing her to find another flight school to continue her training, did she discover that not every flight instructor approached teaching the way that Brad did.

Just as Smith and Ackerman were very different, so too were their instructors. Bob was middle-aged and roughly the same age as Smith. He may well have learned to fly (or worked as an instructor) during World War II and seemed content to make flight instruction his peacetime career. Twenty-five years after Bob taught Smith to fly, Brad was in his 20s, easily a decade younger than Ackerman. Like most pilots his age, he had learned to fly as a civilian, and according to Ackerman, he viewed his current job not as a calling or career but as a stepping-stone to a more lucrative and prestigious position with the airlines. Smith portrayed Bob's abuse as a form of tough love administered in carefully measured doses for maximum effect. Ackerman saw no rhyme or reason in Brad's harangues beyond the fact that she clearly failed to measure up to his version of who a pilot ought to be. But despite these differences, both instructors—and thousands more like them throughout the postwar era—were united by their tough, no-nonsense approach. They modeled themselves after a highly masculinized type of mentor epitomized by military drill instructors and high school football coaches. From their perspective, flying was a demanding and potentially dangerous endeavor, so if a student pilot couldn't take a little heat from his (or her) instructor, then that student had no business behind the controls of an airplane.

Regardless of whether or not this approach to flight instruction had merit, it tended to weed out those who could not—or would not—conform to the masculine image for fliers as defined by the existing community of pilots. Furthermore, the example set by individual flight instructors perpetuated these expectations for decades, as former students trained in this manner went on to enforce the same norms on subsequent generations of aspiring aviators. Postwar flight instructors who had learned to fly during World War II likely took this approach in part because that was how their own instructors had taught them, which in turn almost certainly reflected how *those* instructors originally learned to fly. A joint wartime study by the University of Pennsylvania and Purdue University, commissioned by the CPTP in an attempt to standardize training methods, suggested that the roots of this problem lay in the long-standing assumption that good pilots automatically made good teachers. As a result, according to a 1943 report published by the CAA, "flight instruction . . . remained, for the most part, in the hands of persons *with interest and experience in flying but little or no experience with methods of training.*"[14] The military recognized this problem as well. For instance, a 1944 U.S. Army Air Forces training manual for newly minted instructors (often themselves just out of flight school) listed 20 "Do's and Dont's [*sic*] for Instructors." Rule number three, "Don't be impatient and don't get angry," was accompanied by a cautionary cartoon drawing of an instructor screaming at a mortified student (fig. 2.1).[15]

But even as the military ostensibly discouraged this harsh approach to flight instruction, it also reinforced it, at least indirectly, by emphasizing the masculine nature of flying at every turn. Throughout the war, both the army and navy consciously recruited aviation cadets on the basis of not only stringent physical standards but also psychological makeup. Interviewers asked prospective fliers which parent they favored. Boys who enjoyed hunting and fishing with their fathers were preferred; not surprisingly, there was no room for a "mama's boy" in the ranks of combat pilots. High school or college athletes were considered a good bet, as were young men who liked motorcycles, so applicants improved their chances if they knew, for instance, that the throttle was located on the right handlebar. Bookish, introspective types were unlikely to make the cut. According to the "Naval Aviation Cadet Survey" dated August 1942, respondents' favorite sport was football, and—despite the fact that most had completed at least two years of college—their favorite reading material included Superman comics.[16]

Fig. 2.1. Typical flight instructor–student relationship circa World War II. Accompanying the image was the following advice for newly minted military flight instructors: "Don't be impatient and don't get angry." Contrary to the admonition, this technique was often the norm both during and after World War II. Army Air Forces Training Command, *Instructors' Manual: Basic Flying* (U.S. Army, ca. 1944), 10. Courtesy of the Smithsonian Institution Libraries, Washington, D.C.

There is a certain logic to the premise that combat pilots should be aggressive and that slow learners wouldn't survive long once the shooting started. But many wartime instructors were convinced that harsh treatment during training created better pilots even in peacetime when enemy planes weren't an issue. Such was case with John R. Hoyt. During the war, Hoyt, a U.S. Naval Reserve officer, published a *Manual for Aviation Cadets* meant for a military audience. Then in 1945 he wrote an article for *Flying* magazine directed at civilians who were considering becoming private pilots. His title speaks volumes: "When You Learn to Fly—Get a Tough Instructor." Hoyt opened by blaming a recent rash of avoidable accidents by new pilots on their flight instructors, who he concluded had failed to enforce strict standards during training. He compared such instructors to public school teachers who allowed pupils to "graduate from our school systems who cannot add, multiply, spell or write an elementary essay" because those educators "cling to the theory of the second chance." According to Hoyt, flying was "like any competitive sport, from football to skeet. When you let down, you lose—only losing in the flying game is likely to be pretty permanent." As a result, his ideal flight instructor was the tough, no-nonsense sports coach who wasn't afraid to use a few "red hot remarks" to correct a student's "unairworthiness," not the "the 'soft,' smooth, flattering type" who would grant one more chance after another before ultimately graduating an individual who wasn't fit to fly. He warned, "There are plenty of examples to show that soft instruction fails to produce safe, competent pilots." For this reason, he advised would-be fliers, "It is better to bump into a few harsh words than into the side of a mountain."[17]

Hoyt clarified that "tough instruction does not involve brutality. It certainly does not mean that an instructor should talk too much, too violently, use profanity toward the student or attempt to haze him." Yet a few paragraphs later he related a story that reveals he endorsed inflicting more than just "a few harsh words" to make a point. According to Hoyt, before takeoff, an unnamed instructor noticed that his student forgot to fasten his seatbelt. Once in the air, the instructor deliberately nosed the airplane over into a dive, rendering both occupants momentarily weightless. Unsecured by a seatbelt, "The astonished and frightened student rose two feet out of the cockpit, turning a dull green. Scared? Indeed he was, but he never forgot that lesson nor did he forget the vitriolic lecture he received from the instructor, who took him right back to the field and threatened to wash him out if the incident occurred again."[18] Apparently because there was an important lesson to be learned, this dramatic and potentially dangerous maneuver did not qualify as "hazing" (in a worst-case scenario, the student could have been thrown from the open cockpit plane; fortunately, military fliers wore parachutes, so if he kept his wits about him and pulled the ripcord, he would have at least had a second chance at survival). Hoyt's message was clear: if students couldn't take a little criticism (and maybe a little physical knocking about as well), then they didn't belong in the sky.

Two decades later in the mid-1960s, Dick Weaver, chief pilot for Valley Pilots at Van Nuys Airport in Southern California, agreed with Hoyt that there was nothing wrong with yelling at fledgling fliers. With more than 20,000 hours in the air and 22 years in the military behind him, he proudly declared, "next to a shooting war, I like flight instruction best." Weaver explained that he wore military flight suits when he taught because "I'm wringing wet by sundown" as a result of in-flight confrontations in the cockpit. He also admitted: "Once I kept a punching bag in my garage and I'd paste [the] hell out of it after a day's instructing, just to get the frustrations out of my system. Now, I've bought a P-51 [World War II surplus fighter plane] and, when things get rough, I'll take it out over the desert and go through a full set of aerobatics. Then I can come back and 'be nice' for a while."[19] After likening flight instruction to doing battle with the enemy and gleefully admitting that he still needed a physical outlet for his frustrations after spending all day sweating through his flight suit and yelling at his students, an individual could hardly send a stronger message that flying was an inherently masculine endeavor.

Weaver took pride in his reputation as "the toughest flight instructor in the area," but others questioned whether he was confusing "tough" with senseless (or even damaging) bullying.[20] In response to the magazine article that spotlighted Dick Weaver as a role model for the industry, part-time flight instructor George Glacius declared in a letter to the editor that "this character who enjoys war more than he does flying, and who works himself into such a snit while instructing that he has to wreak his vengeance on a punching bag, should be called on by [the] FAA to pass some sort of psychiatric examination before he is allowed to teach his ways to unsuspecting students."[21] From this comment, it seems that Glacius was not just concerned about the individual student's experience; he was worried as well that those not driven out of flying would accept Weaver's approach as the norm and pass it along to future generations of pilots.

Some instructors openly opposed abusive teaching tactics as counterproductive to training smart, confident pilots. William D. Strohmeier, who published "When You Teach Flying, Understand the Student" in the same 1945 issue of *Flying* that carried Hoyt's advice to "Get a Tough Instructor," advocated a less combative approach. Strohmeier, a civilian instructor with some 3,500 hours flight experience who taught military aviation cadets during the war, hardly qualified as "soft." Yet he insisted that the key to successful teaching involved more empathy than aggression. He observed that "there seems to be the tendency among flying mentors automatically to bestow on any student the stigma of ignorance. This impression is due to the failure of instructors to put themselves in the student's place. They fail to realize that much of this new game is a bewildering complication of procedures which only time and patience can master." After describing how most of his contemporaries believed that yelling and bullying was the best way to get through to their students, Strohmeier lamented, "It has ruined a lot of good pilot material." He went on to explain, "There are many students who react naturally against rough treatment. I'm that type of person myself—and most of the students I've had in the past two years are the same way. They give me much better results when I work with them [rather] than against them."[22]

By the 1980s, few who wrote about flying in books, articles, or even letters to the editor openly advocated the abusive teaching techniques of the past. However, these same sources indicate that at least some believed that such practices remained commonplace. Nearly four decades after Strohmeier penned his negative assessment of fellow instructors, legendary aerobatic pilot and air show performer Duane Cole mused, "There have always been flight instructors who have shouted and cursed and jerked the controls away from students. Such bullying contributes nothing. On the contrary, it confuses and angers students to the point of impairing their ability to comprehend." Cole observed that although "some students can cope with this sort of abuse," he warned that "many give up flying shortly after beginning because of it." He went on to lament, "What a waste. Just a little patience and some encouraging words will keep the sensitive flying and help the slower learner over the rough spots." To Cole, unlike at least some of his counterparts, being "sensitive" or a "slower learner" did not mean that a student automatically deserved to be weeded out.[23]

Others were more optimistic that times had changed. Writing in 1981, aviation columnist and longtime pilot Len Morgan noted that "browbeating too often passed for instructing in the bad old days" but concluded that the field was now populated by "a better breed of instructor."[24] Likewise, in his 1980 book for aspiring aviators, Walter Boyne declared that the era of abusive instructors was long over. Boyne, a retired air force pilot and soon-to-be director of the Smithsonian Institution's National Air and Space Museum, explained that instructors in years past received little training in how to actually teach others to fly. "To hide this deficiency, it became common practice for instructors to bully, berate, and even haze students, criticizing them continuously, banging their knees black and blue with the [control] stick, abusing them . . . and in general behaving in an immature, beastly fashion." But just as Morgan was convinced that "the bad old days" were a thing of the past, Boyne declared with confidence, "Mercifully, most of this went out the window after about 1960, when it finally penetrated the GI [military] psychology that browbeating a student was counterproductive and that intelligent, human-relations-oriented instruction was needed." He concluded brightly, "As a result, both civilian and military flight training offer an essentially pleasant, thrilling experience. If at any time in your training you don't find it to be so, you should immediately look around for another flight instructor."[25] Diane Ackerman, who was busy battling her first instructor, Brad, around the same time that Boyne's book went to press, would have welcomed this last piece of advice.

Although Brad had no firsthand experience in the military, he seemed intent on emulating traditions championed by earlier generations of wartime fliers like Hoyt and Weaver. The persistence of this tough-man approach to indoctrinating new pilots is even more remarkable given the ever increasing attention paid to positive student-instructor relationships since the Army's 1944 manual had simply warned, "Don't be impatient and don't get angry." For instance, the 1956 edition of the federal government's official *Flight Instructors' Handbook* for civilian pilots advised, "The judicious use of praise regularly is the easiest and most potent incentive available to the instructor. Sharp reproof and threats of unhappy consequences are effective for only very short periods, and should be avoided by all competent instructors."[26] The newly revised 1964 edition put even more emphasis on the subject. It cautioned, "Under no circumstances can the professional instructor do anything which implies that he is degrading the student. Acceptance rather than ridicule, and support rather than reproof will encourage learning." Along with ridicule and reproof, the new manual also discouraged harsh words. "In flight instruction, as in other professional activities, the use of profanity and obscene language leads to distrust, or at best a lack of confidence. To many people, such language is actually objectionable to the point of being painful."[27]

As with any advice book or how-to handbook, in describing how instructors *ought* to treat their students, these manuals tacitly acknowledged that many students continued to encounter instructors intent on training civilian pilots using the same harsh techniques used to train pilots for combat during World War II. In 1958 John Pennington wrote an article that began: "If the veterans [by which he clearly meant experienced pilots, not specifically former military fliers] will put up with a bit of impertinence from a rookie pilot, there are a few words that need saying about the modus operandi of some flight instructors." Over the next several pages, the self-described rookie proceeded to take flight instructors, as a group, to task for the various ways that they mistreated their students. George Moloney Jr., who learned to fly more than a decade after Pennington, submitted his own article in the early 1970s with a similar theme.[28] Unlike Diane Ackerman—whose tears and inability to stand up to harsh criticism in the cockpit proved to Brad (and nearly convinced her as well) that she was one of those weak "namby-pambies" who didn't deserve to become a pilot—Pennington and Moloney had stoically endured the indignities of flight instruction like "regular guys" and waited to vent their frustrations only after they had successfully completed their training.[29]

Diane Ackerman was hardly the only woman who found the highly masculinized approach to flight instruction daunting. As noted in the introduction, British aviator Claude Grahame-White declared in 1911, "I have taught many women to fly and I regret it. My experience has taught me that the air is no place for a woman."[30] More than a half a century later, some members of the flying fraternity still seemed determined to keep it that way. At least a few instructors of the immediate postwar era refused to take women students seriously, essentially stonewalling them in a war of attrition apparently in hopes of forcing their undeserving students to give up. For instance, Marion Hart wrote *Air Facts* magazine in 1947 to complain about how she had been passed from one disinterested instructor to another (humorously identified in her article only as Tom, Dick, and Harry). The first was content to ride along (and collect his hourly fee) but taught her nothing as she meandered aimlessly about the sky for lesson after lesson. The second "climbed into the plane with the dejected air of someone who expects only the worst" and constantly badgered her with "'Do this . . . do that . . . don't do this . . . don't do that,' but never a whisper about *why*" until he finally gave up in disgust. Her third instructor was "patient, gentle . . . and inarticulate," and Hart found it impossible to pry any technical information or advice from him. She reported that she finally soloed only when he missed a lesson and a newly hired fourth instructor allowed her to fly on her own (a subsequent article reported that Mrs. Hart had at last earned her private license and purchased her own plane).[31] Other instructors went out of their way to scare off women with what can only be described as deliberate hazing. Such was the case in the mid-1950s when an instructor told retired schoolteacher Nina Stoehr to strap on a parachute before her second lesson. He then proceeded to inflict a series of aerial stunts on his hapless student that had nothing to do with the standard pre-solo curriculum. "That was fun," she recalled grimly. Although Stoehr eventually earned her private license, not surprisingly, she did so with a different instructor.[32] Even when instructors did not intentionally single out women for special hazing, the brusque and even abusive manner that was for so long a staple tactic for all students, regardless of gender, reportedly drove away many women before they completed their training.[33]

Flight instructors were not the only people who shaped a student's introduction to and expectations of private flying. Groups of local pilots—usually men, given their overwhelming statistical preponderance in aviation—often gathered at small and medium-sized municipal airports during their free time. Known as "airport bums" or the "fence crowd" for their habit of lining up along the airport fence to critique other pilots' takeoffs and landings, these aviators played an integral part in educating and indoctrinating students into their elite company. Writing to prospective fliers, Walter Boyne waxed eloquently on this aspect of aviation and enthusiastically predicted that "you'd be surprised at how hospitable airports are if you have the courtesy to drop into the flight operations shack [airport office] and tell them you'd like to look around."[34] But not everyone shared this appraisal. When George Moloney, a successful businessman in his late 40s who had served as a navy officer during World War II, showed up at a local airfield to inquire about taking lessons, he felt like an unwelcome outsider. "Several seedy characters were deeply involved in telling what we used to call 'sea stories' and I suddenly felt that I should be back in the office or someplace other than this place. Pure stubbornness took over and I apologetically interrupted the 'flat-on-my-back-at-40,000' drill, inquiring as to whom I contacted for the introductory flight." Moloney, clearly used to giving the orders by this point in his life, was even less impressed with the storyteller's response to his intrusion. " 'Charlie, give this guy a ride around the patch [airport] in Six Zero November [the airplane's registration number], willya? That'll be five bucks, Mister.' (It was all said in the same breath—like a tower controller at O'Hare [International Airport] at 5 o'clock on a Friday afternoon.)" According to Moloney, " 'Wings' quickly got back into his 'one engine out and a load of ice' story with this other character who was waiting impatiently for him to finish so he could tell about *his* feat—real or imagined—of derring-do."[35]

Boyne and Moloney had very different impressions of airport culture, but both could agree on at least one thing: the way veterans treated newcomers had a significant impact on shaping a new pilot's experiences and expectations of flying. It is entirely possible that Moloney received this callous brushoff because, in his obvious impatience with the storytellers, he failed to follow what Boyne characterized as "the elaborate code of courtesy and conduct of the flying fraternity."[36] It is also possible that the pilots at this particular airport were less welcoming than the norm (Moloney ended up learning to fly somewhere else, so apparently not everyone he met triggered such a negative reaction). Diane Ackerman recorded similar impressions about the overall atmosphere at the first airport where she took flying lessons. Even more telling is what happened when Gordon Baxter, an experienced pilot, prolific aviation writer, and (by his own admission) no namby-pamby, pretended in 1980 to be a nonflier at various airports across the country while working on an article about how well (or poorly) airports treated prospective students. Much to his surprise, few made him feel welcome, and some were downright rude. Invariably, he had to persevere before anyone behind the counter would take him seriously. "Although I sometimes got a very good sales pitch after I broke the ice, my overall impression as an outsider walking onto the flight line is much the same as walking uninvited into a private club." After recounting several examples of this kind of reception, he ruefully admitted, "The painful part of this story deals with that segment of aviation I have always instinctively aligned myself with in the past—the little guy, the romantic grassroots fliers out there in their peaceful meadows. . . . I was unaware of what snobs we are."[37]

The "snobs" of the fence crowd enforced an unwritten code of conduct for pilots and augmented students' formal flight instruction. Boyne advised: "If you are eager, courteous, and obviously ready to learn, you'll be treated like a professional." However, if a student failed to act properly either in the air or on the ground, he cautioned that "you'll find the temperature dropping on all sides."[38] Boyne listed a few of the "rules" that defined acting "properly" but made it clear that no one was likely to gently take the student by the hand to show him or her the ropes. Instead, it was mostly up to newcomers take the initiative to glean this information on their own.

Hangar flying represented one venue where students could learn both technical skills and the pilot code of conduct. Students who hung around the airport in their free time could sit in on "hangar flying" sessions, listening to experienced pilots swap stories about past aerial adventures. If newcomers had the good sense to listen, they could pick up tricks of the trade and learn about potential pitfalls to avoid. Pilots used first- and secondhand accounts of near disasters to reinforce aphorisms like "the two most useless things a pilot can have are runway behind him and altitude above him" in the event of an engine failure shortly after takeoff, and "there are old pilots and there are bold pilots, but there are no old, bold pilots."[39] Students could also hear informal, firsthand assessments of various kinds of airplanes. For instance, in the late 1950s, novice Robert Shaw found other pilots' descriptions of Luscombe aircraft as "tricky [to fly], especially in a crosswind landing," at once "intriguing, challenging, and sobering." He finally decided to buy a used one after hearing someone say that "when you're flying a Luscombe you're really flying, not just herding something around in the sky."[40] Shaw, like thousands of others before and after him, learned early on from fellow fliers what defined "real flying" and, by extension, "real pilots."

Similar to the rites of passage associated with college fraternities, sports teams, and the military throughout the postwar era, students could also expect a certain degree of hazing from the fence crowd as part their entrance dues into the community of pilots. Boyne warned that "pilots' humor usually runs to semi-savage satire" and noted: "You'll know you've become established as a pilot when you find yourself involved in the joke-making." He provided a typical example: "The morning that you've done an elaborate preflight, started the engine, made your radio call, and found that you can't taxi because you've forgotten to remove the tie-down ropes won't be forgotten. When you come back an hour later, hoping that your little faux pas was overlooked, you'll probably find the line boy waiting with a suitable axe, marked 'For Tie-Down Removal.'" And, he warned, a story this good was sure to spread like wildfire. "Afterwards you can count on solicitous inquiries about how low you were planning to fly, or if you need a running start, or whether you wanted to be sure you could find your way back to the parking lot. None of it is very subtle, but it's good-natured, and it means you've arrived."[41]

Some students accepted this sarcastic ribbing from other pilots as just another part of learning to fly. At the start of his first long solo cross-country trip in the mid-1950s, Frank Kingston Smith made the very mistake that Boyne would use decades later as a hypothetical example in his book. "I opened the throttle and the engine roared as the plane moved ahead about three inches and—stopped! I opened the throttle farther, but the plane wouldn't budge. What with all the racket of the wide-open engine on the parking line and my sliding around on my seat looking out of the windows to see what was in front of the wheels, a small group of spectators soon gathered, pointing their fingers and laughing fit to kill themselves." A chagrined Smith allowed the line boy to untie the plane's tailwheel, "and then, to a round of applause, which I didn't particularly relish, I taxied to the end of the runway, checked out the mags [magnetos] and gave her the gun." This custom of laughing at minor mistakes provided an avenue for facing up to one's errors; Smith, for instance, turned this story into a parable about the importance of checklists. It probably helped soothe his bruised ego when Bob, his instructor, informed him after the flight that this type of lapse was "not uncommon to nervous new pilots."[42]

More than two decades later, in the spring of 1979, Smith—by now an accomplished pilot and prolific aviation writer—had this to say from the other side of the airport fence. Perched among "the platoon of rail-birds on the bench out by the line of T-hangars, sometimes known as the Critic's Corner," Smith and the rest of the group eagerly watched as fellow pilots made their first flight of spring since winter's snows had receded. Comparing flying to golf, he noted, "As each airplane waddled out to the practice tee, small wagers were made as to whether the rust-encrusted pilot would duckhook the takeoff and, when an airplane would return and come to gropes [*sic*] with the runway, fingers were pointed and comments heard." He continued, "When a particularly gruesome landing was produced by anyone, the jury of his peers was quick to bring in a verdict. There is more to the fun of flying than just flying. It is still a guild."[43]

Even if this was all meant in fun, be it the early 1950s or the late 1970s, student pilots could not afford to be thin-skinned if they hoped to gain acceptance into this tough-talking "guild" with its "semi-savage satire." Just like those who couldn't put up with verbal abuse from aggressive instructors in the cockpit, anyone who couldn't take a little "good-natured" ribbing on the ground risked being branded as too soft or weak to safely face the very real dangers they might encounter aloft. And if the fence crowd made quips and jokes behind a pilot's back rather than to his or her face, this was a sure sign these critics had branded the person an outsider unworthy of admittance into their ranks.[44]

This culture, in and out of the cockpit, served as a gatekeeper for all potential pilots, but it could be especially unfriendly toward women. For instance, when female fliers assumed the "rather indelicate poses" required in order to check beneath an airplane during a thorough preflight inspection, members of the predominantly male fence crowd sometimes leered or made rude comments, leaving little doubt that they saw these women as sexual objects rather than fellow fliers. These men also had a reputation for carrying the "flying is a man's game" attitude into the local airport's pilots' lounge or coffee shop, where they openly exchanged jokes about women pilots, mostly variations on popular postwar jokes regarding women drivers.[45] This could happen in venues large or small. One woman pilot overhead FAA official Pete Campbell remark at the 1968 AOPA convention, "Some people you just can't teach; I've seen female pilots who would be dangerous as airline passengers."[46] In another case, recounted decades after it happened, a woman student was "just hanging out on the couch in the lobby" of the local flight school while another woman flier settled her account at the counter. As soon as the other woman left the office, the flight school staff started to ridicule her behind her back. The first female student, still sitting in plain view, suddenly realized, "if they were making fun of her like that, they were making fun of me when I wasn't there, too." Like Ackerman, she finished her training elsewhere.[47]

Satirical cartoons and humorous stories published in aviation magazines of the era echoed the banter heard in all-male gatherings of pilots. Many of these cartoons and stories featured women who are comically lost, including one who can't find the runway directly beneath her plane because she's so disoriented that she's flying upside down. Even this is no cause for surprise to the bored-looking male air traffic controller in the airport tower, who wryly informs her via radio, "Madam, you're landing on the sky." Some cartoons suggested that women were lousy at landing, widely regarded as the single most difficult task to master and the benchmark for measuring a good—or bad—pilot. Others joked at their lack of technical prowess in general. One 1967 cartoon featured two otherwise identical personal planes parked side by side: the one in perfect condition is labeled "His"; the other, dented and dinged from incompetent handling, is conspicuously marked "Hers." This last example drew a letter of protest to the magazine's editor from a woman pilot, who complained, "It is hard enough trying to become acceptable in this man's world of flying without things like this. Most thinking men would know this comic for what it is, but there are just enough male types who have no better way to feel superior than by 'standing on the necks' of women." She continued, "I don't expect any favors because I'm a female. On the same hand, I don't expect be run down because I'm a female." In response, the editor offered only a backhanded apology: "Most assuredly, no one thought a female would object to that cartoon, especially the woman [staff member] who selected it for publication."[48] Not surprisingly, cartoons and articles of this nature continued to show up in aviation publications well into the 1970s.[49]

Gender-based assumptions about who would (and would not) make a good pilot were reflected in how the aviation community treated students when they made mistakes more serious than, say, trying to taxi out of the parking area with the plane still secured by tie-down ropes, "duckhooking" a takeoff, or bouncing upon landing. When George Christian took off without a map, got lost, and called for help on the radio, no less than three airplanes from his flight school took off to search for him. Ultimately, a captain for Braniff Airlines joined the rescue effort from his jet cruising at thirty thousand feet, relaying messages by radio between the muddled student and air traffic controllers on the ground. Despite the significant commotion he had caused, Christian discovered: "Nobody scolds. Pilots understand." Indeed, one of those who had joined the search—an off-duty Delta Airlines captain and former World War II bomber pilot—later consoled the embarrassed student: "We've all been there, or we're going to be there. It happens to us all."[50] This story highlights a typical characteristic of the fence crowd in postwar private aviation. For many of these informal "gatekeepers" to the community of pilots, *men* were allowed and even expected to make mistakes in the process of gaining experience and becoming better pilots. So long as this did not happen too often and their errors did not reveal a pattern of fundamentally bad judgment or recklessness, these mistakes could be written off as "live and learn."

Women pilots, on the other hand, often reported that the mostly male fence crowd interpreted their individual errors as evidence that women in general were not cut out to be pilots in the first place. When Patricia Houtz got lost on her first solo cross-country flight in the early 1960s, she acknowledged that this was exactly what fellow fliers would expect to happen. "There has been a lot of kidding about women pilots and cross-country flights mixing like oil and water," she mused. With this weighing on her mind in an already stressful situation, she finally decided, "It's better to be embarrassed and alive than dead and a hero," and rather than press on in hopes of finding the elusive airport, she made a precautionary landing in a cornfield before she ran out of gas.[51] In the end, Houtz consciously chose to face certain ridicule for having lived up (or down) to stereotypes about women pilots always getting lost. Jennifer Ortiz, who more than a decade later became lost during a daunting solo cross-country flight across an inhospitable mountain range, allowed her fear of what fellow pilots might think about her to influence a potential life-or-death decision. "The idea of turning back and admitting that I wasn't as competent a pilot as I had thought was not too appealing. I'd probably be the laughing stock of the [flight] school; I'd confirm male pilots' ideas that women don't belong in aviation." Unlike Houtz, Ortiz decided to press on, all the more dangerous since the rocky terrain offered no place to land should she run out of fuel. Fortunately she managed to find her way to an airport, so the event was recorded in a magazine article rather an accident report.[52]

Some men and women who made their living in the aviation industry decried such practices as being bad for business, much in the same way that some objected to aggressive flight instruction methods. The subtitle to a 1963 article along these lines is descriptive: "Petticoat-operated flying school at Santa Monica, Calif., has successful record in convincing 'reluctant wives' that flying can be done safely." Author Lois C. Philmus argued that women made better instructors for female students because of their ability to combine sound flight instruction with "the patience and understanding that can only exist among women."[53] Another article concerned with widening private aviation's base of potential customers asked, "Where Are All the Gals?" Directed mainly at other men, author Frank Tinker explained how achieving this goal would require male pilots, instructors, and airport operators to change how they behaved in order to make both the airport and the cockpit a less hostile environment for half the nation's population.[54] Despite pleas from individuals like Philmus and Tinker, change came slowly to the postwar aviation community, and women fliers had to develop various strategies to cope with this highly masculine environment and gain acceptance from their male peers.[55]

These examples—along with exceptions to the norm like those voiced by Philmus and Tinker—illustrate how both the instructor in the cockpit and other pilots on the ground introduced students to the masculine world of private flying. Beyond mastering the complex technology of flight, newcomers had to demonstrate through informal hazing that they were mentally tough enough to react calmly and correctly in an emergency situation. They also had to show that they were good company, for hangar flying and hanging out with the fence crowd were important aspects of belonging to the community of pilots. Perhaps most importantly, students learned by example that someday—assuming that they made the grade—they too had a role in educating, acculturating, and vetting future members of their elite ranks.

Acculturation into this masculine community was an ongoing process, and it could be both subtle and overt. But it was probably never more evident than on the day a student completed his or her first solo flight, where a postwar tradition developed that involved the instructor cutting off the student's shirttail immediately afterward. Aside from the fact that the student was assumed to be wearing a shirt and not a dress, there was nothing inherently "male" or masculine about the physical act of clipping shirttails. Nonetheless, postwar pilots found ways to connect this tradition to masculine ideals so that it came to symbolize a rite of passage into the manly world of flying.

In 1997 veteran airline pilot and aviation author Barry Schiff observed, "My curiosity about the custom of clipping the shirttails of new solo pilots began on my sixteenth birthday, June 23, 1954. This is when my instructor destroyed my favorite shirt, adorned it with data, and hung it on the wall of my flight school. I could not have been prouder, but I had difficulty trying to explain the loss of this shirt to my parents, who were unaware that their teenage son had been taking flying lessons." Apparently other pilots at the airport also had difficulty explaining to the teenaged Schiff why they snipped shirttails to celebrate first solos, because he was still looking for answers more than four decades later when he posed this question in the pages of *AOPA Pilot*. Readers rushed to respond with a flood of potential answers. At least one suggested that the tradition dated back to the Roman Empire, reflecting "an honor bestowed upon those who had survived their first battle," which implied that one's first solo was equally fraught with fear and physical danger. Some readers claimed that shirttails represented the flier's first silk scarf, which was supposedly awarded upon successfully soloing back in the days of open cockpits and leather helmets (interestingly, many firsthand accounts of first solos by early aviators make no mention of receiving anything aside from a handshake). Others described how the tradition honored early barnstormers, who would cut scraps of cloth from their clothing to patch holes in the fabric skin of their aircraft caused by rocks or branches, a common problem when using farmers' fields and pastures as makeshift runways. One popular explanation described cash-strapped aviation cadets using their shirttails to create officer's epaulettes for their enlisted uniform shirts during World

War II, when graduation from flight school was accompanied by a commission to the rank of second lieutenant (unless the flier had already earned his commission from West Point, the Naval Academy, or ROTC).[56]

According to Schiff, the most popular explanation submitted by readers (and to him "most plausible") was that the roar of the engine and the rush of the wind in the pioneering years of aviation made it nearly impossible to communicate in open cockpit biplanes, especially since the occupants sat one in front of the other instead of side by side. According to this tradition, the instructor—seated in the rear cockpit—tugged on the student's shirttails to signal when to turn right or left, much like pulling on the reins of a horse. As Schiff put it, "when the student soloed, he no longer needed to receive such instruction, which was symbolized by the ritual removal of his shirttail." After listing a total of 10 potential origins for the shirttail tradition, he concluded: "Which explanation is most accurate? Your guess is as good as mine. Eventually one will survive and become 'fact.' After all, 'history is a myth agreed upon.' "[57] A few years later, Daryl Murphy discovered even more theories regarding how the tradition began, but ultimately he too was forced to conclude, "Many offer their opinions, but no one seems to have solid proof of its origins."[58]

Such "solid proof" remains elusive for good reason, for upon closer examination it appears that none of these explanations qualifies as "most accurate." Scores of early pilots, some famous, others less so, detailed the process of learning to fly in autobiographies published before World War II. Since none mentioned their instructors "steering" them by tugging on their clothing and then clipping their shirttails upon soloing, this explanation seems highly unlikely. Even Schiff, who deemed this explanation the "most plausible" of the 10 he described, admitted a serious flaw: "In all open-cockpit airplanes that I have seen or flown, the instructor sits in front and uses hand signals to direct his student."[59] Similarly, the image of newly soloed aviation cadets trotting down to the tailor shop to have their enlisted uniform converted to reflect their new status upon soloing does not match accounts written by World War II veterans. Instead, their memoirs describe cadets boisterously tossing each other into a swimming pool, pond, or some other convenient body of water to mark this milestone. One former army bomber pilot addressed this question directly in his 1994 memoir: "Today they would cut a piece out of your shirttail when you soloed. However, our shirts were government issued, so they threw you in the goldfish pond in front of headquarters instead."[60]

Buried within a brief 1989 article about a young woman's recent first solo is what may be the closest thing to "proof" that the tradition of clipping shirttails was simply invented, then imbued with meaning after the fact. When the article's youthful author interviewed 90-year-old Paul Garber, the founder of the Smithsonian Institution's National Air and Space Museum and longtime aviator declared with confidence that military aviators didn't start dunking each other until World War II.[61] Garber also disavowed any connection between military flight training and the clipping of shirttails, musing instead that perhaps postwar pilots were just looking for "a 'something unusual' to demonstrate congratulations" from fellow fliers, and that the "shirttail was probably used because there was nothing else handy."[62]

While the origins of this practice are obscure, its symbolic meaning to pilots is not. Although not every individual who learned to fly in the postwar era had his or her shirttail clipped, this became an almost universal ritual to mark the first solo. Adorning the scrap of cloth with the student's name and other particulars about the flight, and then hanging it on the wall for all to see, served as a public announcement to the local community of pilots that another member had completed a momentous milestone in the long process of joining their ranks. But fliers were not content to allow this scrap of cloth simply be a souvenir chosen "because there was nothing else handy." And so, in the absence of a solid historical basis for this tradition, they invented their own creation myths to give it meaning. Even more telling, they chose stories that reinforced the connection between flying and masculinity, from the Roman legionnaire surviving his first battle to the newly independent student finally freed from the backseat badgering of his instructor, from the intrepid barnstormer patching his own plane in a farmer's field to the aspiring aerial warrior moving one step closer to engaging the enemy in heroic combat in the skies over Europe or the Pacific.

Although the legends behind the shirttail tradition were contrived, this in no way denigrates the meaning of the event that it celebrated. Students approached their first solo with a mixture of anxiety and anticipation, and many later remembered the occasion as a rite of passage, not just in the realm of aviation but in life as well. "I'll never forget my first solo" was a common refrain among pilots, even years afterward. In his 1927 autobiography *"We"* published shortly after his historic transatlantic flight, Charles Lindbergh declared, "The first solo flight is one of the events in a pilot's life which forever remains impressed on his memory. It is the culmination of difficult hours of instruction, hard weeks of training and often years of anticipation. To be absolutely alone for the first time in the cockpit of a plane . . . is an experience never to be forgotten."[63] More than half a century later, veteran flier Walter Boyne, in his characteristically enthusiastic and somewhat brazen manner, conjured this analogy in his 1980 book for the benefit of prospective pilots: "In some ways the first solo is very much akin to the loss of virginity: It is an irrevocable act, a dividing line, a boundary, a new life."[64] Private pilot and aviation writer Gordon Baxter made a similar observation in the first chapter of his own 1981 book for aspiring aviators when he asked, "Like the virgin turning from reading romance to the actual experience, could flying possibly be as good as all this heavy-breathing aviation romance that we write about it?" His answer, which plays out over the course of subsequent chapters, can be summed in a single word: "yes."[65] Fliers traditionally mixed talk of sexual conquests on the ground and adventures in the air during hangar flying sessions, described airplanes in feminine terms (often using the pronoun "she" or "her"), and during World War II had painted pictures of women on the noses of warplanes.[66] Therefore, it is not surprising that men would compare their first foray alone into the air with the sexual act that for many marked the dividing line between boyhood and manhood.

This comparison between the first solo flight and achieving full-fledged manhood was so deeply embedded in the male-dominated fraternity of pilots that at least some women felt the need to reassert their femininity when they soloed. For instance, when housewife Janet Davis heard about the tradition of clipping shirttails while learning to fly in the early 1960s, she went home and sewed rows of lace and ruffles and bows onto the tails of her blouse so it would not "look like a man's, when it was hung on the wall."[67] A few years later, Sally Buegeleisen published an article about learning to fly in which she echoed the sentiments of generations of male pilots: "If you have soloed an airplane, you never forget the experience; if you have not, no one can truly tell you about it, but almost every pilot will try." Having just admitted that it was impossible to explain the emotions this event evoked, she gave her best effort anyway: "As you roll out on the runway [after landing], turn off and taxi back to the ramp . . . you may feel a little the way you did when you had your first look at your newborn child—no one has ever managed anything quite so well before."[68] If the first solo marked the passage from boyhood to manhood for men, an event often celebrated in locker rooms as an act of sexual conquest, for women—or at least for Sally Buegeleisen—it compared to the respectable transition from young wife to motherhood.

To Spin or Not to Spin

In most cases, there was a clear dividing line between pilot acculturation and stick-and-rudder flight training. Hangar flying, good-natured hazing, and hanging shirttails on the wall were all part of an informal process intended to indoctrinate new members into the aviation community. Likewise, an instructor's harsh and demanding demeanor simultaneously showed neophytes how "real pilots" were expected to behave and drove off those who couldn't (or wouldn't) put up with the additional stress created by this teaching style. But since most training syllabi followed the same general format, and because the federal government spelled out exactly which skills student pilots had to master in order to qualify for a private license, the student-instructor relationship had little direct impact on *what* was taught, only *how* it was taught. Still, masculine ideals regarding courage, control, and technical competency crept into a few ostensibly objective debates about whether certain skills should, or should not, be required of all private pilots.

Before 1949, all candidates for a private license not only had to practice both stalls and spins during flight training but were also required to demonstrate these maneuvers to a federal flight examiner in order to pass their final check ride.[69] To understand a "spin," one must first understand what it means when an airplane "stalls." When people say that a car "stalled," they mean that the engine quit. In aviation, the term *stall* has a very different definition that has nothing to do with whether or not the engine is running. In the simplest terms, an airplane stalls when its wings quit flying. A slightly more technical explanation is that "a stall is the loss of lift that occurs when a wing is flown at too large an angle of attack."[70] Although postwar aviation texts generally reminded readers that "the airplane will stall at any airspeed, any attitude, any power setting, any configuration, and at any weight," inadvertent stalls usually happen when a pilot accidentally slows an airplane to the point that it ceases to fly. A popular handbook for fliers described what happens next. "When the stall actually occurs, there should be no mistaking the event. Airspeed and control effectiveness will drop sharply. . . . The nose will start to drop, and more back stick [i.e., pulling back on the controls in an attempt to raise the airplane's nose] will serve only to aggravate that condition. Turbulent air from the stalled airfoil [wing] will buffet the airplane structure, causing a shaking vibration that can be felt through the airframe, seat, and pilot's controls." Since the airplane's wings have ceased to fly, it begins sinking toward the ground at a rapid rate of descent.[71]

Stalls are not only dramatic, but proper recovery techniques are counterintuitive, especially to the novice. In the introduction to his widely read 1944 manual *Stick and Rudder: An Explanation of the Art of Flying*, Wolfgang Langewiesche described the many ways in which proper flying techniques were "exactly contrary to common sense." He saved the most dramatic example for last: "And—most spectacular contrariness of all—in emergencies, when the airplane is sinking toward the ground in a 'mush' or falling in a stall or a spin, and you are afraid of crashing into the ground, the only way to keep it from crashing is to point its nose down and dive at the ground, as if you wanted to crash!"[72] By design, most postwar personal planes will recover all by themselves, given sufficient altitude and assuming that the pilot stops interfering and lets go of the controls.[73] Unfortunately, since most inadvertent stalls happen during takeoff or landing, the plane is seldom high enough to correct this condition on its own before it hits the ground. The federal government's 1956 *Flight Instructors' Handbook* summed up the problem nicely: "The most potentially dangerous stall, of course, is one which occurs near the ground where a crash will result from any significant loss of altitude."[74] Pilots can significantly reduce how much altitude they lose by promptly applying proper recovery techniques, but even then, most aircraft can lose several hundred feet before returning to normal flight. And there's a difference between practice stalls conducted in a deliberate manner at a safe altitude and the worst-case unexpected stall that happens close to the ground while the pilot is distracted doing something else. In 1991, nearly a half century after Langewiesche warned that stall recovery was counterintuitive, another expert observed: "Student pilots need to learn that while it's possible to recover from a deliberate stall with no loss of altitude, an accidental stall may involve not only a considerable loss of altitude but also extreme and disorienting nose-down attitudes and secondary stalls."[75]

As Langewiesche explained in his typically down-to-earth manner, "A spin is nothing but a fancy stall; one side of the airplane is stalled, the other is not, and therefore the airplane sinks down twisting."[76] From the pilot's perspective, however, spins are even more disorienting than stalls. Frank Kingston Smith experienced his first intentional spin for training purposes in a normally docile Cessna while he was working toward his private pilot's license in the mid-1950s. He later recalled, "The left wing dropped, and in a flash we flicked over on our back in a twisting, turning dive. For an instant I saw the crazy-quilt pattern of farms through the blue-tinted skylights over my head; then I saw them in front of me, revolving rapidly in a clockwise direction."[77] A more recent article in *Air & Space Smithsonian* magazine, also aimed at a nonpilot audience, described spins no less dramatically: "Imagine you are in a car suspended by a rope around its rear bumper. The other end of the rope is attached to a flagpole jutting out from a very tall building, so you sit staring straight at the concrete far below. Someone starts the car moving in a circle while simultaneously turning on its own axis as the car's front end starts pitching from side to side. Now someone cuts the rope. That's what it feels like to be in a spin."[78]

As with stalls, proper spin recovery techniques are counterintuitive. During a spin, the ailerons—normally used to bank the airplane left or right—become ineffective because the wings (or at least one of them) have stalled. And pulling back on the stick or control wheel, which under normal circumstances raises the plane's nose out of a dive, instead only serves to tighten the nose-down spin. Thus, instead of turning the wheel or moving the control stick in the opposite direction of the spin and pulling back to raise the airplane's nose, proper recovery begins by pushing the controls *forward* into the neutral position, then applying opposite rudder via the foot pedals (e.g., push the left pedal to counteract a right-hand or clockwise spin). Only after the plane stops rotating and exits the stall can the pilot begin to ease the plane out of its nose-down dive. Any attempt to "pull out" prior to this point only serves to tighten the nose-down spin.[79]

Stories involving spins and student pilots were a staple for both hangar flying humor and cautionary tales within the community of pilots, and a steady stream of anecdotes on the subject appeared in aviation periodicals throughout the postwar era. One cartoon published in 1961 depicts an instructor shepherding his student away from their parked airplane following a lesson. As the still-terrified student, teeth clenched and bug-eyed, clutches frantically at an imaginary control wheel, the instructor nonchalantly informs him: "I'll have to flunk you on that spin" (fig. 2.2).[80] Less humorous was Peter Derfner's account of a practice stall gone wrong while he was learning to fly in the late 1970s. Out on a solo flight shortly before taking his flight exam, Derfner accidently put the normally docile Cessna 150 training aircraft into a spin. He panicked, failed to initiate proper recovery procedures, and was saved only when he gave up completely and let go of the controls. True to its designed-in safety features, the Cessna stopped spinning on its own, and Derfner was able to level off with only 800 feet to spare. Fortunately he entered the spin at 2,200 feet; had this happened in the airport traffic pattern (generally at 800 to 1,000 feet above the ground), he would have almost certainly "spun in" and died.[81] Derfner was a novice, but even highly experienced instructors could become victims of spins. A 1985 article warned that "it is well known that student pilots may 'freeze' in an inadvertent spin, pushing on both rudder pedals with equal force." In this worst-case scenario, the instructor might be unable to physically overpower the terrified student in order to apply the full opposite rudder needed to bring the plane out of the spin.[82]

A long-running debate about spin training began during World War II as the CPTP churned out an unprecedented number of new private pilots. In 1944 the Civil Aeronautics Board's Safety Bureau proposed eliminating the requirement for private pilot candidates to demonstrate spins to a federal examiner during

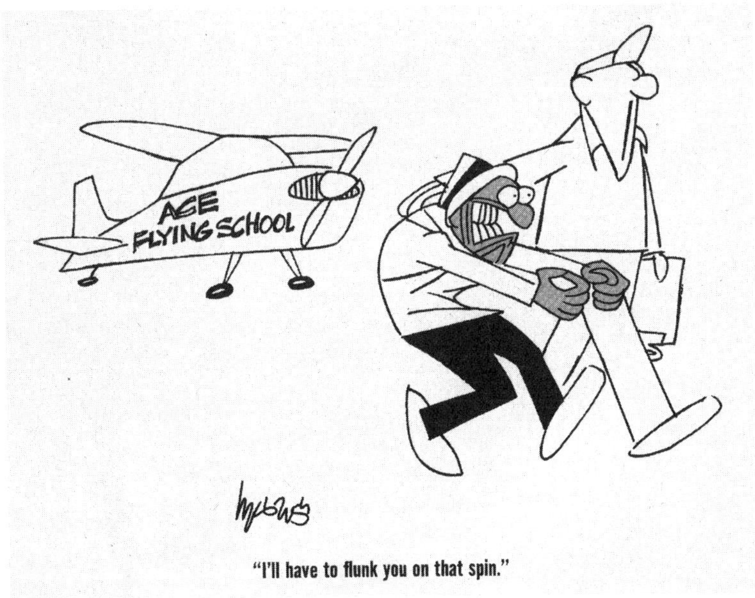

Fig. 2.2. Spin training humor circa early 1960s. Though published more than a decade after the federal government dropped spins from the list of mandatory training for student pilots, this 1961 cartoon depicts the terror this maneuver could inspire. "I'll Have to Flunk You on That Spin" [cartoon], *AOPA Pilot* 4, no. 9 (Sept. 1961): 43. Courtesy of the Hagley Museum & Library.

their final flight exam, arguing that training on this maneuver with a flight instructor was sufficient. However, although the federal government did amend many aspects of the Civil Air Regulations in 1945 to make earning a private license easier, it chose not to eliminate spins.[83]

At the time, stalls and spins were a leading cause of fatal accidents involving small planes, and the crisis continued after the war. For instance, stalls and spins accounted for nearly 50 percent of all fatal nonairline aircraft accidents in 1948 and 1949, and the problem was alarming enough to make the pages of the *New York Times*.[84] While few argued that *nothing* should be done, diverging and sometimes diametrically opposed views quickly emerged within the aviation community regarding how best to solve the problem.

Right from the start, the AOPA vehemently opposed mandatory spin training for postwar private pilot candidates. After polling its members in the spring of 1944, the organization declared: "We believe it is significant that the tremendous number of people opposing the spin requirement were definitely not pilots who had been refused a license for inability to spin, but rather mature pilots," many of whom were flight instructors or flight school operators. "The complaint was universal that the spin requirement was a constant source of their losing both prospective customers as well as enrolled students."[85] As the controversy continued, student pilot Harry Burnett seconded this viewpoint in a letter to the editor of *AOPA Pilot*. "In the fight to drop spin tests from the CAA examination, I may say, as a student pilot (and middle-aged business man) that the idea of having to execute a spin terrifies me already, well in advance of my examination for private pilot."[86] This sentiment was apparently widespread. A 1951 report about factors that had impeded the growth of postwar private flying observed that "many charge that training has emphasized stalls, forced landings and spins and thus built up a tenseness and subconscious fear in the conservative student. The course of study [that includes these maneuvers] appears to be particularly discouraging to the middle-aged business or professional man, who, in the final analysis, is the one who has the ability to pay for an airplane."[87] In short, because spins are scary, spin training is bad for business.

Although the AOPA ardently opposed spin training in the immediate postwar era, this does not mean it took an antisafety stance. Instead of hands-on spin recovery training, the organization promoted technological solutions, including preventative measures such as safer, preferably "spin-proof" new plane designs and mandatory installation of mechanical stall warning devices in all existing aircraft prone to stalls and spins.[88] Although the AOPA also advocated teaching pilots how to avoid stalls, for many years the organization's leadership saw technology as the best solution to this serious safety problem. Others, however, believed that individual pilot skill and judgment were the only way to avoid a deadly stall-spin accident. At the height of a government "anti-buzzing" campaign directed against low-flying private pilots in the immediate postwar era, T. P. Wright, the CAA's chief administrator, actually defended low-altitude training for students. "During their early training, almost all pilots are ground shy. They have a subconscious anxiety or reluctance to bank the aircraft at medium and low altitudes." This fear, he noted, led to incorrect, uncoordinated turns using more rudder than aileron, an error that significantly increases the chance of an inadvertent stall or spin. In Wright's professional opinion, "Correction can be accomplished only by constant repetition of these maneuvers [at low altitude] under an instructor's supervision."[89] Thus, not only did the head of the CAA advocate the importance of training over technology; he also insisted that this training should be conducted at what some (perhaps many) pilots and instructors considered a dangerously low height above the ground.[90]

Wright left the CAA in early 1948. The next year, apparently with little fanfare, the federal government removed spins from the list of required training for private pilot candidates.[91] The AOPA leadership, for whom abolishing spin training was only one part of a multifaceted strategy to make flying easier and safer through better technology, was less than satisfied. The government made no move toward requiring that new aircraft include design features known to reduce the chance of inadvertent stalls and spins, and manufacturers showed little inclination toward introducing such technologies on their own. On the one hand, the AOPA could announce: "New No-Spin Law Eases Requirements for Pilot License."[92] But at the same time, the organization conceded that "it looks as though the elimination of spins is not going to open up a whole new world [of safer airplanes] for the private flyer after all."[93] Worse yet, it appeared that many pilots were not particularly interested in safer technology either. "Are we willing to put up a fight to get such a plane?" one *AOPA Pilot* editorial asked readers, "*Or would we—deep down underneath it all—prefer to continue to be supermen?*"[94]

On the surface, post-1949 arguments in favor of spin training continued to focus on safety. During its anti–spin training campaign in the late 1940s, the AOPA had argued that requiring all students to practice spins, as opposed to introducing safer airplanes, was part of a de facto conspiracy within the aviation community to scare off people who failed to measure up to a "Superman" standard for pilots. At least some advocates of spin training implicitly agreed that spins might screen out certain types of students, but, unlike the AOPA, they saw no problem with this outcome. Nadine G. Tatum complained in 1950: "Everyone learning to fly should have spins, then if the plane goes into a stall the pilot will know how to recover."[95] Flight instructor Truitt Vinson observed in 1958 that "we could have safer pilots if the CAA would let us go back to the flight test we took in the good old C.P.I. [World War II–era CPTP] days" including "two turn on a point spins" and other maneuvers "that had to be just so."[96] And in his 1962 article about modern FAA check rides, Frank Tinker addressed "the most frequently criticized omission in this private pilot check ride: the spin."[97] In these and other cases, explicit arguments about the value of teaching students to recover from spins also implied that this training created better pilots because it weeded out individuals who lacked the discipline or guts (or both) required to control their fears in order to accomplish challenging maneuvers "just so."

These arguments seem mild when compared to what some hard-core members of the pro-spin camp had to say on the issue. In response to Tinker's 1962 article, W. C. Heller penned a letter to the editor that tied high accident rates in private flying to the lack of spin training and then connected this specific issue to the broader "fallacy of so-called liberal progressive educational practices in [public] schools." Heller argued, "Delete the basics in anything and you increase the margin for error." Like John R. Hoyt's 1945 advice to students that "When You Learn to Fly—Get a Tough Instructor," Heller based this seemingly straightforward conclusion on the assumption that the erosion of old, established standards had created an "everyone passes" mentality, which in turn was responsible for most of private aviation's (and American society's) present-day ills.[98] Dick Weaver, who prided his reputation as the "the toughest flight instructor in the area," was even more direct. In a 1964 interview for an article about problems with modern flight instruction, he declared: "I believe that spins should be part of the required curriculum because it puts the student so much at ease after they've completed the maneuver. After all, spins are a very simple exercise, and it's the fear of the unknown that haunts students who don't get this training." Taken at face value, Weaver's statement makes perfect practical sense. But this grizzled military veteran who admitted, "Next to a shooting war, I like flight instruction best," and who used a punching bag to take out his post-flight aggressions, might have just as easily told his students to face their fears like men.[99]

Some pilots and instructors didn't even bother to hide their disdain for those who lacked the courage to handle spins. When 18-year-old Robert Traister started flying lessons in the late 1960s, an instructor called him a "sissy" because he got airsick after doing spins. Traister promptly quit flying, and it took him more than five years to muster up the courage to return to the airport and obtain his private license.[100] Vestiges of this attitude linger into the present day. As recently as 2009, a flight instructor candidate (the only civilian rating for which spin training is still required) complained that his instructor called him "a sissy-man and a stall-baby" for recovering from an incipient spin too quickly rather than allowing it to develop into a full-blown spin.[101] Thus, for some, this maneuver continued to do more than merely teach a pilot how recover from a very specific in-flight event; instead, it served as genuine test of whether someone possessed the character—and the courage—to handle any emergency in the sky.

Hints of this sentiment even appeared in flight manuals of the postwar era. For instance, in 1966, nearly two decades after the federal government removed spin training from the private pilot curriculum, a popular textbook for aspiring aviators included this seemingly neutral statement in its introduction to the section on spins: "Applicants for private pilot licenses no longer are required to demonstrate their proficiency at spin recovery." A few sentences later, however, the author admonished readers, "Yet, the teaching of spin recovery remains a firm part of any pilot training that is intended to do more than get the Sunday pilot around his country airport."[102] The message for readers was clear: whereas a mere "Sunday pilot" might be content to simply learn how to avoid spins, *real pilots* would face their fears, conquer the unknown, and master this maneuver.

Although the legal requirement for spin training ended in 1949, some instructors continued to teach this skill as a matter of course. In the mid-1950s, for instance, Frank Kingston Smith's instructor simply pulled two parachutes from his car and announced: "spins today."[103] Dick Weaver took pride in asserting in the mid-1960s, "Somewhere before they complete their flight program, spin entries and recoveries have been demonstrated to all my students." And in a 1969 letter to the editor to correct another reader's misconception that the FAA had outright *banned* spin training, Spencer F. Houghton noted that "many experienced flight instructors [still] include training in spin recoveries, or spin entry recoveries, in the training of private pilots."[104]

In place of mandatory spins, the federal government required hands-on stall training, on the theory that, by practicing proper stall recovery, pilots learned to avoid spins in the first place and thus would never need to recover from one. Proponents of this approach pointed out that most inadvertent spins occurred in the airport traffic pattern when pilots were flying low and slow and were preoccupied by the extra demands imposed by takeoffs or landings. Under such conditions, they argued, even a pilot who had already mastered spins in training would probably not have time to recover from one before the airplane struck the ground. Some also maintained that the cure was worse than the disease, claiming that more fliers had died in the late 1940s while practicing spins than from crashes involving an accidental spin.[105] The twenty-fourth edition of the *Civil Air Regulations and Flight Standards for Pilots*, published in 1962, left no doubt that fliers should focus on prevention: "THE BEST WAY TO RECOVER FROM A SPIN IS TO AVOID GETTING INTO ONE."[106] And whereas the CAA's 1956 *Flight Instructors' Handbook* had devoted several paragraphs to optional spin training for students "as a safety precaution and as a confidence builder," the FAA manual that replaced it in 1964 contained no mention whatsoever of spin training.[107]

As fewer students learned spins as a routine part of their training, a handful of specialized flight schools stepped in to offer pilots hands-on practice in recovering from spins and other extreme in-flight situations. David Pierce, a private pilot of three years with 114 hours experience, sought out a newly opened aerobatics school in Deer Valley, Arizona, in 1967 for this very reason. Before his first flight with an aerobatic instructor, he grimly reminded himself: "Spins. *You are going to do spins, and before this is over you will be doing them alone. You've got to, or that familiar but never-accepted fear will haunt you as long as you fly—the fear of spinning and of not being able to come out.*" Pierce discovered that he was not alone. Among his fellow students was a corporate pilot who flew Lear Jets for a living. Regarding this experienced professional, Pierce noted, "With 2,500 hours logged, he had never done spins." Another student at the school, a machinist with a brand-new private license and only 60 hours in the air, told Pierce: "On my FAA test they asked me how I recover from a spin . . . all I could give him was what the textbook said. I don't think it's actually right. I won't feel like a complete pilot until I have this aerobatics down." According to Pierce, one of the school's instructors expressed "a profound conviction that a pilot ought at all times to be able to maintain absolute control of his airplane." Another instructor explained, "It's not the spin that gives a guy trouble. . . . It's the psychological effect. Because a guy who gets into one gets into a kind of controlled state of confusion. He doesn't know exactly what to do or when to do it." Since the postwar masculine ideal involved possessing both the "technological skill" and the "considerable amount of toughness in mind and body" needed to master stressful situations, then maintaining "absolute control" at "all times" included having the ability to deliberately enter and exit a spin with one's airplane, senses, and masculinity intact.[108]

By the mid-1970s, the AOPA, which by this time had experienced a changing-of-the-guard among its senior leadership, also reversed its earlier position against spin training. In a 1976 review of an aerobatic training plane, *AOPA Pilot* staff writer Don Downie observed, "It is our opinion that an airplane driver is not really a lightplane pilot until he can recover from a spin and roll off his back if ever tossed inverted" by extreme turbulence or the slipstream of a larger aircraft.[109] And in 1978, after acknowledging that spin training requirements for pilots remained "the subject of considerable debate," William Stanberry, executive vice president of the AOPA Air Safety Foundation, reported that the organization "supports spin training."[110] Nonetheless, the "spin training is bad for business" argument apparently remained a guiding principle for flight instruction. More than four and a half decades after AOPA had first argued against mandatory spin training for this very reason in the mid-1940s, and more than a decade after it had publicly reversed course in the late 1970s, a 1991 *Flying* article on the persistence of stall-spin accidents among private pilots explained that the main reason most flight schools did not offer students hands-on spin training was that it was bad for business: "Selling airplanes and training pilots to fly them go hand in hand; nobody wants to frighten away potential pilots and airplane owners by subjecting them to a dizzying corkscrew ride that most adults in an amusement park would pay to avoid."[111]

After decades of debate, the federal government finally revisited the issue in 1980 when the U.S. House Committee on Science and Technology's Subcommittee on Investigations and Oversight held three days of hearings on the topic. Although the subcommittee recommended reinstating mandatory spin training for private pilots, the FAA disagreed, and the rules remained unchanged. Yet the debate about whether "to spin or not to spin" continued unabated.[112] The fact that pilots and aviation experts continued to recycle terms and arguments from decades of constant controversy indicates that ideals traditionally associated with masculinity—including the ability to face one's fears and maintain full control of both the technology and oneself in any situation, no matter how extreme—remained central to the identity of more than a few who flew for pleasure.

Training Pinch-Hitters, Not Pilots

One of the best ways to illustrate how the process of becoming a pilot was infused with masculine ideals is to examine a deliberate attempt to make flight instruction seem *less* masculine to students. As described briefly in chapter 1, two widely recognized stereotypes for wives of pilots had emerged within the aviation community by the early 1960s: the pro-aviation spouse who enthusiastically participated in her husband's aeronautical activities, and the antiflying wife.[113] In reality, many women existed somewhere between these extremes, but anything less than full support could spell trouble for pilots, who found themselves pressured to spend less time and money on flying, and more on something that their wives could enjoy. Operating under the assumption that many women feared flying in private planes because they mistrusted technology that they could neither control nor understand, the AOPA attempted to address this problem head-on in the early 1960s by teaching fearful wives the basics of their husbands' hobby.

An article published shortly after the organization's 1962 annual convention described how "special programs . . . were arranged for the women, including a session for nonpilot wives, designed to explain to them some of the basic fundamentals of flying, and suggesting ways they might help their husbands while flying." It went on to promise, "This program, presided over by Joyce Case of Beech[craft Corp.], one of the country's most accomplished women pilots, proved so popular and informative that it is planned to expand the idea to a short course for nonflyers, to help them better understand general aviation and to alleviate some of their unfounded fears."[114] True to its word, the AOPA Foundation partnered with the Ohio State University's School of Aviation to create a completely new flight training syllabus, which debuted the next year as the "Pinch-Hitter Course" at the AOPA's 1963 annual convention.[115]

To teach the first course, organizers assembled 76 highly experienced flight instructors, 5 of them women (this ratio of 1 in 15 is noteworthy given that only 332 of the nation's nearly 29,000 instructors, or 1 in 87, were women).[116] Enrollment for the inaugural class was capped at 150 students, 142 of whom graduated. An article lauding the program noted that although one woman quit partway through the course, she was an exception. "Dropouts occurred mainly for two reasons: (1) inability to attend [the convention], (2) flat refusal of a few of the wives to take the training after they had been entered in the course [by their husbands] without their knowledge." There was only one male student, 17-year-old Bruce Cooper, who sat in at the last minute for his mother when she was unable to attend the convention for unspecified reasons. Given that he was the lone male in a class full of fearful female fliers, apparently young Mr. Cooper felt the need to reestablish his masculinity, for he confidently told an interviewer that he "thought" he could land his father's plane safely even before he completed Pinch-Hitter training.[117]

The creators of the Pinch-Hitter Course were up front about the fact that they deliberately simplified flight instruction to make it seem less intimidating to women. The instructor's guidebook emphasized that "only the absolutely essential items concerned with the operation of the airplane should be used" and went on to stress, "It must be remembered that these students should be handled differently than the ordinary flight student since the basic mission of this course is entirely different from that which the ordinary flight student would pursue."[118] Unlike the "ordinary flight student," Pinch-Hitter participants did not need to worry about soloing. Instead, their "final exam" involved taking over the controls in flight, then guiding the plane to the airport for a successful—if rough—landing. All the while, the instructor was inches away in the left-side pilot's seat (Pinch-Hitter students flew from the right-side front passenger's seat, where they would most likely be seated in a real emergency situation). From there, the instructor could always take over if necessary. Pretending to be a ground-based air traffic controller speaking to the student by radio, the instructor also provided a steady stream of advice, encouragement, and instructions so that the student did not have to remember or figure everything out by herself. Nearly 40 years later, an *AOPA Pilot* retrospective attempted to explain why the Pinch-Hitter Course was so popular among the wives of pilots. "The key to the program's success was its nonthreatening approach to flying. There were no heavy discussions about aerodynamics and no complex procedural advice—just the bare essentials. . . . Instructors were chosen for their affability . . . and the main ground rule was that no one was to flunk the course. The pressure to succeed was removed, so real, unfettered learning could begin."[119]

Recounting his wartime observations as an army psychiatrist serving with the Eighth Air Force, psychiatry professor Douglas M. Bond observed in 1952, "Men do not like to fail conspicuously, and any failure in flying is a conspicuous one. Especially was this true when most flyers had 'talked up' to their friends and families the glamour of flying and their own part in it."[120] Perhaps for this very reason, flunking or "washing out" of military flight school was considered to be a tacit failure of manhood. Plenty of male students failed to solo or earn their civilian pilot's license in the postwar era, but their stories are almost entirely absent from the pages of aviation magazines. Within the world of flying, a man who failed at flying was not a man worth talking, writing, or reading about. Removing the supposedly masculine "pressure to succeed" marks one important way that the creators of the Pinch-Hitter Course attempted to reshape traditional approaches to flight instruction according to their assumptions about the proper gender identities, roles, and inherent characteristics of women versus men. There would be no yelling in the cockpit, no students sweating through spins or fretting about their first solo, no admonitions of "C'mon . . . fly it *right*!" from irate or exasperated instructors.

Indeed, unlike the masculine culture of pilots where failure was met with ridicule or exclusion, instructors and pilots seemed more than willing to laugh off potentially serious mistakes by the nonfliers who signed up for the Pinch-Hitter program. Maxine Whetsler, who took part in the inaugural Pinch-Hitter course at the 1963 AOPA convention, forgot to lower the landing gear on the Beech Bonanza in which she was taking her training, and as a result made a wheels-up landing on the runway. Instead of being chastised or ridiculed for this expensive and embarrassing mistake, Whetsler was awarded "a battered flyer's mug—with one wing missing" at a dinner party later during the convention. The master of ceremonies also congratulated her on joining "the ranks of some of the world's greatest flyers, who also had made belly landings." Most important, even though she had wrecked the airplane, she passed the course without retaking it. To be fair, since she was not a real pilot, her instructor was ultimately responsible for ensuring the safe completion of the flight. Still, that's the point: no "real pilot" in training could ever hope to pass any check ride that ended in a crash.[121] At the very least, the student would have to not only endure the shame of failure and the derision of the fence crowd but also go back for remedial training before taking the test again and having an opportunity regain the respect of fellow fliers.

Not only did the Pinch-Hitter program attempt to remove all pressure to succeed; it even left the door open for what was popularly assumed to be the least masculine way to respond to a challenge: quitting. In a 1966 interview, Agnes

Streech admitted that before she registered for the course, "I thought of a hundred reasons and excuses why I shouldn't. . . . But because flying means so much to my husband, I finally signed up. I figured that if the going got too rough, I could always drop out." She was further reassured when course supervisor Don Sundin "stressed at the outset that we could withdraw whenever we wished." As intended by course organizers, Streech noted that being given this potential escape clause actually gave her "an added incentive to complete the course."[122]

Concentrating a large group of women students in the same time and space, and insulating them from the rest of the community of pilots (and aspiring pilots) by the very nature of the program, also meant that Pinch-Hitter students enjoyed a different experience on the ground. There were no catcalls or comments from a critical fence crowd if someone made a mistake or botched a landing. Like all fliers, Pinch-Hitters congregated between their ground school classes and flights with instructors for their own informal, all-woman hangar flying sessions. One 1966 article included a photograph of "Pinch-Hitters row," showing how course participants had set up folding chairs outside a hangar so they could "swap notes, await their turn at the aircraft controls, or just . . . bask in the warm California sun." But these informal meetings apparently bore little resemblance to the typical gathering of male pilots where talk of women, technology, and tales of derring-do or near-disaster dominated the conversation. Instead, a 1964 letter from Mrs. Bernie Shapiro suggests that Pinch-Hitter classes resembled "group therapy" meetings where women could openly share their fears and frustrations and expect to receive support and encouragement (instead of ridicule) in return.[123]

Even the student manual reassured potential Pinch-Hitters that they were not signing up for regular flight training. In fact, one did not need to open the manual to get this message; a statement printed on the front cover declared that this was "a curriculum designed for people who are not now pilots, and (we assume, for the purposes of this course) have no intention of becoming pilots." The cover went on to explain: "Purely and simply, the AOPA Pinch-Hitter Course has been designed to enable the nonpilot in the right-hand seat to take over the controls in case the pilot is incapacitated. The objective is only to train the nonpilot in just enough of the rudiments so that he or she can keep the plane under control, in level flight, then locate a nearby airport, and land safely."[124]

Despite the inclusive use of "he or she" on the cover, the rest of the manual was obviously and exclusively directed at women. All of the illustrations and photographs of Pinch-Hitter training in progress depict male instructors and female students. The intentional feminizing of flight instruction is also written into the text itself. As the cadre of instructors who developed the Pinch-Hitter Course sought to simplify flying for their intended audience—whom they viewed as inherently timid and technically inept—they fell back on widely accepted assumptions of the era regarding gender and technology. Rather than highlight the significant differences between operating airplanes and automobiles, a crucial early step in becoming a *real* pilot, the course's architects instead made a point of comparing flying to driving whenever possible. The introduction to the student manual reassured nonfliers, "If you know how to drive an automobile—and who doesn't these days?—it will be relatively easy to learn how to drive an airplane." And the next section, labeled "How to Control the Plane," compared aircraft to cars six different times.[125] This approach may indeed represent the quickest and easiest way to teach a neophyte the basic skills needed to "keep the plane under control, in level flight," and then muddle through the sky long enough to make an emergency landing. However, it also instilled poor piloting habits right from the start, habits that Pinch-Hitters would have to later unlearn if they ever decided to actually pursue a pilot's license.[126]

The course relied on the same approach to teach participants how to use the radio. As one instructor put it, "women spend enough time on the phone. . . . Just have them use the mike [microphone] like it was a telephone and don't worry about the technical language. Tell the ground station who you are and ask for what you need to know."[127] The student manual offered a sample radio conversation and then concluded in a tone that could be interpreted as either reassuring or condescending, depending on the reader's perspective: "Don't you feel better, more confident, now that you have talked to someone? It's like telling your problems on the phone to your closest friend."[128] This informal approach to radio communication worked well enough in an emergency situation, when air traffic controllers would clear the airwaves of other pilots to devote their full attention to the Pinch-Hitter in distress. However, this advice to use a radio "like it was a telephone" had nothing in common what actual student pilots were taught, and it had no place during routine flight operations when pilots and controllers exchanged information in a rapid-fire staccato of technical jargon to avoid monopolizing a single channel that served many users.

In addition to comparing the complex systems of an airplane to relatively easy-to-use, everyday technologies that were familiar to housewives, the instructors who created the Pinch-Hitter Course sought to demystify the cockpit by deliberately ignoring technical details. "All those dials and gauges on the instrument panel are not really as formidable as they appear to be," the manual reassured readers. "Perhaps you never paid much attention to them while flying as a passenger. Perhaps you felt they were mysterious mechanical gadgets which only a trained pilot could understand. Actually those instruments will be your guide and helper to a safe landing, and there are only a few of them that you will need to use." Those friendly "helpers" and "guides" (as opposed to "mechanical gadgets") included the airspeed indicator, engine tachometer, altimeter, and artificial horizon indicator. A bit later, the manual introduced three more instruments that could help the Pinch-Hitter navigate to the nearest airport: the compass, the directional gyro, and the VOR (very high frequency omnidirectional radio range) receiver.[129]

Rather than attempt to explain the complexities of each instrument and the information it provided to the pilot, the Pinch-Hitter Course focused on teaching participants just enough to get by. The section about the airspeed indicator, for instance, neglected to introduce concepts vital to routine flight operations, including optimal speeds for liftoff and clearing obstacles during takeoffs; the crucially important "best glide" speed that would help a pilot reach a safe landing spot in case the engine quit; or the many issues affecting the accuracy of airspeed readouts such as pitot tube icing, outside air temperature, and pressure altitude. Instead, Pinch-Hitters simply learned that, in addition to numbers, most airspeed indicators were marked with white, green, yellow, and red shaded arcs that signified specific speed ranges. As long as they always kept the airspeed needle somewhere within the green arc and avoided steep turns, the airplane would not stall and fall from the sky.[130]

The course organizers employed the same simplified approach to explaining aerial navigation. Although the manual acknowledged that a compass was tougher to use than a directional gyro and could be "difficult to read under certain conditions," it quickly added (without further explanation), "But don't be alarmed if you have only a magnetic compass; it is a reliable instrument." Any mention of the factors that made a compass "difficult to read" in a moving airplane—including magnetic dip, a confusing tendency for the needle to "lead" or "lag" during turns, and other issues like magnetic variation and deviation—were deliberately left out. And instead of having readers learn formulas and practice how to compute compass headings in their head, the manual instead advised a practical exercise that clearly assumed the reader was a housewife: "Stand with arms outstretched in the middle of your kitchen and consider that the sink is North. Now practice turns to various headings, turning your body as you take each new heading. If anyone asks what you're doing, simply explain that you're practicing turns in an airplane. They won't bother you any longer."[131] Postwar manuals for pilots-in-training assumed that students—most of whom were men—at least knew the difference between north, south, east, and west. The instructors who put together the Pinch-Hitter manual were not so sure this would be true of the wives of pilots, hence their advice for the Pinch-Hitter to stand in front of her sink, stretch out her arms to pretend she was an airplane, and slowly turn in circles as her first step to learning the basics of navigation.

It is possible to take the stark contrast between Pinch-Hitter training and regular flight instruction too far. Despite enthusiastic claims that the Pinch-Hitter Program might create "a new air age . . . of women!" the course was never actually intended to overcome the gender imbalance in the postwar population of pilots.[132] And, gender considerations aside, its intended audience, male or female, was made up of people who had no desire to become pilots but rather were taking flight training owing to extenuating circumstances (in this case, because they had a flying spouse). Nonetheless, it serves as a useful counterexample to traditional pilot training.

Flight instruction in the postwar era was such a highly masculinized affair that the AOPA—the largest organization for private pilots—believed that ordinary flight lessons would be ineffective in helping pilots' wives overcome their fears of flying. At the same time, its leaders were convinced that hands-on training was the best way help people overcome these fears. Thus, the AOPA invented a completely new approach, one that simultaneously imparted fundamental technical information and yet trivialized that same knowledge to make it seem less complicated. In addition, by removing the milestone of soloing—universally recognized as the first true measure of success for an "ordinary flight student"—the course creators further deemphasized the importance of individual skill and technical prowess. In doing so, they aimed to make flight instruction seem less threatening to women, a decision that reflected their own assumptions about gender and technical competence. Last, but by no means least, the fact that Pinch-Hitters were *not* training to be pilots made it permissible for instructors to treat them differently. Since full admission into the flying fraternity was not on the table, there was no need to put Pinch-Hitters through the informal hazing, in the cockpit or on the ground, that would help prove whether or not they had the "right stuff" to hack it as pilots.

Learning to fly encompassed far more than mastering the myriad technical tasks required by federal regulations and methodically laid out in various private pilot course syllabi. It also involved the complex dynamics of the student-instructor relationship, hazing from the fence crowd, lessons gleaned from hangar flying, and the symbolic traditions surrounding the first solo. Disguised as a normal part of flight instruction or as business-as-usual at the local airport, these informal aspects of flying educated students in practical matters of flying as well as how to behave as pilots. And, as the long-running debate about spin training reveals, sometimes the line between stick-and-rudder flight instruction and acculturation could become blurred. Whereas advocates of spin training believed that it made better and safer pilots, opponents described it as a throwback to the not-so-good-old-days of flying when only "supermen" were allowed to take to the air. These rites of passage and practical aspects of flight training served as a gatekeeper that demanded applicants pay their dues before acceptance into the highly masculine community of pilots. And just as important, if not more so, they helped drive away those who, for one reason or another, failed to fit in.

CHAPTER THREE

The Family Car of the Air versus the Pilot's Airplane

Technology as Gatekeeper to the Sky

One Sunday morning in November 1942, Brazil A. Freed decided to take his wife along on a routine flight from Moline, Illinois, to the nearby town of Galesburg. When they were ready to return home later that day, there was no one around to help Mr. Freed start his plane, which lacked an electric starter. He left his wife seated in the passenger seat and stepped around front to "hand prop" the engine himself. He spun the propeller manually until the motor sputtered to life, then shouted for his wife to ease the throttle back to idle, but she misunderstood his instructions and gave it full power instead. Worse yet, when she tried to correct her mistake, she accidently released the parking brake. Engine roaring, the airplane surged forward, knocking Mr. Freed to the ground as it accelerated across the airfield. Shaken but not seriously injured, he watched in horror as the pilotless plane lifted off into the sky, carrying his wife with it. Fearing the worst, Mr. Freed called for a doctor and an ambulance to come to the field. The airport manager, also fearing the worst, called the fire department and then moved the airplanes parked near the runway out of harm's way, just in case Mrs. Freed found her way back and tried to land. No one on the ground had any reason to believe this flight would end happily.

Meanwhile, since the plane lacked a radio, Mrs. Freed, who had no previous aeronautical experience beyond riding as a passenger with her husband a few times, had no choice but to teach herself to fly through trial and error. She discovered that when she turned what looked like a steering wheel slightly to the right or left, the plane turned right or left, just like the family car. She also found when she pushed forward on the wheel, the plane entered a shallow dive, and when she pulled back, it began to climb. Though different from a car, this too seemed intuitive enough. The throttle, which had caused all the trouble in the first place, turned out to be fairly straightforward as well: when she pulled the knob all the way out, the engine grew quiet; when she pushed it all the way forward, it roared to full power, just like it had during takeoff. Having figured this out, Mrs. Freed did the rest of her flying with the throttle positioned somewhere in between. After about 30 minutes of these experiments, she managed to return to the airport and, on her second attempt, land on the runway, bending the front landing gear and breaking the propeller when she touched down too hard but otherwise escaping "with nothing but a bit lip and a couple of bruises."[1]

Mrs. Freed deserves considerable credit for this feat of aeronautical prowess. It is impressive enough that she maintained the presence of mind to recover from the terror of finding herself alone in the air. In addition, whereas most student pilots have the benefit of 8 to 10 hours of hands-on training with a professional instructor before making their first solo flight, she managed to make it back alive (and nearly unscathed) after just 30 minutes of do-it-yourself lessons. But the plane deserves its fair share of credit, too. Thanks to its simplified control system, Mrs. Freed did not have to worry about rudder pedals as she learned how to steer through the sky. More importantly, by design the plane would not stall or spin, no matter how much she slowed it down or pulled back on the controls. If ever there was a flying machine that promised to "democratize" the sky by allowing anyone to take off into the not-so-wild blue yonder, this had to be it.

The airplane that carried Mrs. Freed aloft was a prewar design with its origins dating to the 1930s. Known as the Ercoupe, this "foolproof" personal plane entered limited production in 1939 and enjoyed a spectacular burst of popularity and sales immediately after World War II. Early accolades proved short-lived, however, and even before the personal plane industry was rocked by the downturn in private aviation in the late 1940s, fliers began denigrating the Ercoupe, leaving it with a tarnished reputation it would never overcome. Instead of an "Airplane for Everyman," members of the flying fraternity wanted—and deliberately chose—a "Pilot's Airplane" that required considerably more skill to operate than an Ercoupe. Ironically, the challenge facing the Ercoupe was that it delivered all too well on its promise of built-in safety and ease of use.[2] Most postwar private fliers were by no means "daredevils"; however, they built their identity as pilots around their mastery of complex machines and reliance on skill to overcome the dangers of the sky. As such, they had little interest in flying a "foolproof" plane, for not only did this take away the challenges that made flying enjoyable, but the term *foolproof* also implicitly labeled the individual who chose to fly one as a "fool." Just as bad, if not worse, the Ercoupe's unprecedented array of built-in safety features threatened to open the sky—and therefore the fraternity of pilots—to everyone, a prospect that many postwar private pilots seemed loath to accept.

Making Flying Foolproof

Historically, most efforts to create an airplane that almost anyone could fly were motivated by the longtime desire to use private planes as a practical form of personal transportation. Serious efforts to design and manufacture a foolproof airplane predated the Ercoupe by at least a decade, when flying was still limited to a just few thousand "intrepid birdmen" and private aviation was practically nonexistent. For instance, on the eve of America's entry into World War I, aviation pioneer Glenn Curtiss designed and built what he called an "Autoplane." Curtiss gave up after his machine—a hybrid between an automobile and an airplane—reportedly made a few short straight-ahead hops but failed to achieve sustained flight.[3] In the mid-1920s, Henry Ford announced that he would apply his principles of mass production to create an aerial Model T for the masses, but he scrapped his plans after his chief pilot, Harry Brooks, died testing a prototype of this so-called flying flivver.[4] Several years later, the U.S. Department of Commerce sponsored a contest to encourage entrepreneurs to design and build a $700 Airplane for Everyman that would cost no more than a typical family car. Bureau of Air Commerce director Eugene L. Vidal devised this competition to simultaneously unlock what he believed was private aviation's great untapped potential and bolster the nation's flagging Depression-era aircraft industry. Using New Deal funds, the federal government invested in several private research projects to accomplish these goals. The production version of the winner, the two-place Stearman-Hammond Y, incorporated several innovations that make flying somewhat safer and easier, but at $3,200 it cost nearly five times the target price. Other submissions were even more expensive (two exceeded $12,000), delivered lackluster performance, and at least one was reportedly downright dangerous to fly.[5] In terms of putting an airplane in every garage, the government's Airplane for Everyman program was a failure.

The genesis for the Ercoupe came after Ford's failed foray into private aviation but before Vidal announced his contest. In the early 1930s, Fred Weick, a promising young aeronautical engineer employed by the federal government's National Advisory Committee for Aeronautics (NACA), started working in his spare time to design a personal plane that was safe and easy to fly. Using his own funds and with help from friends, Weick built a working prototype called the W-1 in his garage.[6] In 1936 Weick left the NACA to join the Engineering and Research Corporation (ERCO). There, he went to work on a production model based on lessons learned from his earlier experiments with the W-1. Dubbed the Ercoupe (generally pronounced "air-coop," a combination of the company's initials and the word *coupe*, a common label for a two-seat automobile), ERCO built just over 100 of these small two-seat airplanes—including the one in which Mrs. Freed taught herself to fly—before World War II interrupted production. Nonfliers and aviation community insiders alike heralded the Ercoupe as the private plane of the future. In December 1940 Richard Thruelsen, an editor for the *Saturday Evening Post*, announced, "Get ready today, groundlings, for tomorrow we fly" after previewing Fred Weick's revolutionary plane. He went on to conclude that thanks to the Ercoupe, "The eagle is changing into the sparrow and uncounted thousands of us will be flying soon."[7] Postwar prospects for the new design seemed bright.

Even inexperienced observers could see at first glance that the Ercoupe was different. All other aircraft of the era were equipped with two main wheels under the wings and a smaller wheel located beneath the tail. This "conventional" landing gear configuration was inherently unstable during landings and gave all such "taildraggers" an unnerving tendency to swerve out of control on the runway in a violent and occasionally disastrous maneuver called a *groundloop*. Weick equipped his new design with a nosewheel instead of the conventional tailwheel, making the Ercoupe the first widely produced aircraft to employ tricycle landing gear, which remains the industry standard to this day. This difference made the plane inherently stable on the ground, which in turn made it considerably easier for pilots to execute a safe landing (fig. 3.1).[8]

Weick also simplified the Ercoupe's controls. One of the many differences between flying and driving is that changing direction in the sky normally requires a pilot to coordinate three separate controls (rudder, ailerons, and elevator), whereas turning a car on the ground involves just one (the steering wheel). In most aircraft designed before World War II (and many prewar designs built during the immediate postwar period), simply pushing the stick or moving the control wheel in the direction one wished to turn, without also applying the proper amount of rudder via a pair of foot pedals, resulted in a condition known as "adverse yaw." This meant that although the plane banked (tilted its wings) in the direction that the pilot wanted to turn, the plane continued to travel straight ahead or, in extreme cases, entered a sloppy, wallowing turn in the opposite direction. Even if the pilot used enough rudder to get the plane going in the correct direction, failure to continuously adjust the proper combination of aileron and rudder input throughout the maneuver resulted in an "uncoordinated turn" and caused the plane to "slip" or "skid" through the sky. At the very least, this maneuver was inefficient, clumsy, and uncomfortable for the pilot and passengers to the point that it could induce airsickness. At worst, an uncoordinated

Fig. 3.1. ERCO Ercoupe. The first (and only) supposedly "foolproof" personal plane to enter production, the ERCO Ercoupe is shown here circa 1940s with its distinctive twin tail and revolutionary tricycle landing gear clearly visible. Hans Groenhoff Photographic Collection, Smithsonian National Air and Space Museum (NASM HGC-804).

turn executed at low speed and low altitude, such as in the airport traffic pattern, could cause the aircraft to enter a potentially deadly stall or spin.[9]

Weick overcame the problems of adverse yaw and uncoordinated turns by connecting the Ercoupe's wing-mounted ailerons to its tail-mounted rudder so that simply turning the plane's control wheel automatically moved both the ailerons and the rudder the right amount.[10] This, in turn, allowed him to dispense with the floor-mounted rudder pedals altogether. Just as Mrs. Freed had figured out on her own, except for pulling back to raise the plane's nose or pushing forward to dive toward the ground, pilots used the Ercoupe's control wheel to steer through the sky exactly as they would use a car's steering wheel to drive down the road. Weick employed the same concept to make the Ercoupe easier to handle on the ground. In an airplane equipped with conventional controls, the control wheel or stick (which operates only the elevators and ailerons) does nothing to turn the plane while it is taxiing. Instead, pilots must learn to steer using their feet on the rudder pedals to turn the tailwheel or nosewheel, combined with "differential braking" using two separate floor-mounted brake pedals (one for each main wheel). This arrangement differed greatly from what drivers used in a car: to stop, pilots had to carefully apply equal pressure to both brake pedals, since pushing on just one would cause the plane to swerve violently off the runway. Weick simplified ground handling considerably by connecting the Ercoupe's control wheel to its steerable nosewheel so that pilots could "drive" the plane on the ground just like a car, and he replaced differential braking with a single automotive-style brake pedal that applied equal stopping power to both main wheels.[11]

Early articles and advertisements about the Ercoupe boasted that, thanks to these simplified controls, flying was now even easier than driving, a claim based on the fact that most cars of the era were equipped with manual transmissions. One ERCO ad declared: "When you drive a car, you have to coordinate footwork and handwork [with the clutch pedal and shift lever]. When you fly your Ercoupe, your hand on the wheel is enough."[12] Maxwell Hamilton, who had never been in an airplane of any kind before he went aloft with an instructor to review the Ercoupe for *Mechanix Illustrated*, exclaimed: "I Flew My First Time Up!" Hamilton's opening lines—"Now it's easier to fly a plane than it is to drive an automobile. You needn't spend months learning—you just climb in and take off!"—describes both the content and tone of the rest of his 1941 article.[13] Another reviewer, this one an experienced pilot writing in 1946 to an audience that included both aviation enthusiasts and fellow fliers, compared flying the Ercoupe favorably to driving an automobile no less than four times in a single page.[14]

Weick's third innovation, although invisible to the naked eye, was even more significant than the plane's revolutionary landing gear or its simplified controls. At the time, around half of all fatal accidents involving private aviation occurred when pilots inadvertently allowed their airspeed to drop below the minimum needed to maintain flight, causing their aircraft to stall or enter an uncontrolled spin. This type of mishap usually occurred shortly after takeoff or just before landing, when the plane was already flying slowly and the pilot, busy with multiple tasks, could easily lose track of the airspeed. To make matters worse, if a plane stalled under these conditions, it was often too low for the pilot to regain control before hitting the ground.[15] Weick eliminated this problem by designing the controls such that a pilot could not force the wing to exceed the critical angle of attack and stall. Instead, the worst that a heavy-handed pilot could do was put the airplane in a relatively flat, "mushing" glide toward the ground.[16] The CAA rewarded Weick's efforts by certifying the Ercoupe as the first (and, as it turned out, only) "spin-proof" production aircraft, and even created a new type of private pilot's license with reduced training requirements for students who learned to fly in this kind of airplane.[17]

ERCO resumed production of the Ercoupe at the end of World War II, and orders in 1945 and 1946 far exceeded capacity. The company also started work on two larger, faster models that met the same safety standards as the Ercoupe: the four-seat, 165 horsepower "Ercoupe Four," and an easy-to-fly twin-engine plane with five seats named the "Ercoach."[18] At least one other established aircraft company, Aeronca Aircraft Corporation, took steps to develop its own plane, the Chum, with the same safety features as the Ercoupe.[19] For a short time, Macy's Department Store in New York City offered Ercoupes for sale, and Bamberger's Department Store of Newark, New Jersey, advertised that it had an Ercoupe on display. But even as ERCO expanded its factory in College Park, Maryland, the demand for personal planes fell through the floor. Despite optimistic prewar and wartime predictions to the contrary, offering a safer, easier-to-fly airplane wasn't enough to turn the public loose in the sky. For one thing, with a price of $3,150 in 1946, a new two-seat Ercoupe cost about 35 percent more than a brand-new Cadillac four-door sedan.[20] And even if a family could afford to purchase and maintain both an airplane and an automobile in an era when two-car ownership was still rare, practicality proved to be a problem. Whereas automobiles could safely navigate all but the worst weather, unless a pilot could afford costly blind flying instruments and was willing to complete the complex (and expensive) additional training required to use them, airplanes were grounded by conditions that wouldn't even slow highway traffic.[21]

With the market in a nosedive, ERCO executives gave up on the fickle field of private aviation in the late 1940s, and the company moved on to other, less volatile ventures.[22] Production rights for the Ercoupe changed hands several times over the next two decades as first Sanders, then Forney, and finally Alon tried to cash in on a perceived demand for a super safe, user-friendly airplane.[23] Alon also attempted to introduce a larger four-seat version called the Model A-4, but this project, like the ill-fated Ercoupe Four before it, was shelved when the company went out of business in the mid-1960s.[24] Mooney Aircraft purchased production rights later that decade in hopes of marketing a modified version of the Ercoupe as an entry-level trainer, but the company soon abandoned this project after selling fewer than 120 planes. Most of the nearly 5,700 Ercoupes and follow-on variants ever produced were built prior to 1947, an astounding 4,300 of these in 1946 alone. Although the design still has a faithful following to this day, this safe, easy-to-fly plane never carved out a sustainable niche in the mainstream market for personal planes.[25]

In contrast, Cessna Aircraft Company met with success in the personal plane market by shifting its focus *away* from easy-to-fly aircraft. During World War II, the company had wholeheartedly embraced the same concepts of simpler and safer flying by design that Fred Weick engineered into the prewar Ercoupe. When Cessna started marketing its own "Family Car of the Air" in 1942, advertisements promised future buyers that its engineers had already drawn up plans and production would begin as soon as the war was won. Endorsements by experts such as Captain Edward A. Stewart, United Airlines senior pilot, who described how he had examined the blueprints in person and came away convinced that this was "an airplane that the average person can buy and fly," added credence to the company's claims.[26] Later ads were even more explicit, promising that anyone who could drive a car would be able to fly.[27]

Then in October 1945, just weeks after World War II ended, without warning Cessna officials publicly executed an about-face. After several years of promising consumers a postwar Family Car of the Air, the company introduced a new sales pitch touting "Cessna, The Pilot's Airplane." The first advertisement with this new theme declared: "If you are a pilot now you want a pilot's airplane as all statements from pilots indicate. If you are not a pilot now, you will be when you start flying. Then, you, too, will want a pilot's airplane."[28] Later ads were slightly more specific: "Every Cessna will be a 'Pilot's Airplane,'—not in the old sense of a hot, hard-to-fly plane. That's as old-fashioned as the open cockpit. Today a pilot's airplane means safety, high performance, speed, and full, easy, positive control."[29] Although Cessna's new marketing campaign never mentioned the Ercoupe by name, remarks about pilots wanting "full, easy, positive control" were clearly directed against Weick's simplified control system, which threatened to replace pilot skill with technology. This new marketing campaign acknowledged that while any airplane that promised to make flying as easy as driving might sound great to members of the nonflying public, no real "pilot" would actually want such a plane once he (or she) had learned to fly.

Cessna explained that it derived the specifications for its postwar Pilot's Airplane from "thousands of letters and interviews" collected between 1942 and 1945.[30] This suggests that despite claims in dozens of previous advertisements for its Family Car of the Air, the company did not in fact have a production-ready design on the drawing board. This possibility is reinforced by Weick's recollection that Cessna's sales manager Don Flower "was thoroughly familiar with my work on the Ercoupe" and that Cessna had purchased two Ercoupes "in order to study the design carefully" during the war. Even more damning is Weick's description of a phone call he received in early 1945 from Cessna's president and general manager, Dwayne Wallace, just months before the company suddenly switched to promoting its more challenging Pilot's Airplane: "I was surprised to hear Wallace invite me to come with Cessna for the purpose of designing the Family Car of the Air. This was an offer I hated to turn down, but at the time I felt that I had to."[31]

In the end, Weick decided to remain with ERCO, and Cessna was on its own to come up with a postwar design. Whether it was an in-house engineering team that was unwilling (or unable) to incorporate Ercoupe-style safety technology without Weick at the helm, fears that licensing fees for Weick's patented innovations would drive up costs too much to be competitive, or, as the company publicly claimed, a direct response to input from thousands of pilots about which features they did and did not want, Cessna's first generation of postwar personal planes in no way resembled its much touted Family Car of the Air. Instead of producing an airplane that anyone could fly with minimal training, company engineers stuck with conventional (taildragger) landing gear and conventional (nonsimplified) controls. Furthermore, unlike the stall-proof, spin-proof Ercoupe, Cessna's so-called Pilot's Airplane was fully capable of stalling and spinning.[32] And although sales never reached the numbers that overly optimistic wartime public opinion polls and industry analysts had predicted, they sold better than anything else on the market. Between 1946 and 1956, Cessna found customers for 7,600 two-seat model 120s and 140s and more than 5,100 four-seat model 170s (essentially a stretched version of the 140 with a larger engine and bigger wings to carry more weight), for a combined total of roughly 12,700 aircraft.[33] Thus, in just 10 years, the company's entry-level personal planes outsold the entire production run of easier-to-fly Ercoupes by a ratio of more than 2:1.[34] On the basis of sales alone, postwar pilots chose Cessna's so-called Pilot's Airplane over the Family Car of the Air ideal embodied in Fred Weick's revolutionary design.

Not a Real Airplane

Some Ercoupe fans insist that the plane failed to catch on because it was "too advanced for its time," and even its detractors found few purely technical faults with the sturdy little plane.[35] But Richard Thruelsen's December 1940 *Saturday Evening Post* article that had optimistically proclaimed, "Get ready today, groundlings, for tomorrow we fly," contained thinly veiled hints that a significant gulf existed between the Airplane for Everyman ideal that inspired the Ercoupe and the kind of airplane that existing pilots actually wanted. After the *Saturday Evening Post* editor, a former army aviator, settled into the pilot's seat, ERCO president and CEO Henry A. Berliner warned him, "The first thing I want you to do is to forget, if you can, everything you know about flying." As the flight progressed, Thruelsen admitted that the plane was so easy to handle, "I began to feel a little foolish. Was this flying? Was this the art which required what the Army called, in my training days, 'inherent flying ability'?" Later, as he watched the plane that he had just flown depart into the setting sun, he reflected nostalgically that it seemed like "the mystery of the air and the glamour of flying were disappearing over the horizon. . . . It's getting too simple."[36]

Throughout his article, Thruelsen focused on a not-so-distant future where "The eagle is changing into the sparrow and uncounted thousands of us will be flying soon," an idea that many of the so-called eagles did not relish. By the time ERCO resumed production after World War II, many of these eagles—that is, experienced pilots who had learned to fly in traditional airplanes during the war and who apparently reveled in the "mystery" and "glamour" of aviation—were already denouncing the Ercoupe. When Max Karant reviewed the plane for *Flying* magazine in 1946, his opening remarks described how a schism had developed within the flying fraternity: "No one ever talks about the Ercoupe in calm, dispassionate terms. Pilots and plane owners either are violently pro-Ercoupe or they pound tables for hours on end trying to convince you that it is one of the world's most dangerous aircraft." Karant, an experienced flier whose positive comments about the plane placed him squarely in the pro-Ercoupe camp, summed up the main points of contention in a single statement: "Tricycle gear and two-control system are most often condemned by conventional-plane pilots."[37]

One complaint was that the Ercoupe sacrificed pilot control in the name of safety. Weick's ingenious method of linking the plane's rudder to the ailerons had significantly reduced the amount of skill needed to execute a properly coordinated turn in flight. But at the same time, by removing the rudder pedals, he also reduced pilots' ability to execute certain maneuvers. Many early postwar personal planes (including the Ercoupe) lacked wing flaps, which are used during landing to slow the plane by creating more aerodynamic drag. In planes without flaps, pilots employ a maneuver called a sideslip to quickly lose altitude during final approach to the runway. To execute a sideslip, a pilot banks the wings in one direction and applies opposite rudder with the foot pedals, creating adverse yaw on purpose. By the 1950s, most new planes came equipped with flaps, so pilots no longer had to use a sideslip during normal landings. However, virtually all postwar planes, with or without flaps, required pilots to master a variation of this maneuver to land in a crosswind (i.e., when the wind blows from the side rather than directly down the runway). Correcting for a crosswind involves slipping the plane into the wind and applying exactly the right amount of opposite rudder to keep the nose aligned with the runway throughout the final approach and landing. Called a "cross-controlled slip" because the pilot uses the stick and rudder in opposition rather than together (e.g., banking to the left and applying the right rudder to counter a left crosswind), this procedure can be difficult and demanding, especially in high winds or gusty conditions. Even when executed properly, this type of landing is dramatic to watch: instead of coming in with wings level and settling to the ground on both main wheels, the airplane completes its final approach banking into the wind and touches down on just one wheel. Thus, the rudder—and proper use of the rudder pedals by the pilot—plays a pivotal role in the success or failure of the single most demanding task in flying: the landing.[38]

Lacking rudder pedals, the Ercoupe cannot be slipped. Instead, its pilots must use alternate techniques that demand far less skill than a properly executed slip but are also far less elegant in execution. To descend rapidly without picking up unwanted airspeed, the Ercoupe owner's manual instructs pilots to pull back on the controls to place the plane in a mushing stall-like condition on final approach—a deadly dangerous trick in any conventional plane because of the risk of stalling and spinning in—and then to simply push the nose down to level off just before touchdown. To land an Ercoupe in a stiff crosswind, the pilot "crabs" into the wind with the aircraft's wings level and its nose pointed several degrees off the actual direction of flight over the ground (some pilots reportedly boasted of landing in a direct crosswind of 40 miles per hour with the plane's nose cocked 30 degrees to the right or left of the runway centerline). When the plane touches down in this manner, the result is screeching tires and a significant jolt to the occupants as the plane abruptly straightens itself out (with no input from the pilot) to roll straight down the runway. Weick deliberately designed the Ercoupe's heavy-duty landing gear to absorb this extra punishment, and relied on the inherently stable tricycle gear configuration—not pilot skill—to correct the plane's path after it hit the runway.[39]

These techniques were so unconventional that Ercoupe fans cautioned conventionally trained pilots to "forget, if you can, everything you know about flying" before setting foot in the little plane.[40] Max Karant went even further when he declared in 1946 that an experienced aviator "must learn to fly the airplane in the manner to which it, not he, is accustomed. And until he makes up his mind to that, the old-timer is the most dangerous of all Ercoupe pilots."[41] Critics focused instead on the negative effect that the Ercoupe had on pilot skill and warned, "It'll just plain ruin your flying" under the presumption that anyone who became habituated to these methods would forget how to land properly and thus become dangerous behind the controls of any other airplane.[42] When Karant, still a dedicated Ercoupe supporter, reviewed the larger and faster Ryan Navion in 1948, he observed that "the Navion is an ideal modern airplane for the ham-handed amateur. Ercoupe pilots, particularly, take to this airplane instantly."[43] While he meant this as a compliment for the Navion's relatively docile behavior compared to other planes in its class, this statement could easily serve as further proof to critics that the typical Ercoupe pilot was, by definition, a "ham-handed amateur."

Not only did critics attack the Ercoupe, but they also disparaged, to varying degrees, the people who wished to fly them or any other aircraft with similar features. In a 1943 letter to *Air Pilot and Technician*, private pilot W. L. Jones of Park Ridge, Illinois, described how he hoped someone would offer a postwar personal plane that was faster and carried more than the diminutive two-seat Ercoupe but still offered the same patented safety features, including "coordinated rudder and aileron control," "tricycle landing gear," and a design that would render it "non-spinnable." According to Jones, his personal preference to "rely on good sense and better airplanes" instead of luck and skill to avoid trouble in the sky had led fellow pilots to "catalog me as a sissy flyer."[44] In 1945, as production of personal planes resumed, longtime pilot Swanee Taylor wrote a scathing critique of simplified controls as part of an Ercoupe debate published in *National Aeronautics.* Capping off numerous examples intended to prove that simplifying aircraft controls was a misguided idea, Taylor argued that removing the rudder pedals "is nothing short of wanton murder of a pilot's best friend" because "a rudder never hurts anybody, unless it is jammed, and it has pulled pilots out of hundreds of thousands of tight spots," especially through the use of sideslips. He concluded by declaring unequivocally that simplified controls were appropriate only for "oldsters, simpletons and sissy-britches aeronauts who won't fly otherwise."[45] At least one Ercoupe pilot (and perhaps many more) took offense, as evidenced by A. Naughton Lane's letter to the journal's editor published two months later. "My main purpose in writing is to express a small amount of resentment at being branded as an 'oldster' (I'm 34) 'a simpleton' (I'm vice president and general sales manager of my company and a college graduate) and 'a sissy-britches aeronaut' (I don't know what he means by that)." Lane continued, "If he means that *he* and all other flyers should be the bravado, stunt flying heroes of the air, I'm afraid my flying days are over."[46] A decade later, private-pilot-turned-author Frank Kingston Smith took a milder tone than Taylor by damning the plane (and its pilots) with faint praise. After informing readers that the Ercoupe was "a good airplane for cloud loafers," he immediately clarified that *he* was no cloud loafer by adding, "I like standard controls better."[47]

Patronizing remarks like "cloud loafers" in postwar reviews of the Ercoupe outnumbered outright attacks on character like "sissy flyer" or "sissy-britches aeronauts," but pilots still found ways to question the manliness of those who were willing to surrender so much control over their airplanes to built-in safety features. Such was the case in 1958 when Ray Withman, aviation editor for *Science and Mechanics* magazine, flight-tested the Forney Aircoupe (one of several attempts to revive Weick's design). He began his article by explaining the underlying attraction of private flying to his readers, most of whom were not pilots. "Precision flying is like making love. From the pilot, it requires the same highly tuned sensitivity to the peculiarities of his plane and the forces acting on it. And it yields something, more than the simple business of getting from one place to another, that is almost impossible to describe to the non-flyer." This, he noted, "is why flying is habit forming to a degree that [it] almost becomes a vice for many of us nonprofessionals when it strains our budgets." According to Withman, because the Aircoupe did not require this level of intimacy with the aircraft, flying it was inherently less satisfying for real pilots. "In the Aircoupe, you have to sacrifice some of these sensations. In making a turn, you don't have to do any of the feather-touch coordinating between stick or wheel and rudder pedals that you are accustomed to. The Aircoupe does the coordinating for you mechanically, and there is nothing whatever to do with your feet—no pedals." After noting that the plane's interconnected controls did "a reasonably good job," he observed: "If you are willing to forego the precise coordination that's such a joy to sense in a skilled pilot, this little plane will get you where you want to go efficiently, neatly and safely." Withman reported that Forney was "tooled up to make as many as four or five a day of the new, significantly better, Aircoupes" and predicted—wrongly, it turns out—that they would sell readily. However, he was careful to qualify who would and would not want these planes. In his assessment most buyers would not be "skilled pilots" like himself. Rather, the type of people who would want this airplane were interested in flying only as a means of getting from point A to point B. Although he never used Cessna's trademarked motto of the immediate postwar era, in his opinion an Aircoupe did *not* qualify as a Pilot's Airplane.[48]

Another review of the Forney Aircoupe, this one written in 1957 by Alice S. Fuchs for *Flying* magazine, highlighted the connections between gender identity and pilots' desire to feel that they were in control of the aircraft. Fuchs herself believed that the Aircoupe was a perfectly acceptable design, but she noted that many of her fellow fliers disagreed. "Pilots are a strange, proud breed and there is a feeling that man should be master of his fate—at least to the point of pushing his own rudder pedals. Without admitting to it themselves, these pilots feel there is something rather unsporting about making pilotage too easy, that a real he-man airplane should be more difficult to handle."[49] This tongue-in-cheek observation, meant to poke fun at pilots she described as "the young-in-heart who dream of the helmet and goggles days and want their flying rugged," helps explain where derogatory labels like "cloud loafer" and "sissy-britches aeronaut" came from. In a society that considered "autonomy, mastery, [and] technological skill" to be hallmarks of masculinity, any technical innovations that so completely devalued these character traits in flying automatically threatened a pilot's individual masculine identity.[50]

Consciously or otherwise, even the aircraft's most devoted fans employed terms and anecdotes that challenged time-honored connections between flight, skill, and masculinity. Lloyd Mangrum admitted in the late 1950s: "I believe that any idiot could fly an Ercoupe. That's why I bought one."[51] Fred Weick preferred to use stories involving women to emphasize the user-friendly features he had designed into the plane. For instance, in 1957 he sold his own Ercoupe to Tex Anderson, a friend and fellow pilot who wanted his wife to learn to fly. As Weick recalled in his autobiography, Mrs. Anderson "was having difficulty learning to fly with a taildragger" but quickly caught on and earned her private license after she switched to the docile Ercoupe.[52] He also loved to describe the story used to open this chapter, which he called "the unexpected flight of Mrs. Freed," as further evidence that the plane was so user-friendly that it almost flew itself.

Even more amazing than Mrs. Freed's totally unplanned solo were cases involving children who took off in stolen Ercoupes and lived to tell the tale. For instance, in 1948, two boys, aged 11 and 12, piloted a pilfered Ercoupe 120 miles from Oklahoma City to Cheyenne, Oklahoma. After making "a perfect landing" in a field near Cheyenne, they caught a horse and rode it to the outskirts of town. The boys later explained that they figured out how to fly the Ercoupe using information gleaned from comic books. According to State Highway Patrolman Arch Hamilton, "They thought we were silly not to know how."[53] A few years later and half a continent away in Long Island, New York, 13-year old Eddie Cates took his 11-year-old friend Roy Brosseau up for a ride in a "borrowed" Ercoupe. The older boy had flown once before as a passenger and claimed to have learned everything he needed to know just by watching the pilot. According to the newspaper account of their misadventure, which made the front page of the *New York Times*, "They flew the craft out over Great South Bay, waggled its wings over Center Moriches and Eastport, and buzzed their own homes not far from the airport at treetop height. They also flew the length of Fire Island and performed a bit over Suffolk Air Force Base at Westhampton. After an hour of adventure, Eddie brought the plane down on its tricycle gear for a neat landing." The reporter seemed impressed, even envious. "Two small boys have done what every boy dreams of doing. They have gone up in an airplane alone, and thereby hangs a tale that has this community buzzing with excitement today."[54] Stories like these of hapless housewives and mischievous children who literally taught themselves to fly an Ercoupe may have sold newspapers, but they also eroded pilots' respect for the aircraft and anyone who chose to fly them.

Although Fred Weick was proud that his design was inherently safe and easy to use, he was never comfortable with widespread claims that it was completely foolproof. In a paper titled "Development of the ERCOUPE . . . an Airplane for Simplified Private Flying" that he presented at the annual meeting of the Society of Automotive Engineers in March 1941, he described his designed-in safety features in detail but then cautioned, "This does not mean, however, that the airplane is foolproof. Safe operation requires that the pilot know the capabilities of the airplane and operate within them."[55] Nearly five decades later, and with the advantage of hindsight, Weick expanded on these reservations about the plane's reputation. "Looking back with today's perspective . . . I believe the CAA's recognition of the Ercoupe as an extraordinarily safe airplane did more harm than good." He followed this up by joking, "It's the old story of the most important 'nut' in the airplane being the 'nut' at the controls!" Weick concluded by musing that technology alone could not make flying less dangerous and observed, "It is safer to fly with a very competent pilot in a tricky and hard-to-fly airplane than it is to fly with an incompetent pilot in an airplane that is easy to fly safely."[56]

As Weick's reflections suggest, the Ercoupe's forgiving nature enabled people who lacked skill, sound judgment, or both to gain access to the sky in the immediate postwar era, thus setting the stage for accidents caused by blatant pilot error. One pilot crashed near Manhattan Beach, California, after flying a severely overloaded Ercoupe into fog, leading the AOPA to remark that "even the Ercoupe can't overcome that type of pilot judgment."[57] Another pilot, on his way home after picking up his newly purchased Ercoupe, spotted his wife driving below. A little too eager to show off to "the missus," he knocked the nosewheel off on a wire across the road as he buzzed her car. To make matters worse, he failed to realize what had happened until he crash-landed his new plane—minus its nosewheel—on the runway at the local airport.[58]

Apparently fooled by the Ercoupe's mild manners, some aspiring aviators never even bothered to finish their training before attempting to show off their aerial prowess. One student pilot celebrated his high school graduation with a night flight in an Ercoupe, despite the fact that he had not yet soloed, even in daylight. Fifteen minutes after takeoff, he flew into the ground. The airplane was destroyed; fortunately, he and his companion sustained only minor injuries.[59] Another case, ready-made for a modern-day morality tale, indicates that such antics were not limited to the young. As the rightful owner of an Ercoupe reportedly sat in church in Kansas, an inebriated 40-year-old former student pilot (his license had expired years earlier, and it was doubtful that he had ever soloed) stole the plane, took off, and disappeared. Two days later, a fence rider found the thief—deceased—in the wreckage some four miles from the airport.[60] One accident after another seemed to prove that even if talented engineers like Fred Weick could design a plane loaded with foolproof features, they could not build one that was, as some pundits put it tongue in cheek, "'damn-fool' proof."[61] These and other stories about "damn-fool" pilots who managed to mishandle what early advertisements had touted as "the world's safest airplane" provided ample proof to members of the anti-Ercoupe camp that even the most ingenious technology could not completely deskill flying.[62]

It should come as no surprise that not everyone who chose to fly an Ercoupe qualified as a "damn fool." A 1953 article described Dr. Harold C. Hand, a college professor who learned to fly in 1948 at the unusually advanced age of 46, just as the GI Bill flight training frenzy peaked (there is no indication that he was a veteran). By 1953, Hand—now a 51-year-old grandfather—regularly used his Ercoupe to fly from his home in Illinois to speaking engagements across the country, logging an impressive 1,350 hours and an estimated 130,000 miles in his Ercoupe on multiple trips to the West Coast, Florida, New York, and points in between. Far from foolhardy, he prided himself in being a safe pilot—"I don't take chances with the weather"—and he reportedly planned all trips with an alternate means of travel in mind in case marginal conditions left him grounded. In addition, he steadfastly refused to fly on moonless nights. Instead, according to the article, "When he reaches the airport where he plans to spend the night, he obtains permission to sleep in the operations office." Professor Hand apparently had a good sense of humor about the impression he made when he showed up in his Ercoupe with a bedroll tucked under his arm, for he told the article's author, "The boys around the hangar never fail to chuckle at the sight of an old grandpappy like me breaking out a sleeping bag."[63]

The same year that *Flying* carried the article about Professor Hand, Frank Fulkersin of Flint, Michigan, organized the Michigan Ercoupers club. According to an article published a year later in 1954, he obtained the names and addresses of all 150 Ercoupe owners in his home state from Michigan's Department of Aeronautics, then sent each a letter proposing his idea of "banding together for fun, frolic and flight." The six photographs accompanying this article suggest that the typical "Michigan Ercouper" was considerably older than the average private pilot of the era. Many of the men have thinning hair and expanding waistlines and look as if they could be in their 50s or older; a like number of the women (most of them identified as wives of the pilots) resemble white-haired grandmothers.[64]

If Hand, Fulkersin, and the rest of Fulkersin's Michigan Ercoupers represent the "typical" Ercoupe owner throughout much of the postwar era, even though they didn't come across as "damn fools," there was little about their staid image to dispel common criticisms of the plane. Instead, Ercoupe owners appeared to be older than the average flier—senior citizens, even—which played into original marketing claims that the plane was so easy to operate that *anyone* could take to the air. Furthermore, hints that some had chosen the plane because they lacked the skill and confidence to handle the challenging conditions inherent in aviation—take, for example, Hand's adamant refusal to fly at night—further reinforced preconceived notions within the aviation community that Ercoupe drivers were less skilled than "real pilots."

This attitude lingered in private flying circles long after the last Ercoupe variant was produced. A 1979 article on used planes observed that if one purchased an older "classic" to save money, in most cases this would not result in being stigmatized by owners of newer, flashier aircraft at the airport. "The odds are you will be known as the Stinson pilot. Or the Luscombe pilot. Or the one with the Aeronca or the T-Craft [Taylorcraft]. From there on down, you may be looked upon as the Tri-Pacer owner. Or the Ercoupe pilot." Here the author paused to explain. "Wonder why the Ercoupe is last on the list? In many ways, an Ercoupe is a better flying machine than a [Globe] Swift, yet you can gather up bundles of fair to really good Ercoupes at $3,500 or $4,500. Yet a Swift of the same vintage, with its tricky reputation and P-40 silhouette wings, is worth about $12,000. If you can find one."[65] And this was no isolated opinion that pilots generally held Ercoupes (and the people who flew them) in low esteem. A used-plane review published in 1980 observed that "bringing up the Ercoupe idea to any group of pilots is a call for an animated and opinionated debate." After explaining that since the plane had "no flaps, no rudder inputs, trouble-free crosswind landings and inherently coordinated flight, there simply wasn't much for a pilot to do," the author noted that critics still complained that the Ercoupe encouraged sloppy flying habits and made for "an incomplete pilot experience." The author of yet another review, this one published a decade later in 1990, admitted up front that Weick's design was frequently "dismissed as 'not a real airplane.'" Even more recently, Ercoupe owners complained how some fellow fliers still used the word "sissy"—probably behind their backs—to describe anyone who flew a plane that lacks rudder pedals.[66] Thus, defining a "real airplane" hinged upon how much skill it required, determining who qualified as "real pilots" depended upon whether or not they flew a "real airplane," and anything (or anyone) that failed to measure up to these standards in the manly world of postwar private aviation risked being described in unflatteringly unmasculine terms.[67]

Safety Doesn't Sell

Although the Ercoupe provides a ready-made case study of masculinity-in-action within the community of pilots, debates regarding individual skill versus built-in safety features extended far beyond Fred Weick's revolutionary design. Throughout the postwar era, a handful of aviation commentators complained that private fliers consistently resisted or rejected even relatively minor design modifications and accessories precisely *because* they promised to make flying less dangerous. In 1955, 14 years after Fred Weick had described his revolutionary Ercoupe before an attentive audience at a meeting of the Society of Automotive Engineers, Arthur L. Klein declared in a speech delivered during the same group's Golden Anniversary convention, "The problem of safety for aircraft is largely psychological." Klein, a professor of aeronautics at the highly regarded California Institute of Technology, went on to assert (possibly in direct reference to the Ercoupe's recent demise): "It is notable that 'safe' small aircraft do not sell. People do not buy small aircraft because they are safe; they buy them to inflate their ego and therefore the more dangerous, the better." Although he had no professional qualifications as a psychologist, and interviews later in life revealed that he also had little interest in (and less sympathy toward) private aviation, Professor Klein had no qualms about offering his "expert" opinion on why private pilots and other alleged risk-takers behaved the way they did. "The psychological attitude [of private pilots] is similar to that of mountain climbers and hot rodders, and if everyone was properly adjusted, none of these [traditionally masculine hobbies] would exist."[68]

One year after this speech, *AOPA Pilot*, which was far more supportive of private aviation than Professor Klein, published an editorial as part of a long-running campaign to convince manufacturers and pilots to voluntarily adopt technologies that promised to improve the safety record for private flying. The editor listed several such devices, ranging from an instrument designed to warn pilots of an impending stall or spin, to something as seemingly straightforward as installing shoulder belts to protect occupants from injury in the event of a crash. His four word title for this brief piece, "You Can't Sell Safety," succinctly summarized his grim conclusion after more than decade of such efforts.[69] Although the AOPA's leadership was clearly disappointed in the lack of consumer interest in safety-related technology for personal planes, pilot attitudes of the era reflected broader public views about automobiles. Most automakers did not start offering seatbelts as optional equipment until the 1950s, and even then, according to the National Highway Traffic Safety Administration only 10 to 15 percent of drivers and passengers consented to actually use this simple lifesaving device. It would take Ralph Nader's 1965 controversial exposé, *Unsafe at Any Speed: The Designed-In Dangers of the American Automobile*, to raise public interest in automotive safety to the point that the federal government took substantive action. In 1967, following high-profile congressional hearings, the U.S. Department of Transportation finally required seatbelts in all new cars, the first regulation in what would become a long list of mandatory Federal Motor Vehicle Safety Standards.[70]

In the summer of 1969, nearly a decade and a half after Professor Klein and the AOPA's leadership had agreed that private pilots didn't care enough about safety-related technology, two young aeronautical engineers took the aviation industry to task for continuing to build and sell unsafe products. Modeling their efforts after Nader's *Unsafe at Any Speed* and working as members of a consumers' rights activist group known informally as "Nader's Raiders," James T. Bruce III and John B. Draper detailed their findings in a 196-page publication titled *Crash Safety in General Aviation Aircraft*. Their report, excerpts of which were reprinted in *AOPA Pilot*, contended that "travel by light plane is the most lethal of the major forms of transportation in the United States." Drawing on reams of preexisting crash-safety reports and accident investigations, Bruce and Draper observed that numerous deadly accidents in postwar personal planes need not have been fatal. According to their analysis, many pilots and passengers died when poorly designed cockpits failed to provide adequate protection in otherwise survivable situations. As a result of these and other design deficiencies, Nader's aeronautical raiders concluded that "flying in most light aircraft today is similar to mailing a couple of eggs and a bag of nails together in a cardboard box."[71]

The two engineers spread the blame for what they saw as a deplorable lack of safety in postwar private flying. In their estimation, "profit-orientated lightplane manufacturers" were designing and building aircraft to comply with the absolute minimum safety standards set by a complacent federal government (which, unlike its long-standing hands-off tradition regarding automobiles, had started regulating certain aspects of aircraft design in the mid-1920s).[72] But they also faulted consumers—that is, the private pilots who bought, flew, and sometimes died in these personal planes—for accepting a fundamentally dangerous product. As evidence that "safety doesn't sell," Bruce and Draper cited an ill-fated advertising campaign by Beechcraft that highlighted crash-survival features of the company's aircraft. "Beech's sales apparently lost ground and the sales department called a halt to the safety pitch as a result." According to the report, this decision had industry-wide repercussions: "The cry, 'Remember Beech,' has been an effective excuse against the progress of crash safety ever since." Echoing Nader's book on automobiles, the authors concluded that because neither manufacturers nor pilots seemed interested in adopting technology that could reduce the number of deadly crashes, the only sure path to safer flying was for the federal government to pass new laws mandating this technology in personal planes.[73]

Despite accusations to the contrary, most private fliers were not simply ego-driven daredevils who reflexively spurned any technology that promised to make flying easier and safer. Nor were all of them "the young-in-heart who dream of the helmet and goggles days and want their flying rugged" whom Alice Fuchs had jokingly described in her review of the Forney Aircoupe. Instead, most postwar private pilots shunned the Ercoupe ideal and preferred to rent, purchase, and fly what Bruce and Draper described as inherently "unsafe" airplanes for two related reasons: they placed a high value on individual technical skill and therefore did not want flying to become too easy; and certain built-in safety features such as simplified controls and the stall/spin-proof design required them to sacrifice a significant degree of control over the airplane. Throughout the postwar era, the same community that consistently rejected the Ercoupe ideal would come to accept or even embrace stall warning devices, tricycle landing gear, and other innovations that made the use of rudder pedals optional most of the time. Not only did these technologies leave the pilot in full control, but, just as important if not more so, they did not decrease the level of skill needed to fly a plane to the point that anyone and everyone could, in the words of *Mechanix Illustrated*'s Maxwell Hamilton, "just climb in and take off!"[74]

For instance, although the Ercoupe's stall-proof design never caught on, after some initial reluctance or indifference, most pilots ultimately came to rely on a different technological approach to avoiding potentially deadly stalls and spins: the stall warning device. Introduced in the late 1940s by Safe Flight Instrument Corporation, the system employed a wing-mounted sensor and an audible alarm in the cockpit to warn the pilot of an impending stall.[75] This alarm was not as effective in preventing stalls and spins as Weick's built-in safety features, which worked in part because he had limited how steeply pilots could climb, dive, and bank the Ercoupe. But Safe Flight's solution to the stall/spin problem did provide some measure of increased safety without sacrificing any of the pilot's control over the airplane. In an activity where skill mattered to the participants, this approach left the ultimate responsibility of preventing potentially deadly stalls with the individual pilot, not some anonymous engineer.

Some experienced pilots initially resisted adopting stall warning devices because they believed that all aviators should be able to sense and prevent an impending stall on their own. This premise relied on the fact that most airplanes shudder and buffet as airflow over the wings is disrupted at the start of a stall, something that astute pilots can literally feel through both the controls and the aircraft's seat (hence the phrase "flying by the seat of your pants"). In a letter to the editor of *Flying* published in July 1948, Clarence H. Oskey complained, "I don't think much of the stall-warning indicators," which by this time the magazine was actively promoting. "I hope this gives you some idea how many persons feel about so-called safety-devices. Lots of us fly by the seat of our pants, and can fly better that way than by depending on a panel-full of gadgets."[76] William A. Scott echoed this sentiment in another letter published in the same issue, in which he warned that overreliance on technology would allow underskilled individuals to take to the air. "I feel sure pilots would be safer were they not taught to depend upon an instrument which could, at a most inopportune moment, be unserviceable."[77]

In the July 1948 issue of *Flying*, letters against the stall warning device outnumbered those supporting it by a ratio of two to one. But then, it seems, pilots began to change their minds. Such was the case with Ed Hoadley, who wrote in a 1951 article, "The controversy rages with pilots about evenly divided on the subject." He went on to explain, "My thinking on this subject started on the SOP (Seat-of-the-Pants) side, and since the war has veered in favor of the SWD (Stall Warning Devices) thesis." He was not, however, quite ready to discard pilot skill and place full reliance on this newfangled technology, and declared, "It is my belief that both are necessary to all who have soloed." Furthermore, he advised that pilots should rely first and foremost on individual skill and conjured up a common gendered stereotype to help make his point: "The warning devices should be but like a cautious wife—to backseat drive for you."[78]

The debate over stall warning indicators continued into the mid-1950s, after which it all but disappeared from the pages of aviation periodicals. By this time, more and more pilots were relying on this technology. For instance, in an article describing his decidedly masculine hobby of flying his floatplane into remote lakes to go fly fishing, Edwin D. Merry wrote in 1958, "Frankly, I wouldn't be without my alert stall warning instrument. I'm not so darned smart—having talked with a better pilot who 'went in' [crashed] and now can't fly."[79] Some fliers even complained that certain stall warning devices were not intrusive enough. For instance, when Norman Jacobshagen penned his 1960 review of the Beechcraft Debonair he complained: "One thing I did not like. The stall warning has only a red light. It seems to me that at those times when a stall is most likely to occur, the pilot's eyes are apt to be diverted from the instrument panel. I, personally, would vote for the stall warning horn along with the red light."[80] Jacobshagen, at least, wanted his backseat driver to have a louder voice. By 1969, an article about the federal government's aircraft certification standards devoted just two sentences to the fact that a handful of pilots still opposed putting stall warning devices in certain kinds of airplanes.[81]

Pilots embraced the stall warning indicators so completely that, with the encouragement of the AOPA, they even retrofitted older planes that did not come with these as original equipment. In mid-1949, *Flying* magazine noted that "stall warning indicators were collecting dust on [airport] operators' shelves until AOPA stepped in" and convinced its members to start buying these devices as aftermarket upgrades for their planes.[82] A year later, the AOPA proudly reported that "by the end of 1949 over 3,000 stall warning indicators had been installed on AOPA members' aircraft."[83] While only a drop in the bucket compared to the roughly 92,000 general aviation aircraft in service at the time, this marked a significant increase given the aviation community's initial negative reaction to this safety feature.[84] By 1974, the Safe Flight Instrument Corporation reported that more than 200,000 general aviation aircraft had been equipped with its stall warning devices over the past quarter century. Since U.S. general aviation manufacturers had produced approximately 208,000 planes during the same time period, this suggests near-universal adoption of the technology.[85] Eventually, and with little ado, stall warning devices became standard factory equipment on nearly all new aircraft built in the United States.[86] Thus, unlike Fred

Weick's airplanes that would not stall or spin by design, postwar private pilots could get on board with technology that warned them of a potentially dangerous problem but then left it to the individual to understand and properly react to the situation.

Just as pilots accepted stall warning devices in lieu of Weick's stall/spin-proof design, they also seemed more than willing to adopt technologies that produced the same result as the Ercoupe's simplified controls—acceptable turns without rudder input from the pilot—as long as this didn't force them to give up rudder pedals altogether. When Piper Aircraft introduced the PA-22 Tri-Pacer in the early 1950s, the company stopped short of installing simplified flight controls in this entry-level, four-seat, single-engine plane. Instead, it employed a feature that accomplished the same ends without sacrificing a pilot's control over the plane. The Tri-Pacer came equipped with a standard three-control system (foot-operated rudder pedals and a control wheel that turned left or right to bank the plane and moved fore and aft to raise or lower the nose). However, the control wheel and rudder pedals were connected by means of a spring-loaded cable. As a result, if the pilot turned the wheel but neglected the foot pedals in flight, the ailerons and rudder automatically worked in concert to produce a properly coordinated turn, just as they did in an Ercoupe. But unlike the Ercoupe, the Tri-Pacer still gave pilots the option to use all three controls, and if they exerted enough pressure on the wheel and pedals to overpower the spring, they could temporarily override the interconnected controls to execute slips as well. On the basis of positive consumer response, Piper decided to incorporate this feature into some of its later designs including the PA-24 Comanche (introduced in 1957), a high-performance single-engine plane equipped with a variable-pitch propeller and retractable landing gear that was never in any danger of being labeled "not a real airplane." Just in case anyone had doubts, in a 1965 advertisement, the company boasted, "There isn't a sissy in the Piper family."[87]

Cessna did not go so far as to connect the rudder and ailerons in its planes; instead, its engineers used another technological refinement, the Frise-style aileron, which creates extra drag on the low wing in a bank (e.g., the right wing during a right turn) to counter adverse yaw.[88] Piper eventually abandoned interconnected controls when it introduced a new two-seat trainer in the late 1970s, the PA-38 Tomahawk, and instead relied on the "differential aileron" system, which reduces adverse yaw by decreasing the drag created by the downward deflected aileron (in other words, by decreasing the drag on the high wing in a bank).[89] As with the interconnected controls used in earlier postwar Piper aircraft, both technologies—the Frise aileron and the differential aileron—allowed pilots to make passable turns without much skill. When private pilot and aviation writer Gordon Baxter went undercover as a neophyte to write an article about learning to fly, he reported that during one introductory flight lesson in a Tomahawk, the instructor "never mentioned rudders. We flew the whole ride [with our] feet on the floor. So did the others, but the good ones pointed out that rudders were important and that I would learn more about coordinated flight later." Baxter added, almost as an afterthought, that this oversight didn't really matter: "Both that Piper [Tomahawk] and the Cessna 152 fly along just as happy with dead and ignorant feet."[90] However, although pilots could get away with "dead and ignorant feet" in flight, they still had to master the challenging balancing act of slipping the plane down to the runway for a safe crosswind landing. Thus, unlike the Ercoupe, both Piper's and Cessna's approaches made flying easier without completely deskilling the pilot's job. And unlike the two-control Ercoupe, the various Piper, Cessna, and other manufacturers' models equipped with these features sold well throughout the postwar era. When it came to rudder pedals, it seems that many pilots were content to do without—just as long as they had the *option* to use them.

In an interesting twist, some Ercoupe owners, dissatisfied with inelegant crosswind landings and the other limitations imposed by simplified controls, retrofitted their planes with aftermarket rudder pedal kits, and later variants of the plane—only a few of which were produced—included factory-installed rudder pedals as standard equipment. But this was only a partial fix to the "problem." Reviewers and owners alike made negative comments (e.g., "Even with the rudder-pedal conversion there's very little rudder authority") and suggested that a pilot's best bet was to rely on the original, clumsy but effective method of landing Ercoupe in a crosswind. As a result, even an Ercoupe with rudder pedals still didn't count as a "real airplane" in the eyes of many conventional pilots.[91]

Beyond doubt, Fred Weick's most popular safety innovation was the tricycle landing gear. He pioneered this concept in his W-1 prototype and then incorporated it into the Ercoupe, making it the first true production airplane to feature this type of landing gear. Aircraft that are equipped with tricycle landing gear—that is, one wheel under the nose and two main wheels under the wings—are inherently stable during landings and tend to travel nose-first down the runway with little input from the pilot. In contrast, conventional gear airplanes (those with two wheels up front and a small wheel under the tail), are inherently unstable on the ground and require constant attention from the pilot to prevent the airplane from swerving into an out-of-control groundloop that can drag a wing-tip on the ground, collapse the landing gear, or even flip the plane onto its back. Aside from the Ercoupe, only a few of the early postwar personal planes were built with tricycle landing gear. By the mid-1950s, however, practically every new personal plane designed, built, and sold in America (including Cessna's second generation of postwar personal planes) was equipped with Fred Weick's patented innovation. When it came to making landings safer and easier, pilots, acting as consumers, quickly embraced tricycle gear, even though landing a plane so equipped required less skill than landing a tailwheel airplane. Such behavior complicates sweeping conclusions like "you can't sell safety" (*AOPA Pilot*) and "safety doesn't sell" (Nader's Raiders) or claims that private pilots bought airplanes only "to inflate their ego and therefore the more dangerous, the better" (Professor Klein).

The question of "how easy is *too* easy?" in private aviation was seldom straightforward, and the community of pilots continued to revisit this issue throughout the postwar era. One airport manager complained in 1957, "I find that people are more and more trying to substitute 'fool-proof' design and electronic gadgets for good flying habits, knowledge, skill, common sense, and basic navigation procedure." As proof, he noted that more than half of the licensed pilots who wanted to rent his planes failed his mandatory flight test.[92] Yet manufacturers continued to report that pilots and instructors alike wanted flying to deliver a certain degree of challenge. After Mooney Aircraft purchased production rights to the Ercoupe from Alon in 1967, the company removed the interconnected controls, replaced the plane's original H-shaped double tail with the trademark "forward-swept" single vertical tail common to all Mooney aircraft, and gave it a new name as well: the Mooney Cadet. According to one reviewer, these and other changes were not merely a marketing ploy (i.e., to rebrand the old Ercoupe by making it look more like other Mooney aircraft) but also in response to demands from the field. "More and more instructors are insisting that students be introduced to spins," the reviewer reported, and he was favorably impressed with the results. "The gentle old rocking-horse Ercoupe has been moved to bronco land, no fooling."[93] Mooney stopped building Cadets in 1970 after the company went bankrupt, so it is impossible to predict with certainty how pilots and flight schools might have responded if larger numbers of this more aggressive version of Weick's formerly docile design had entered the marketplace.

When Piper decided to develop a new two-seat trainer a decade after Mooney's failed attempt to reinvent the Ercoupe, it too chose to move further away from Fred Weick's foolproof, user-friendly philosophy. Responses to a questionnaire the company mailed randomly to 10,000 flight instructors across the country led

Piper's engineers to deliberately design the PA-38 Tomahawk to be less docile than its immediate predecessor, the Piper Cherokee (which Weick had designed in the late 1950s during his post-ERCO career as director and chief engineer of the Piper Aircraft Development Center), as well as its main competitor, the Cessna 150/152 model line, all of which were already more difficult to fly than the original Ercoupe.[94] Piper succeeded perhaps a bit too well, and its new trainer quickly earned the unfortunate nickname of "Traumahawk" for its dramatic and "unconventional" behavior during stalls and spins.[95] The Tomahawk was also more complicated to operate than most other trainers of the postwar era. For instance, its fuel system requires the pilot to regularly switch back and forth between separate tanks to keep the plane properly balanced (not to mention to prevent running one tank dry and crashing when the engine quits, even if the other tank is completely full), and also to use an auxiliary fuel pump when switching tanks and during takeoffs and landings. By comparison, the Cessna 150 is equipped only with a simple "on/off" fuel valve and has no extra fuel pumps.[96] These and other features in the Tomahawk increased the workload for students, but proponents argued that this taught valuable cockpit management skills that would make it easier for pilots to transition into larger, faster, more complex aircraft after they had earned their private license. Although the plane had its share of detractors, and in spite of its negative nickname, it still proved popular with both flight schools and private owners.

In 1991 the editor in chief of *Flying* magazine declared, "There are as many reasons to become pilots as there are pilots, but there is a common thread. All of us want to know how to master a complex machine operating in an ever-changing environment."[97] Though sweeping in nature, this broad statement accurately describes the attitudes and actions of several generations of postwar private pilots. Fliers took pride in mastering the technology and skills required to earn the title of pilot, a fact reflected in comments about what kind of plane they learned to fly in. Among pilots, a "hot" plane was one that is particularly demanding to fly. Frank Kingston Smith reported that his Cessna 140 taildragger had a reputation for being a "warm" airplane (few within the aviation community would have taken him seriously had he claimed that the little two-seat Cessna was "hot"). In a form of self-congratulation masked as self-deprecating humor, Smith observed that fortunately he didn't hear about the Cessna 140's reputation until he had already learned to fly in one.[98] Diane Ackerman, who started out a quarter century later in a famously docile Piper Cherokee, had this to say about the Cessna 150 in which she completed her training: "Now I know why people say that it's such a good trainer: if you can learn to fly a little bronco like a 150, you can fly anything."[99] Later still, a flier identified only as "Harty" responded to perennial complaints about the Cherokee's replacement, the Piper Tomahawk, by declaring: "It is a demanding plane and that's hard to admit, since it is only a small trainer. But if you learned flying in a Tomahawk, you can almost fly anything. Say that from a Cessna."[100] Smith, Ackerman, and "Harty" all learned to fly in different aircraft, but they all talked up their planes as being more difficult to fly than other models available on the flight line as a way to highlight their own hard-won skills, and title, as "pilots." By comparison, the historical record is noticeably lacking of any such boasting by individuals who earned their wings in an Ercoupe.

The story of the much-maligned Ercoupe—and of designed-in-safety features in postwar personal planes—would have almost certainly played out in another way under different conditions. If Americans had come to rely on personal planes in the postwar era to the same degree that they depended upon private automobiles, more private fliers would have accepted (or even demanded) aircraft that demanded less skill in order to operate safely. Such was the case with cars. Electric starters, automatic transmissions, and other technical innovations that made driving less arduous were adopted in no small part because women took up driving during the interwar era as an extension of domestic duties such as running errands and transporting family members to and from work, school, and appointments. Despite popular jokes to the contrary, women proved to be perfectly capable drivers, but relatively few based their gender identity on their mastery over complex technology. As driving became an everyday task, automakers found a willing market among both men and women for accessories that made driving easier. In an era when most families had only one car, thus precluding the purchase of "his" and "her" automobiles with significantly different features, men could maintain their masculine pride by claiming that they had acquiesced to these "feminine accessories" for the convenience of having two willing drivers in the household.[101] Because personal planes offered limited utility and remained too expensive for most families, no similar situation emerged for private aviation in which women felt compelled to fly for practical reasons and men felt obliged to buy an aircraft that their wives might find more user-friendly.

The initial demographics of postwar private flying also played a significant role in determining what did—and did not—qualify as a Pilot's Airplane. Most postwar private fliers were young men who had served in the military during World War II. Everyone who learned to fly through the military or the CPTP was familiar with conventional aircraft controls and had already mastered stalls, spins, and landing an airplane equipped with a tailwheel. This group had no compelling reason to "step down" to a simpler airplane like the Ercoupe, especially since doing so would invite ridicule from fellow pilots. Meanwhile, most novices to the air at the time were young veterans learning to fly under the GI Bill. While the Ercoupe might make their first foray into the air easier, all who earned their license in this plane were legally restricted to flying aircraft with similar safety features. Since ERCO never put its larger Ercoupe Four (or the larger-still Ercoach) into production, and no other manufacturer adopted Fred Weick's designed-in safety features, this meant that the Ercoupe proved to be a de facto dead end: Ercoupe-trained pilots could only fly Ercoupes until they completed additional government-mandated training to upgrade to a conventional airplane. Faced with this limitation, and subject to the peer pressure to fly the same planes as those who learned to fly before or during the war, most of these new pilots also ultimately rejected built-in safety features in order to continue flying.

By refusing to buy enough ERCO Ercoupes (or later, Forney and Alon Aircoupes) to keep the design on the market, private fliers inadvertently limited how many of these planes would be available for future generations of people who wanted flying to be easy. And by insisting that the Ercoupe's designed-in safety features actually created unsafe pilots, the private aviation community called into question the wisdom of training students in this kind of airplane, which explains Mooney's decision in the late 1960s to radically modify Fred Weick's docile "rocking-horse" into a bucking "bronco" that would stall and spin. At the same time, the widespread acceptance of numerous innovations, especially those that left the individual pilot instead of the designer ultimately responsible for the safe operation of the aircraft, indicates that complaints like "you can't sell safety" and "safety doesn't sell" fail to fully describe consumer attitudes toward the technology of private flying. Most pilots did not want planes that were downright dangerous to fly, but at the same time they found satisfaction in mastering an activity that was considerably more challenging than driving a car.

The vehemence with which some fliers argued that the Ercoupe's simplified design and built-in safety features posed a new kind of safety hazard reveals another aspect of postwar private aviation. Private pilots not only enjoyed mastering the complex technology and arcane skills associated with aviation, but they also liked the fact that their technical prowess set them apart from those who lacked this esoteric knowledge and ability. For pilots, deskilling flight too much took a good deal of the joy out of flying and allowed anyone, even the timorous sparrows, to join the ranks of eagles. In his 1940 *Saturday Evening Post* article about the Ercoupe, Richard Thruelsen had lamented that, thanks to Fred Weick's design, "the mystery of the air and the glamour of flying were disappearing over the horizon. . . . It's getting too simple."[102] But in the end, influenced by a pervasive culture of masculinity that placed a high value on individual skill, the postwar community of pilots made deliberate decisions to define what was and was not considered a Pilot's Airplane and, as a result, helped preserve private aviation's eagles from extinction.

The "Right Stuff" Syndrome

Risk, Skill, and Identity within the Community of Pilots

In 1981 Richard B. Weeghman, editor and executive director of *Aviation Consumer*, observed that pilots possessed a "natural pride in their accomplishments and skills." This pride, he noted, led some of the better pilots to embrace complicated, accident-prone airplanes *because* these planes had reputations for requiring more skill from those who flew them. As evidence, Weeghman singled out several specific aircraft that fit this profile. For instance, the Beechcraft Bonanza, long considered the epitome of excellence for high-end personal planes, had developed a reputation for fatal in-flight structural failures that some attributed to its distinctive, V-shaped tail. Weeghman remarked that when *Aviation Consumer* first reported on this matter in the late 1970s, loyal Bonanza owners had responded with "angry howls" as they protested that these accidents were the result of "'klutzy' pilots," not poor aircraft design. This "survival of the fittest philosophy," he noted, extended beyond specific models to encompass entire groups of aircraft, including *taildraggers*, a common term for planes equipped with conventional landing gear rather than the safer tricycle landing gear. To prove his point, Weeghman asked readers, "Ever heard a tail-dragger pilot taunt his tricycle-gear flying friends (who overall have a much better safety record) for their 'training wheels'?" These and other cases, he concluded, led to a sobering question: "When you come right down to it, how often does the 'Right Stuff' syndrome intrude upon our common sense?"[1] The title of Weeghman's editorial—"The Wrong Stuff"—left readers no room to doubt where he stood on this issue. Had he carried this brief essay further, Weeghman might have noted that just as some pilots chose to fly more-difficult-than-average aircraft to prove their superior skill, many others avoided using planes that would brand them as inferior fliers. Aside from the entry-level Ercoupe, nowhere is this more evident than in the history of a revolutionary multi-engine plane—the Cessna Skymaster—that promised additional safety at the expense of the technical challenges (and thus the panache) associated with traditional multi-engine designs.

Taken together, these three case studies—the Beechcraft Bonanza, taildragger aircraft in general, and the Cessna Skymaster—reveal that some postwar private pilots derived personal satisfaction from mastering the additional challenges posed by high-performance aircraft, old-fashioned landing gear, or multi-engine airplanes. In many ways, their behavior resembles that of the entire flying fraternity when it collectively chose to reject the Ercoupe, except that this smaller group of pilots embraced the ideal of technical competency to an even greater degree. Not content with simply differentiating themselves from the non-flying public, they instead sought new ways to further sort themselves into a skill-based hierarchy *within* the community of pilots. Individuals who chose to fly a particularly challenging aircraft consistently argued that the quality of the pilot, not the aircraft's design, was the crucial factor in determining the safe outcome of every flight. In the process, they reinforced their own self-image as competent pilots who were willing—and, more importantly, able—to take on this increased risk in a safe and responsible manner.

A Tail of Controversy

In a 1995 textbook on aircraft design, an engineer with decades of experience in postwar aeronautics observed that "aircraft designed for the same use are becoming so similar in appearance that it is extremely difficult for even an experienced observer to tell them apart." To make his point, he chose three common personal planes: "Of course, the high-wing designs can be separated from the low-wing models, but try to distinguish a Beech Musketeer from a Piper Cherokee in the landing pattern. And even all the high-wing Cessnas [regardless of size] look much alike because they were derived from the same basic design."[2] His comment reflects one of many aspects of postwar private flying that separated it from the world of driving. In an era when automobile companies relied on cosmetic changes to set their products apart not only from the competition but also from last year's model of the same car, aircraft manufacturers found it difficult to differentiate their offerings on the basis of either performance or appearance.[3] Each new aircraft design had to undergo extensive and expensive testing before receiving government approval, a process that tended to stymie technical innovation. Furthermore, purely cosmetic features routinely used in cars such as tail fins or chrome trim added too much weight and aerodynamic drag to an airplane, where every extra pound of superfluous frills reduced its overall carrying capacity—in either baggage or fuel—by equal measure. Costly government regulations, combined with the unbending laws of physics and aerodynamics, helped ensure that most postwar personal planes were designed and built along the same general lines. But there were a few planes that stood out from their peers.

One of these was Beechcraft's model 35 V-tail Bonanza, which became an icon to many and an anathema to others during a development and production history that spanned 1945 to 1984.[4] Thanks to its looks, but also because of its excellent performance coupled with a decidedly mixed reputation with regard to safety, the Bonanza was well known among pilots. As the name implies, the plane had two tail fins joined in the shape of a V instead of a conventional-looking tail with three fins joined at right angles in the shape of a cross or inverted T. Because it had only two instead of three tail surfaces, the V-shaped tail supposedly contributed to the Bonanza's undeniably impressive performance by reducing both structural weight and aerodynamic drag.[5] And the Bonanza was fast. Depending on model year (later versions had more powerful engines), it boasted a cruising speed ranging from 170 to 203 miles per hour (224 mph for a turbocharged version), leading to claims like "no single [engine personal plane] is faster below 12,000 feet."[6] By comparison, a World War II–era Piper J-3 Cub puttered along at about 75 mph on its 65 horsepower (hp) engine, the docile Ercoupe cruised at 100 mph, and pilots could expect little more from later two-seat trainers like the Cessna 150 or Piper Tomahawk.[7] Despite some complaints that the Bonanza's lack of a vertical tail fin caused it to imperceptibly weave back and forth or "fishtail" though the sky (a tendency that could induce airsickness), many pilots considered the airplane to be exceptionally stable, light on the controls, and "delightful to fly."[8] Just as important, if not more so, since it was the only V-tail personal plane produced in any numbers, its distinctive look made the Bonanza instantly recognizable on the ground or overhead—even to people who knew next to nothing about private flying (fig. 4.1).[9]

The Beechcraft Bonanza was pricy, even within the notoriously expensive world of private aviation. A new one cost several times more than a basic four-seat Cessna 172. Even a used model in decent condition could command far more than many other brand-new but less impressive (and capable) personal planes.[10] As a result, simply by economic necessity, Bonanza owners tended to be wealthy, upper-middle-class professionals such as doctors, lawyers, and business executives.[11] According to a well-known truism within aviation circles, the only people who could afford such an expensive plane for personal use also tended to be driven, self-reliant, and extremely confident (perhaps overly so) in their

Fig. 4.1. Beechcraft model 35 Bonanza. Shown here circa late 1940s with its distinctive and eventually controversial V-shaped tail, the Beechcraft model 35 Bonanza was widely regarded as the "Cadillac" of postwar personal planes. Hans Groenhoff Photographic Collection, Smithsonian National Air and Space Museum (NASM HGC-203).

own abilities and judgment. At the same time, such individuals also tended to be too busy with their professional lives to stay proficient in a demanding, high-performance aircraft like the Bonanza.[12] Some members of the aviation community concluded that this combination of plenty of money, high self-confidence, and lack of proficiency explained a rash of fatal mishaps involving V-tail Bonanzas and began referring to the plane as the "fork-tailed doctor killer."[13]

By the 1970s, pilots fell into two broad groups when it came to the V-tail Bonanza: those who believed that the design was sound and blamed pilot error for most accidents, and those who argued exactly the opposite and claimed that the plane's unique tail, not poor pilots, was the root of the problem. In a 1982 *Flying* article simply titled "V-Tail" (a nod to the design's ubiquity within aviation circles), Richard L. Collins succinctly summed up the situation: "Nobody picks on an airplane like the [Piper] Super Cub. It's like a loveable puppy. But the Bonanza is sleek and fast; it's the kind of machine that you either love or hate. The airplane has always had its detractors, yet even while being sniped at, the V-tail Bonanza has inspired a cultish devotion."[14]

Regardless of whether pilot or plane was at fault, a common accident profile emerged in which pieces of the tail broke off in flight, sending the airplane into an unrecoverable and invariably fatal dive. These accidents generally happened when the aircraft encountered severe turbulence, or when the pilot lost control after becoming disoriented in the clouds and then overstressed the airframe by pulling up too vigorously. The low margin between the Bonanza's maximum cruise speed and its certified "never exceed speed," combined with the sleek design's low aerodynamic drag, meant that even a relatively shallow dive, entered during a few moments of inattention by a distracted pilot, could lead to this potentially deadly situation.[15]

After years of growing controversy and speculation surrounding the V-tail Bonanza, the FAA commissioned an independent study in 1978 to examine in-flight breakup accident rates for all personal planes. The results were mixed. On the one hand, the study found that the V-tail Bonanza had the best overall safety record for any personal plane in the general aviation fleet. On the other hand, the same report indicated that despite its enviable safety record, the V-tail Bonanza was 24 times more likely suffer an in-flight structural failure than Beechcraft's model 33 Debonair (also known as the "straight-tail Bonanza" since it was essentially identical to the model 35 except for its conventional-looking tail).[16] Supporters of the V-tail argued that because the sturdy, straight-tail Debonair experienced such a remarkably low number of in-flight airframe failures compared to any other airplane in the study, any direct comparison between the Debonair and the V-tail Bonanza was unfair, especially since, aside from the Debonair, the so-called doctor killer had an in-flight failure record comparable to most other planes in its class.[17] Others, however, focused on the dramatic 24:1 accident ratio between these nearly identical designs to argue, according to one postmortem of the controversy, "that something was not quite right with the V-tail Bonanza."[18] A television exposé by *60 Minutes* in 1980, combined with a series of highly critical articles published in *Aviation Consumer* around the same time, fanned the long-smoldering debate into a full-blown argument within the aviation community. All sides weighed in with equal vigor, and aviation periodicals printed articles by experts and letters from individual pilots on the subject for several years.

More than a few former fans were converted to the idea that the V-tail design—that is, the technology itself—posed a significant danger. For instance,

Bert Shivers confessed in 1980: "I loved the V-tail Bonanza until *The Aviation Consumer* and other publications educated me."[19] Some owners even resorted to what one article described as "radical surgery," spending thousands of dollars to install an aftermarket kit that transformed their V-tail Bonanzas into "straight-tail" planes.[20] Others sold or traded in their Bonanzas for other aircraft. Bill Prymak was one such pilot who, after "fish-tailing through the skies in a V-tail" for 12 years and 2,000 hours, gave in and bought a used Debonair. He had nothing but praise for his newly acquired plane: "The ride is better, speed numbers are the same," and the center of gravity "extends further aft," which allowed the plane to carry more weight in its baggage compartment. In trading in his V-tail, Prymak had consciously removed himself from a thinly veiled competition among fellow pilots to fly the most challenging personal plane on the market. At the same time, he ruefully acknowledged that many fliers still sought to earn the respect of their peers by taking on the Bonanza and predicted negative consequences to his own pocketbook as a result. "I must face up to one ironic fact: when it comes time to sell my F33A [Debonair], it will be worth $5,000 less than a comparable V-tail. This goes to prove that Bonanza pilots still buy with stardust in their eyes, and that the ego-macho trip bears more weight than the fish-tailing or inherent structural dangers."[21]

One recent historical case study relies on used-plane prices to argue that members of the private flying community behaved as "rational agents" before, during, and after the Bonanza "crisis" of the late postwar era, paying more for the V-tail than the straight-tail version when they believed that the design was perfectly safe and less when they were wary about its reputation. And the numbers suggest that for a few years, the controversy swirling around the contested V-tail design outweighed the "ego-macho trip" of owning this distinctive airplane. Contrary to Prymak's gloomy prediction, by the early 1980s, not long after he had traded in his V-tail Bonanza, the average price of a used model 35 V-tail Bonanza dropped below that of an otherwise identical straight-tail Debonair. V-tails remained less expensive until 1991, by which time Beechcraft had installed (at its own expense) a mandatory, FAA-approved technical fix to reinforce the tail where it meets the fuselage.[22] One independent safety review later noted, "After the installation of the kit, the in-flight breakups decreased dramatically." And as accidents decreased, prices for used V-tail Bonanzas rebounded.[23]

While appealing in its simplicity, this argument that the price for V-tail Bonanzas fluctuated solely on its perceived safety record (i.e., that pilots decided to stop flying the plane because they thought it was "unsafe") overlooks other potential explanations for consumer behavior. For instance, it may be that the price of V-tail Bonanzas dropped during the debates about their safety because owners and prospective buyers feared that expensive mandatory modifications loomed on the horizon, regardless of whether they personally believed that these were actually necessary (Beechcraft's eventual decision to modify the planes at its own expense was unprecedented in general aviation, where individual aircraft owners—not the original manufacturer—legally bear the full financial burden of government mandated repairs or technical upgrades). Decades before the controversy erupted into front-page news and prices of used planes began to drop, the V-tail Bonanza was known as a "hot," unforgiving, even dangerous airplane within the private aviation community. Given this reputation, had most V-tail owners and prospective owners truly acted as "rational agents" from an economic perspective, the plane likely would not have remained in production for nearly as long as it did. Instead of trying to analyze the V-tail controversy in purely technical and economic terms, it is instead useful to consider cultural reasons that pilots may have gravitated to these planes in the first place, along with how they justified flying such a supposedly dangerous aircraft.[24]

Few of the plane's proponents publicly insisted that the Bonanza was easy to fly, especially under adverse conditions. Instead, most seemed to agree that that this model required a skilled pilot to operate safely, a belief that dates back to the plane's introduction. In November 1948, not long after the V-tail Bonanza entered production, *AOPA Pilot* republished the details of a fatal crash involving one of the new planes. Instead of blaming the Bonanza, which the article described as the type of "high-performance airplane" that "every private flyer has wanted for years," it cited pilot error. It also hinted that this was not the first accident of its kind when it observed "another case of a Bonanza being flown in weather, getting in too deep for safety, then having its controls mishandled at excessive speed." As the article concluded, "This type of accident is common to the Bonanza because pilots simply will not take the trouble to recognize certain basic truths about high speeds in airplanes."[25]

Six years (and several accidents) later, the September 1954 issue of *AOPA Pilot* included two accident reports that described in-flight structural failures involving V-tail Bonanzas. Both took place in poor weather conditions, with the pilots relying totally on instruments to keep their planes level and on course as they flew through clouds. In one case, "residents of Leipers Fork, Tenn., heard a loud explosion and observed two bodies (the pilot and his passenger) and parts of wreckage as they fell to the ground." In the other instance, "Persons on the ground at Roselle, Ill. heard a loud explosion and saw the wreckage fall to the ground." The editor's comments about the second accident suggest a growing perception within the aviation community that the Bonanza was too much airplane for some pilots to handle safely: "This appears to be another of those unfortunate weather accidents to which Bonanzas seem to be prone. The pilot 'pushed it' into weather and conditions beyond his ability."[26]

Despite the Bonanza's growing reputation for being unforgiving in inept or careless hands, many pilots considered it the paragon of personal aircraft design. Writing in the mid-1950s, Frank Kingston Smith summed up the positive side of the Bonanza's reputation by comparing it to a familiar postwar cultural icon: "What Cadillac has stood for, since World War II, in the automobile industry, the Beechcraft Bonanza has represented as a personal plane; it has 'class.' " He dismissed the plane's negative reputation by declaring naysayers' criticisms as "Hogwash! All of it." The lawyer-turned-writer then took on the list of complaints one by one almost as if he were addressing a jury in the courtroom, declaring that if the plane got going too fast in instrument conditions and a "tense pilot" panicked and pulled up too quickly—"a bad technique at excessively high speed"—the result would be "Tough luck all around." He went on to explain, "But, for a proficient pilot, the nose-down attitude [dive] can't be too dangerous." Smith personally owned a secondhand two-seat Cessna 140 at the time, but he left no doubt that he admired both the V-tail Bonanza and those who flew them competently. He also made it clear that he could see himself as one of those pilots. "If I were rich, or if I had a business where I needed fast transportation, it would be a Bonanza."[27]

By this time, even the legal system was shaping pilot opinion about the plane. In October 1953, Nathan Prashker, a 27-year-old private pilot and president of his family's business, took off in the company-owned V-tail Bonanza, flew into the clouds over Pittsburgh, lost control, and crashed. According to court documents, the left wing came off shortly before the deadly accident and was found near the wreckage, but parts of the tail were missing, indicating that it had separated at higher altitude. Prashker's estate sued Beechcraft, claiming that the plane was improperly designed and that the manufacturer had received prior notice of this flaw (presumably as a result of previous crashes) but failed to take any action. A lower court held that blatant pilot error caused the crash; specifically, a non-instrument-rated pilot had flown into instrument conditions, where loss of control by an untrained individual was almost guaranteed. When the plaintiffs appealed, a higher court upheld the original ruling, helping to set a legal precedent that would stand for years. Publicized in *AOPA Pilot*, this case, with its detailed expert testimony about the role of pilot skill (or lack thereof) in safely operating a "fast, 'clean' airplane" like the Bonanza, reinforced readers' opinions that the plane demanded an experienced hand at the controls.[28]

Throughout most of the 1950s, knowledge about the V-tail Bonanza's accident rate was confined to the community of pilots, but the airplane entered the public limelight on February 3, 1959, in an event memorialized to this day in pop culture as "the day the music died." Early that morning, a Beechcraft Bonanza carrying Charles Hardin "Buddy" Holly, Ritchie Valens, and J. P. Richardson (aka "the Big Bopper") took off from Mason City, Iowa, disappeared into the predawn darkness and swirling snow, and then crashed in an open field five miles from the runway. Everyone aboard was killed. Federal investigators determined that the inexperienced, non-instrument-rated charter pilot, unable to discern the outside horizon in a snowstorm at night, lost control shortly after takeoff and flew the Bonanza into the ground. Although this particular accident did not involve an in-flight structural failure, the investigation revealed that the plane hit the ground at high speed in a 90 degree bank (i.e., with the wings perpendicular to instead of level with the ground), a sure sign of an incipient "spiral dive" that often ended with the tail coming off when a panicked pilot pulled up too quickly. Regardless of whether the V-tail design had anything to do with this particular crash, the event further cemented the airplane's reputation for demanding much and forgiving little. Yet far from having a negative impact on sales, these and other stories over the next several decades only seemed to enhance the Bonanza's image as a plane that tested—and proved, in the ultimate courtroom of life or death—a pilot's skill and proficiency.[29]

Both the FAA and its predecessor, the CAA, were well aware of the Bonanza's accident record long before the federal government commissioned the 1978 study on in-flight structural failures. According to former FAA structures engineer Vern Ballenger, one of his co-workers attempted to convince Beechcraft to reinforce the Bonanza's tail as early as 1957. As Ballenger recalled nearly three decades later in 1985, "Since the V-tails all clearly met the minimum requirements" of federal regulations for aircraft certification, his colleague was unsuccessful. Ballenger also recalled that he had personally "sketched an automatic self-powered speed-limiter door, which might limit the speed of the airplane," but lamented that nothing came of this effort either.[30] Perhaps prompted by these earlier suggestions by Ballenger and his colleague, in the early 1960s the FAA contracted with Beechcraft's H. A. Slingsby to explore automatic methods "of reducing speed buildup which occurs in a dive." Using a Bonanza as the test aircraft, researchers experimented with four different systems and reported that while all were effective in preventing excessive speed buildup, they were "more complicated than manual systems . . . and therefore would cost more." Once again, nothing further came of this project.[31]

Two decades later in the early to mid-1980s, by which time the controversy about the Bonanza's V-tail was in full swing, fans of the embattled design echoed and expanded upon Frank Kingston Smith's earlier assessment that the Bonanza was the Cadillac of the air, placing it atop the hierarchy of personal planes for its appearance, performance, *and* the challenges that it posed to pilots. For instance, in a *Flying* magazine article titled "Prized Possessions," Nigel Moll extolled the plane's virtues: "Fast, stylish, and delightful to fly, the Model 35 is a success symbol for many a discerning businessman, doctor, lawyer and heir to the family fortune." He explained why the Bonanza enjoyed such lasting appeal. "Perhaps the most elusive quality to design into an airplane is the intangible something that makes it the prized possession of those who possess it and coveted by those who don't. Most airplanes meet the first goal, but few the second."[32] According to Moll, Bonanza owners weren't the plane's only fans; instead, just about everyone who flew wanted one.

Interestingly, nowhere in his article did Moll address the darker side to the plane's reputation. But other fans of the embattled design were less reticent. In 1981 *AOPA Pilot* editor Edward G. Tripp penned an article suggestively titled "The Bonanza Mystique." After singing praises about the aircraft, from its solid construction to its clean lines to the aesthetic appeal of its unique "butterfly" tail ("it still stands out"), he confidently declared that "most pilots find the Bonanza delightful to fly." However, after acknowledging that this venerable design was "still an object of desire, and for some it is a cult object," unlike Moll, Tripp cautioned readers that even if they could afford its exorbitant price tag, not everyone was cut out to fly this airplane. True, its "light, responsive controls" helped "build confidence rapidly that the pilot truly is in control," but, he warned, "There is a trap in that" because "there are conditions in which the Bonanza can turn from Dr. Jekyll to Mr. Hyde." For instance, "Fly a Bonanza in relatively smooth air, and you will feel like Captain Midnight. Wander into turbulence, and the character of the machine changes."[33]

Tripp repeatedly admonished current and prospective Bonanza pilots that they—and not the plane—were ultimately responsible for the safety of themselves and their passengers. This was especially true regarding the "infamous spiral dive," which, he admitted, was bound to develop under certain conditions if an "inattentive, inexperienced or nervous pilot" failed to properly cope with this "hands-on airplane." Regarding its two-faced "Dr. Jekyll to Mr. Hyde" personality,

Tripp insisted, "This does not make the Bonanza a dangerous airplane, despite the inferences or outright claims of others." Instead, he observed matter-of-factly, "the spiral instability and the turbulent instability are characteristics that anyone who wants to fly . . . [a V-tail Bonanza] must be aware of, competent to deal with or decisive enough to leave the airplane in the barn [hangar] if they are not."[34] Without actually using the pejorative nickname, Tripp left no doubt that in his professional opinion, these so-called doctor killers did not kill their pilots; rather, pilots were killing themselves by taking off in a plane that was beyond their abilities.

Although Tripp described the conditions that led to in-flight structural breakups and admitted that the Bonanza had "a besmirched reputation," he did not discuss the role that the aircraft's distinctive tail supposedly had in this kind of accident.[35] Several years later, another experienced pilot unflinchingly addressed the question of tails breaking off in flight. In the opening lines of his 1985 article, "V-Tail Bonanza: A Tail of Controversy," *Flying* magazine's technical editor J. Mac McClellan described what happened when he announced his intention to trade in his old reliable Mooney 201 four-seat airplane for a larger (and more prestigious) used V-tail Bonanza. In response to this news, a fellow flier chided McClellan: "When the tail breaks off you'll be dead from embarrassment before you hit the ground. You should know better."[36]

McClellan vehemently disagreed with his friend's assessment of the plane, and throughout the article he carefully deconstructed standard accusations that the V-tail Bonanza was an inherently unsafe design. After dealing with the statistics (by this time, a well-rehearsed argument among members of the pro-V-tail camp), he acknowledged that the plane was indeed challenging to fly safely, at least for inexperienced or inept individuals. For instance, under instrument conditions, he observed that "a pilot of marginal instrument-flying ability needs his full attention to fly the Bonanza well. Throw in thunderstorm turbulence, ice or some other problem and the marginal IFR [instrument-rated] pilot may be in over his head." He added, somewhat ominously, "I don't see how a VFR [non-instrument-rated] pilot could last long in the clouds in a Bonanza." However, he noted that none of this should come as a surprise: "The Bonanza has been a demanding airplane to fly for 37 years."

Eventually McClellan got around to answering his friend's accusation that he would "die from embarrassment" before he hit the ground. "Am I less safe in my V35B [Bonanza] than I was in my Mooney 201? Obviously, I don't think so. My insurance company doesn't think so, either; it charges the same premium for identical coverage." After summarizing a typical accident report about an in-flight airframe failure associated with the V-tail Bonanza, including a staccato list of contributing factors that all pointed to pilot inexperience and error, he concluded: "These sound like the epitaphs of some pilots, not like indictments of an airplane."[37] To McClellan, mastering his plane's technical challenges provided an opportunity to establish himself as a member of an elite group. With this in mind, he may well have died "from embarrassment" if the tail ever came off his Bonanza—not because he should have known better than to buy it in the first place, but because this would have proved that he didn't have the right stuff to handle the plane, after all.

The history of the American Bonanza Society (ABS) reveals that individuals like Frank Kingston Smith, Nigel Moll, Edward G. Tripp, and J. Mac McClellan were far from alone in their views regarding the embattled design. In 1970, just three years after it was founded as a venue for Bonanza owners to share tips for maintaining and flying their airplanes, the ABS boasted 1,900 members. Given that there were some 6,300 registered V-Tail Bonanzas in the United States plus an additional 1,100 straight-tail Debonairs at the time (pilots of both planes could join), this suggests that as many as a quarter of all Bonanza or Debonair owners belonged to the ABS. Many were experienced fliers, too: 60 percent of the organization's members reportedly held an instrument rating in 1970, compared to the national average of 32 percent for all pilots.[38] By 1974, there were 4,300 ABS members, indicating that up to half of those who owned or regularly flew a Bonanza or Debonair (now totaling 8,100 registered aircraft) belonged to the society. Membership continued to climb even as the debate over the V-tail's integrity heated up. By 1985 the organization had more than 7,500 members; a decade later its membership topped 10,000. A 2009 article reported that more than 44,000 pilots had belonged to the ABS over its four-decade history, with the average membership lasting 10 years, which makes sense given that many Bonanza pilots acquired their planes later in life and probably left the ABS when they bought a different kind of airplane or finally gave up flying for good.[39]

Dr. B. J. "Mac" McClanahan, M.D., ABS cofounder and the organization's first president (not to be confused with *Flying* magazine's technical editor J. Mac McClellan, quoted earlier), told the author of a 1974 article about the Bonanza, "I feel we need a two- or three-day check-out school" for everyone who purchased one of these aircraft, either new or used. When asked about the plane's reputation for in-flight breakups, however, he confidently declared: "An old wives' tale, unless you ignore red lines [the official maximum allowable airspeed, marked with a red line on an airplane's airspeed indicator dial]." He also explained away growing concerns about the Bonanza by referring to an accident that was more likely related to an early wing design issue rather than the airplane's distinctive tail: "Way back in 1946, a prototype test aircraft did lose a wing during a dive test. Somehow, people contrive to distort and perpetuate."[40] To McClanahan, it was up to the properly trained and highly proficient individual pilot, not designers sitting behind drafting tables, to ensure the safe operation of a demanding, high-performance machine like the Bonanza.

Long after "Mac" handed over the reins of the ABS, its members continued to defend the V-tail Bonanza. As the controversy about in-flight breakups reached fever pitch, the organization's monthly newsletter carried letter after letter with titles like "A Response to the *Aviation Consumer* Article" and "To Speak Up *for* the V-Tail."[41] Although the newsletter also carried information about optional structural "beef-up" kits that promised to "dramatically reduce the chance of an in-flight tail breakup," the editor was careful to include disclaimers that this did "not imply that the tail is unsafe. . . . [A]s long as the aircraft is operated within the operating limitations."[42] As always, according to fans of the V-tail Bonanza, the key to safety lay in proper pilot technique.

True to its founder's original vision, in 1983 the ABS introduced a rigorous Bonanza Pilot Proficiency Program (BPPP), featuring 10 hours of classroom instruction on everything from aircraft systems to emergency procedures, followed by 4 hours of hands-on flight training, "to help improve the safety record of Bonanza pilots through the standardization of piloting techniques for all models of the Beechcraft Bonanza."[43] By 1997, Scott Mathews of Sikeston, Missouri, could report that he had attended six BPPP clinics over the past 11 years, and some ABS members claimed that they repeated the course on an annual basis. But many other members (not to mention those Bonanza owners who decided against joining the organization) apparently believed that they already possessed the necessary skills to fly these demanding airplanes safely. Such was the case, at least for a time, with Ray L. Leadabrand. Before purchasing a used turbocharged Bonanza in 1986, Leadabrand had flown his 1949 Bonanza model A35 for 15 years. He later recalled, "Although I had been hearing about the ABS BPPP for several years, I didn't think I needed it for flying my A35 model." However, with his newer, more complicated airplane, he decided that he "needed help and the BPPP seemed to offer a good way to get that help." Leadabrand reported that "as a result of the absolutely outstanding learning process" he experienced during his first encounter with the program, he "continued to sign up for the course each year" (he also became president of the ABS in 1990–91). But many ABS members didn't share Leadabrand's conversion experience. Jim Slavik, a self-described "three time survivor" of the BPPP, expressed dismay in 1990—

some seven years after the course was introduced—when he learned "that only six percent of the membership of the American Bonanza Society has had the opportunity to attend these outstanding programs."[44]

The actions and attitudes of ABS members summarize what legions of Bonanza loyalists like McClellan found appealing about the plane. It was fast, flashy, and expensive. But perhaps even more important, because it was challenging to fly, it provided an opportunity for pilots to demonstrate their technical prowess as well as their willingness to tolerate (and hopefully control) risk. One had to be an above-average pilot in order to fly the Bonanza and survive, not just a well-heeled amateur who could purchase instant status in the form of a fancy airplane. The fact that someone like J. Mac McClellan could simultaneously agree with a skeptical friend that the plane could be dangerous (at least in the wrong hands), yet also insist that he was not putting himself or his passengers at any increased risk by flying it, conveys the message that he considered himself one of those pilots who had earned the right, through superior skill and experience, to take on the V-tail Bonanza. The same might also be said regarding many Bonanza pilots who refused to take the BPPP. These individuals almost certainly did not view themselves as reckless daredevils; rather, they—like Leadabrand—probably felt that although the course might be a good idea for new owners, it was unnecessary for experienced pilots like themselves who had already been flying a Bonanza for some time. And, since everyone in the aviation community—critics and champions alike—instantly recognized the distinctive silhouette of this V-tail plane, they could draw their own conclusions about the pilot behind the controls. The Bonanza's lasting appeal—thanks both to its appearance and because of, not despite, its reputation as a potentially dangerous plane—highlights the ways in which some pilots used technology and skill (or the presumption of skill) to set themselves apart from, and above, their fellow fliers.

Taming the Taildragger

Whereas Bonanza pilots proved their competence within the flying fraternity by mastering this fast, high-performance airplane with an unforgiving nature, there emerged midway through the postwar era another group of fliers who chose to reject tricycle landing gear in favor of an earlier technology that challenged their skills as pilots in different ways. Like the Bonanza's distinctive V-tail, taildragger landing gear was visually apparent, making it easy for other fliers to identify these planes. Taildraggers also earned a reputation for being tough to operate safely, albeit in a different phase of flight than the Bonanza: the landing. And like the Bonanza, this combination of a distinct physical appearance, plus a near mythic reputation regarding the additional skill required to handle them on the ground, meant that those who flew taildraggers could rest assured that other pilots would instantly recognize them—and their accomplishments—wherever they landed.

As described in chapter 3, taildraggers were equipped with "conventional landing gear," which meant that they had two main wheels under the wings and a much smaller wheel located beneath the tail.[45] As the nickname suggests, these planes seem to drag their tails on the ground as they taxi across the airport. Airplanes equipped with tricycle landing gear, with the third wheel located under the plane's nose instead of its tail, first entered production in the late 1930s. Although this new style of landing gear was far easier to control during landings than the conventional tailwheel configuration, it did not become the industry standard for personal planes until the 1950s. Over the next two decades, a special, skill-based identity emerged for fliers who chose airplanes equipped with an old-fashioned tailwheel. As a result, by the early to mid-1970s, members of the aviation community, regardless of what kind of plane they flew, could find humor in the cartoon that depicted a judge informing a defendant in the courtroom: "No, a jury of your peers does not mean you can select tail-dragger airplane pilots."[46]

In the earliest airplanes, landing gear was a decidedly ad hoc affair. Professor Samuel P. Langley's Smithsonian-sponsored "aerodrome"—which crashed into the Potomac River in failure just before the Wright brothers' successful first flights in December 1903—had no real provisions for landing. Langley's first priority was achieving powered human flight; he viewed controlled landings as a separate problem to be dealt with later.[47] The Wright Flier took off down a short track balanced on a wheeled trolley that remained on the ground when the plane lifted off; Wilbur and Orville then landed their slow moving craft on simple skids built into the airplane's structure. Later designs by the Wrights and other aviation pioneers incorporated wheels in various configurations to permit planes to take off without a track and land safely at higher speeds than skids allowed, but taxiing to and from the runway still required a ground crew to guide the wingtips and push on the tail to steer these clumsy craft.

By World War I, most military aircraft were true taildraggers, with two main wheels located under the wings and a wooden skid fixed beneath the tail. Although friction from the tailskid helped stop the plane after landing (most early planes had no brakes), ground handling was still difficult. Pilots had to "blast" the engine at full power in short bursts to create enough airflow over the rudder to change direction on the ground; often even this increased airflow was not enough, and it took additional helpers to maneuver the plane. In the 1920s, aircraft designers and builders began replacing the tailskid with a small wheel. In some planes this tailwheel swiveled freely; in others it was connected to the rudder pedals to provide additional control when taxiing. These innovations, combined with brakes on the main wheels, made aircraft far easier to maneuver on the ground. As gravel and paved runways replaced open grass airfields, tailwheels also allowed aircraft to operate from these harder surfaces that proved unsuitable for wooden skids. Pilots viewed the tailwheel as an improvement over earlier tailskids, but these planes were still inherently difficult to control on the ground, especially at the critical moments during landing when an aircraft transitioned from flying to rolling.

When a plane is moving, its center of gravity (CG)—the theoretical point upon which an object will balance, like the pivot of a see-saw—wants to lead the way.[48] Designers position the wings, tail, and other aerodynamic components in relation to the CG so that an aircraft tends to travel nose-first during flight. However, for much of the first half century of powered flight, designers ignored this principle when it came to ground handling. On a taildragger, the main landing gear is located in front of the airplane's CG; otherwise the plane would tip onto its unsupported nose when sitting on the ground. But because the aircraft's CG always wants to lead the way, and because the taildragger's CG is located *behind* the main landing gear, this means that taildraggers have a disconcerting tendency to suddenly whip around and roll tail-first down the runway or tip over as they attempt to do so. This unintentional maneuver, known as a groundloop, is at best embarrassing. At worst, it can cause a serious accident, demolishing the plane and possibly leading to a fatal post-crash fire. One modern flight instruction manual compares landing a taildragger to balancing a broom vertically with the end of its handle cupped in the palm of one's hand: while possible, skill and constant effort are required to keep it from tipping over. Thus, by their very design, taildraggers are inherently unstable during the most difficult phase of flight: landing.[49]

Before World War II, aviators generally accepted difficult ground handling as just another of the many challenges inherent in flying. When designers bothered to think about landing gear, they focused on developing retractable undercarriage; their goal was to streamline planes to attain higher cruising speeds by reducing drag, not to create an aircraft that was easier to land.[50] Likewise, flight instruction manuals of the era wasted little space on the subject of landing; this was an "art" best learned firsthand from one's instructor. Groundloops received even less attention in print: these were merely the result of poor piloting (or perhaps bad luck) at the crucial moment of touchdown. Even Roscoe Turner's 1941 manual for aspiring aviators, *Win Your Wings*, which devotes a generous 30 out of the book's 649 pages to the subject of learning how to land, includes only one rather vague paragraph on the subject of groundloops. This dearth of discussion of groundloops suggests that many pilots and even designers did not fully grasp the reasons why airplanes were so difficult to control during and immediately after touchdown.[51]

Fred Weick was the first engineer to fully describe this problem in print and identify a potential solution. While employed by the NACA, Weick used his own funds and worked out of his garage during his spare time to develop, test, and then patent what he called "tricycle landing gear" in the mid-1930s. With two main wheels under the wing and a steerable wheel under the nose, tricycle landing gear eliminated the stability problems inherent in conventional taildragger designs. Because the main wheels are located *behind* the plane's CG (otherwise the plane would tip backward onto its tail like a taildragger), a tricycle gear airplane, left to its own devices and with no input from the pilot, tends to roll nose-first down the runway with no inclination to groundloop (fig. 4.2).[52]

Weick traveled to professional conferences throughout the 1930s with a simple device to illustrate the difference between conventional taildraggers and his new tricycle gear. Setting up a long, shallow ramp to simulate a runway, he would release a small model of a taildragger—with its CG located behind the main wheels—to roll down the slope. He reported, "Of course, the [model equipped with] tail-wheel gear turned ever more sharply into a turn and ground looped just about every time; it was obviously unstable in taxiing." Next, Weick would unveil his tricycle gear version, which was identical to the tailwheel model except for the arrangement of its wheels in relation to the CG. This one, he reported, followed its CG (located in front of the main landing gear) and "went straight down the runway every time," simultaneously demonstrating the stability of his invention and reinforcing the inherent instability of the conventional taildragger arrangement he had shown moments earlier.[53]

Weick's demonstration impressed audiences, and he incorporated his innovation into the prewar Ercoupe, which entered production in 1939. However, several factors impeded widespread adoption of tricycle gear before World War II. Aviation was still widely viewed as the domain of "intrepid birdmen" made up primarily of wealthy playboys, daredevil barnstormers, and a small but growing number of military and professional airline pilots.[54] This elite fraternity of fliers was more interested in high performance in the sky than ease of handling on the ground. Meanwhile, the American economy, still in the throes of the Great Depression, was less than favorable for selling personal planes to those who were

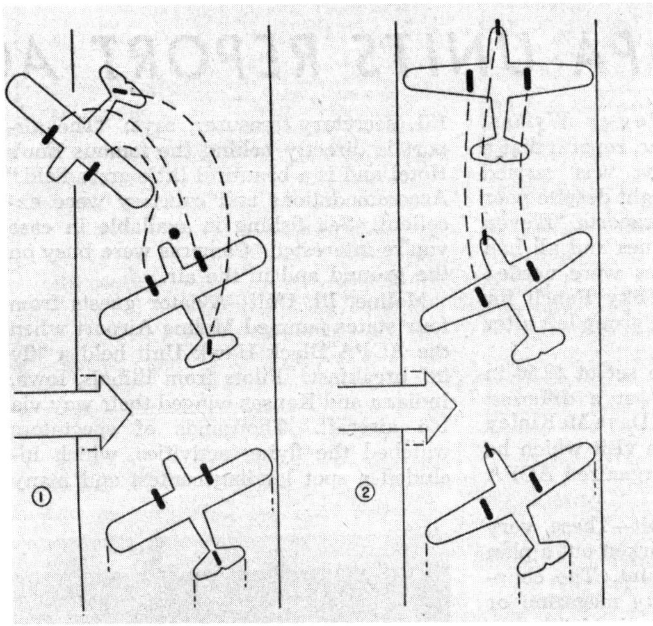

Fig. 4.2. Unstable taildragger versus self-correcting tricycle landing gear. Detail of a 1946 illustration shows two aircraft landing in a crosswind blowing from the left (as indicated by arrow). The left aircraft has inherently unstable taildragger landing gear, leading to a disastrous groundloop; the right one is equipped with self-correcting tricycle landing gear, resulting in a successful landing. "Self Castering [*sic*] Landing Gear," *AOPA Pilot* (Aug. 1946): 50f. Courtesy of AOPA and the Hagley Museum & Library.

interested in flying for recreation. When the United States began mobilizing for war, the government chose tried-and-true conventional-gear aircraft like the Piper J-3 Cub to teach future military pilots how to fly rather than investing in trainers featuring Weick's new tricycle gear.[55]

Although the postwar community of pilots collectively rejected Weick's Ercoupe at least in part because it was too easy to fly, they slowly began to embrace other personal planes equipped with tricycle gear. At first, only high-priced, high-performance planes like the Beechcraft Bonanza came equipped with tricycle gear. Cessna's two seat model 140 and its bare bones sibling, the model 120—the company's answer to its wartime promise to produce a low-priced, easy-to-fly Family Car of the Air—were equipped with a conventional, groundloop-prone tailwheel. So too were the thousands of Piper Cubs that flooded the postwar market, along with larger four-seat versions of that classic design, including the Piper PA-20 Pacer. In 1951 Piper finally introduced the not-so-new concept of

Fig. 4.3. Piper PA-20 Pacer (*left*) and PA-22 Tri-Pacer (*right*), circa early to mid-1950s. These airplanes are essentially the same, aside from landing gear configuration. The center of gravity (CG) on both is located roughly where the two wing struts meet at the bottom of the fuselage. On the taildragger Pacer, the main landing gear is attached in front of the aircraft's CG; on the tricycle gear Tri-Pacer, the main landing gear is mounted behind the CG. Smithsonian National Air and Space Museum (NASM 00081200).

tricycle gear to entry-level personal planes with its PA-22 Tri-Pacer, which was essentially identical to the PA-20 Pacer taildragger except for its landing gear (fig. 4.3). The Tri-Pacer enjoyed immediate commercial success. An early review published by the AOPA's Max Karant described the new plane with cautious optimism, predicting that pilots unfamiliar with tricycle gear might suffer "an unwarranted rash of ground accidents."[56] But five years later, when the organization offered a new Tri-Pacer as a prize to the member who submitted the best new slogan to replace its long-standing motto "AOPA makes your flying safer, less expensive, and more useful," *AOPA Pilot* gushed praise for the plane, declaring that "its tricycle landing gear takes the skill out of landings, even with windy conditions."[57]

By January 1953, Piper reported that the Tri-Pacer was outselling the taildragger Pacer five to one, prompting the author of one article to note: "At long last, the tricycle landing gear has found a permanent niche in the Piper world. [Company president] Bill Piper himself is now so enthusiastic over the tricycle gear that I doubt whether Piper will ever again design a new airplane without it."[58] Two years later, the firm stopped producing the taildragger Pacer altogether, citing a 20-to-1 customer preference for the Tri-Pacer. By this time, Cessna was nearly ready to introduce its own tricycle gear plane, the four-seat model 172. Touted as an updated version of the popular model 170 taildragger, the 1956 Cessna 172, with its patented Land-o-Matic spring steel landing gear, inspired aviation writer Frank Kingston Smith to describe tricycle gear as "the greatest innovation since the wired brassiere."[59] Cessna did not even bother to market its taildragger and tricycle gear models simultaneously but instead shifted all production efforts to the new design.[60] Accident statistics compiled in the late 1970s by the National Transportation Safety Board (NTSB) indicate good reason for all the excitement about tricycle gear. For example, the Cessna 170 experienced 9.91 groundloop accidents per 100,000 hours of operation, compared to 1.00 for its successor, the Cessna 172. In short, the taildragger version was nearly 10 times more likely to groundloop than its tricycle gear counterpart. The rugged Cessna 180, often flown by professional bush pilots, fared little better, with a groundloop rate of 6.49 accidents per 100,000 hours compared to 1.06 in the nosewheel-equipped Cessna 182.[61]

With the widespread introduction of tricycle gear planes in the mid-1950s, taildraggers seemed doomed to eventual extinction save for a few kept by antique airplane aficionados or longtime pilots too thrifty or stubborn to buy newer models. More and more flight schools switched to tricycle gear planes for training students, so fewer and fewer new pilots learned to fly with conventional landing gear. In the used-plane market, prices for taildraggers dropped or stagnated relative to tricycle gear models, indicating consumer preference for the newer technology. Although this made used taildraggers a potentially attractive "budget buy," cost-conscious pilots also had to factor in the reason that these planes were cheaper: they were tougher to land and more prone to accidents, especially on windy days.[62]

Many pilots eagerly traded in their taildraggers for newer tricycle gear planes. Speeding the eventual demise of taildraggers, some owners modernized their tailwheel planes by retrofitting them with tricycle landing gear. A conversion kit developed by Piper aircraft dealer Tom E. Smyer of Ponca City, Oklahoma, represents an early example of an aftermarket upgrade for postwar personal planes.

According to a 1949 article on Smyer, "With a fairly simple little conversion job—and $260 for the conversion kit—you can transform your meek little fair-weather Cub into an even more useful airplane." Because landing a Cub with tricycle gear was so much easier, "operators find their students prefer a Smyer-ized Cub to the conventional model." And by adopting tricycle gear planes that are less affected by windy conditions, pilots could fly the modified Cubs more frequently, "which is the same as cash in the bank." Smyer reportedly sold several such kits to flight school operators and private owners between 1949 and 1951, when Piper introduced its Tri-Pacer.[63] Another company, Met-Co-Aire, based at Municipal Airport in Fullerton, California, touted its kits to convert Cessna taildraggers with selling points like: "Modernize Your Aircraft. Increase Safety. Increase Value!"[64]

Through the 1960s, there is little evidence to indicate widespread nostalgia for the good old days of groundlooping taildraggers. Instead, in the aviation press, taildraggers came to epitomize obsolete or worn-out airplanes. The drawing accompanying a 1961 article entitled "The Value of Speed" shows a family of four standing next to their old, slow, "puddle jumper" airplane. The father and two sons gaze "starry-eyed" as a sleek modern "speedster" zooms by overhead ("Mama," on the other hand, looks on with obvious dismay while clutching her purse, which plainly represents the family's finances). The old-fashioned family plane is unmistakably a taildragger; the new and desirable "speedster" definitely is not.[65] One 1964 cartoon depicted a decrepit-looking taildragger parked in the grass near the runway. From its cockpit, the pilot snaps angrily at a passerby, "What do you mean 'Have I filed an accident report?' I haven't taken off yet."[66] And in a 1972 cartoon, an airport manager nonchalantly warns the pilot of a dilapidated taildragger that is vibrating violently, belching smoke, and dripping oil as it idles on the tarmac: "Don't park it there. We have rubbish pickup in the morning."[67] In these and numerous other examples, illustrators and aviation humorists chose to reinforce the theme of obsolete, worn-out airplanes by tapping into pilots' common perceptions of taildraggers at the time.

At least initially, a few pilots quietly continued to fly their "old-fashioned" machines. Some of those who simply could not afford to "step up" to a new plane eventually bought a tricycle gear model once such planes became available on the used-plane market. Others had no interest in changing: they had mastered taildragger technique back in the days before they had any choice and saw no reason to get rid of their perfectly good airplanes. According to this logic, taildraggers should have disappeared after a slow but steady decline as these pilots eventually retired from private flying. Articles in the aviation press through the late 1960s gave no indication that the community of pilots valued flying a taildragger as a specialized skill, but the turning point seems to be sometime around 1970 when *Plane and Pilot* published, "Some Can, Some Can't: Tame a Taildragger." Lest readers miss the not-so-subtle point, the subtitle left no doubt: "Who's Boss When You Fly One with the Little Rear Wheel? Or Are You Afraid to Try?"[68]

Five years later, Don Downie's article, "Taming the Taildragger," described learning to fly a tailwheel-equipped plane as an exercise that helped an individual become "a better all-around pilot." Downie, who owned a Cessna 170, agreed with the author of "Some Can, Some Can't" that the additional skill required to land a taildragger safely deserved admiration. "I continue to be amazed that so many competent pilots who fly with me just can't handle this ship on the ground." No amount of tricycle gear experience, he argued, granted individuals entry into the exclusive club of taildragger pilots until they had paid their dues of "toil, tears, and sweat." And Downie made this sound like a club worth joining. He quoted one veteran FAA official who divided those who flew into three skill-based categories—"airplane drivers, pilots and aviators"—and who was thoroughly convinced that flying a taildragger required a "plain seat-of-the-pants sensibility" that was "not really needed in a nosewheel airplane." In this hierarchy, it took an "aviator" to master a taildragger, in contrast to the mere "airplane driver" at the opposite end of the spectrum who got by thanks to the forgiving nature of the nosewheel. Downie also pointed out, like a boxer or warrior for whom battle scars signified courage instead of defeat, "With almost any high-time taildragger pilot, the question isn't whether or not he's groundlooped, but when did he do it last?"[69] Although the consequences for making a mistake in a taildragger were likely less serious than losing control of a V-tail Bonanza in the clouds, taildragger pilots, like their V-tail counterparts, believed that their higher technical proficiency could offset any additional risks they incurred by flying these tougher-than-average airplanes.

Downie's article unleashed a flurry of reader responses. The main point of contention centered on whether it was proper technique, as Downie had indicated, to touch down on the tailwheel first, followed almost immediately by the main landing gear, or whether all three wheels should touch the runway simultaneously. The debate about how to correctly execute a three-point landing raged back and forth for several months in the form of letters to the editor, each insisting that they had the "right" answer and citing their credentials to back this up. But none of this correspondence challenged the underlying premise that "taming the taildragger" deserved respect from fellow fliers; that, it seems, was a foregone conclusion.[70]

Part of the bragging rights that came with taming the taildragger involved making inside jokes about how difficult it was. Given the old adage that there are two kinds of taildragger pilots, "those who have groundlooped . . . and those who will," this macabre humor allowed fliers to explain in advance the inevitable botched landing.[71] It also served to inform the majority of pilots who flew only tricycle gear planes—occasionally derided as "nosedraggers" or chided for using "training wheels" to help them land—that taildragger pilots deserved respect for their special skills.[72] In 1976 aviation writer and humorist K. O. Eckland took up the topic in his regular column, "The Old Aerodrome." He began by describing how "Roger Over" had learned to fly, soloed, and "got his commercial and instrument tickets" in a Cessna model 210, one of the most modern, high-performance, single-engine, tricycle gear airplanes on the market. This was certainly not the kind of plane in which most pilots learned to fly, suggesting that "Roger Over" was either a very competent pilot to have achieved so much so early in his flying career or, more likely, one who desperately *wished* to be viewed by others as an accomplished aviator. Eckland continued, "With this exciting background, he decided that it was high time he got into taildraggers and for the past three months, he has been trying to prove his mastery over those funny airplanes with their wheels on backwards." Eckland showed neither mercy nor pity for his imaginary subject, who was evidently meant to parody a particular type of pilot common at most general aviation airports across the country (fig. 4.4). "So far, the plane is ahead 35 to 6, and it's only the first quarter. Once Roger gets airborne, there isn't much of a problem until it comes time to land. . . . He has become known to the fence crowd as 'more-bounce-per-ounce' Roger. Crash crews suit up and stand by the pumper when they hear his call over the radio."[73] Although obviously intended to entertain readers, this brief essay also demonstrated to other pilots—whether they aspired to fly taildraggers or not—that landing these planes was a difficult skill that was worthy of admiration. Appropriately, the author left the reader in doubt as to whether Roger ever actually mastered this skill and earned the respect he craved.

A generation after Piper introduced its Tri-Pacer, several companies perceived enough consumer demand for "those funny airplanes with their wheels on backwards" to revive classic taildragger designs. In 1968 Charles Feris resurrected Taylorcraft Aviation, which had declared bankruptcy years earlier. By 1972 he was producing the classic side-by-side two-seater F-19 model, mostly unchanged except for a slightly larger engine equipped with an electric starter. Meanwhile, in the early 1970s, Douglas Champlin reintroduced the Great Lakes model T-2, a two-seat sport biplane first built in 1929.[74] Although these companies produced

Fig. 4.4. Taildragger humor circa mid-1970s. K. O. Eckland's fictional character "Roger Over" attempts to learn to fly a taildragger in order to gain status among his fellow pilots. The humorous essay accompanying this illustration highlights the additional skill required to master landing an airplane equipped with conventional (taildragger) landing gear. K. O. Eckland, "The Old Aerodrome," *AOPA Pilot* 19, no. 8 (Aug. 1976): 60. Courtesy of K. O. Eckland's family and the Hagley Museum & Library.

planes for a small niche market that reveled in nostalgia, the aerobatic Citabria (Airbatic spelled backward) and Decathlon aircraft were another story. Though also based on a classic design—in this case the Aeronca Champ, a popular 1940s rival to the famed Piper J-3 Cub—these versions were significantly updated with stronger wings and airframes to serve a small but growing demand among private pilots for aircraft that could safely execute loops, rolls, and other basic aerobatic maneuvers. With a more modern-looking angular tail in place of the old-fashioned rounded one, new spring steel main landing gear, and up to three times more power depending on model (with 180 hp in the Super Decathlon, compared to 65 hp in the original 1940s Champ), these planes combined the status associated with "taming a taildragger" with the added prestige, even mystique, conferred upon those who were not afraid to depart straight-and-level flight. By the time Bellanca discontinued production in the early 1980s, several thousand Citabria and Decathlon taildraggers were flying.[75] These planes, in addition to the 10,000-plus Piper PA-18 Super Cubs built between 1949 and 1982, meant that pilots did not necessarily have to search the used-plane market for decades-old classics if they wanted to fly a taildragger.[76]

Throughout the 1970s and into the 1980s, a burgeoning homebuilt aircraft movement provided pilots with another opportunity to choose between tailwheel and nosewheel designs. Early on, nearly all homebuilt designs featured conventional landing gear. One observer noted that "during the 1950s and early 1960s the goal was to get in the air as economically as possible," and homebuilders used materials, designs, and building techniques reminiscent of the 1930s.[77] But starting in the 1970s, several companies introduced more sophisticated designs using professionally prefabricated metal parts or modern composite materials. With high-performance engines, retractable landing gear, and the latest avionics in the cockpit, homebuilders could produce aircraft that cost less than—and yet outperformed—anything on the factory-built new plane market. While it is not surprising that "low-and-slow" models like the fabric-covered Kitfox (first produced in 1984) were taildraggers, it is noteworthy that sleek, high-performance planes like the composite Glasair (which made its public debut in 1980) and the all-metal Van's RV-6 (introduced in 1986) offered customers the option of building their planes with either tailwheel or nosewheel configuration.[78]

By the late 1970s, pilots were even converting newer tricycle gear airplanes into taildraggers. Some planes, like the Piper PA-22 Tri-Pacer and its two-seat derivative, the Colt, were relatively easy to modify. After all, these fabric-covered aircraft with welded steel tube frames were based on the PA-20 Pacer taildragger of the immediate post–World War II era. According to the Supplemental Type Certificate issued by the FAA, the main modifications needed to convert a PA-22 into a taildragger involved installing a metal bracket to support the tailwheel, welding new attachment points to the frame to move the main landing gear forward of the plane's CG, and replacing the single brake lever in the cockpit with individual brakes on each rudder pedal to facilitate ground handing.[79] Pilots who desired a more modern, all-metal taildragger could pay specialty shop owners like Ralph Bolen to modify a newer Cessna model 150, 152, or 172. Much like the "unibody" automobiles produced today, these aircraft were built from sheet metal and lacked traditional internal frames, which made the conversion from tricycle gear to taildragger more complicated than it was in the Piper Tri-Pacer. According to a 1978 article by Donald Chase, modifying a Cessna in this manner required installing 62 parts, approximately 50 to 60 hours of shop time, and no small expense. A new landing gear support structure installed under the cabin floor, plus reinforcement for the tail section, negated any weight saved by removing the heavy nosewheel and its shock-absorbing strut. But because the smaller tailwheel produced less aerodynamic drag than a large nosewheel, pilots of these so-called Texas Taildraggers reported slightly higher cruise speeds and, according to the article, fairly docile ground handling as well. "This is not to say that the 150 T/G [as this variant was labeled] could not be groundlooped," Chase noted, "but you'd really have to work at it." Thus, pilots of these converted planes could enjoy the practical perks of flying a taildragger without having to deal with the same degree of inherent instability that made older tailwheel aircraft notorious.[80]

After Chase dutifully recited the standard litany of technical advantages of conventional landing gear—better rough-field performance, plus higher cruising speed due to less drag—he admitted that there was another benefit as well: "the prestige of flying a taildragger." He observed that more than just a few pilots were interested in earning this distinction: "Many flight schools are supplementing standard 150s with the T/G. Several report that 'everyone wants to learn in the taildragger.'" When the article went to press, Bolen already had more than 40 conversion kits in the field and was tooling up to produce another 200 over the next 12 months. Furthermore, he reported that he had received 1,200 inquiries regarding conversion kits for the two-seat Cessna models 150 and 152 and at least twice that many for a soon-to-be-certified kit for the four-seat Cessna 172. The article concluded with a comment remarkably reminiscent of Frank Kingston Smith's comparison of tricycle landing gear to wired brassieres some two decades earlier: "Just like women's fashions, if you wait long enough, the trend seems to return. So from tailwheel to nosewheel it's now back to tailwheel."[81]

Simmering just below the surface of this comparison to women's fashions lies the reality that many pilots responsible for the tailwheel revival wanted to prove both to themselves and to others that they had what it took to "tame the taildragger." Writing in 1985 about why members of a new generation of pilots chose to fly old-fashioned planes despite the extra skill and risk involved, Mark Twombly explicitly described this motivation: "Part of the pleasure of owning a tailwheel airplane is the added degree of difficulty in landing it. You have to fly a little harder than in a tricycle-gear airplane." He cautioned, however, that "some tailwheel aircraft have reputations for trampling pilots' egos like spring grass in the touchdown zone."[82] On the surface, Twombly's choice to compare flying a taildragger to the manly sport of football reflects the masculine culture of private aviation. But his analysis also articulates how taildragger pilots used technical expertise, and their willingness to take on additional risk, as both a source of personal pride and a chance to set themselves apart from fellow fliers in a manner that meant little, if anything, to those outside the aviation community.

Trouble with Twins

Just as some pilots chose to fly a taildragger or a V-tail Bonanza in order to demonstrate their competence and ability to tolerate and mitigate higher risk, others who could afford to do so flew twin-engine personal planes. Also known as "light twins" to differentiate them from larger multi-engine commercial aircraft, these planes were unquestionably more complex and difficult to operate safely than single-engine models. When Cessna attempted to revolutionize multi-engine personal planes by making them inherently safer by design, pilots who found conventional light twins attractive because of their speed, prestige, and association with military and commercial aircraft ultimately rejected Cessna's new offering. But this wasn't all. Many who flew conventional twins—and more than a few who did not, but aspired to—saddled Cessna's new design with a negative reputation that carried over to its pilots. In the short term, this made it seem like a plane that any self-respecting pilot should steer clear of; in the long term, it helped to seal its fate.

At first glance, the question of whether two engines are better than one on an airplane seems elementary. Throughout the postwar era, one of the greatest fears nonfliers had about personal planes—and a common concern of many experienced pilots as well—was the specter of engine failure. A single-engine plane becomes a glider when the motor quits, leaving the pilot little choice but to find an open field, a clear stretch of road, or some other suitable surface on which to attempt an emergency landing.[83] Under the best conditions, success is far from guaranteed. Flying over open water, dense forests, mountains, and congested urban areas increases the likelihood of a catastrophic crash, as does flying at night or in the clouds when a pilot has little hope of spotting a suitable landing site. One obvious solution to this problem is to add a second engine so that the plane can continue flying even if one motor quits. The first aircraft equipped with multiple motors predate World War I, and by World War II fleets of multi-engine planes dominated military and commercial aviation. The image of the damaged Allied bomber limping home safely with one or more engines shut down came to represent the increased margin of safety that extra engines afforded. Introduced into the private aviation market in the early 1950s, light twins had other attractions as well. They were fast, expensive (which conveyed instant status upon the owner), and looked far more like a modern airliner than any single-engine plane, even a Cadillac of the sky like the Beechcraft Bonanza. As one aviation writer summed it up in 1971, "Multiple-engine airplanes have an aura about them; most pilots can easily imagine themselves involved in the adventure of commanding a dozen pistons with fistfuls of throttle."[84]

In a 1928 article titled "Lindbergh on Flying: Aircraft for Private Owners and How to Choose One," the world-renowned aviator declared, "There is no doubt that multi-motored planes have a much greater degree of safety when they are properly designed and operated." But, he warned, simply adding another engine was no cure-all. "If they are not so designed and operated, then they not only lose their advantage over the single-motored machine, but they have only a fraction of its reliability."[85] Nearly six decades later, Dr. Paul Nichols, another highly qualified (albeit far less famous) aviation expert, observed: "Many people, aviators included, think twins *are* safer than single engine aircraft"; he then proceeded to debunk this notion.[86] Although safety was a major selling point for multi-engine personal airplanes, statistics indicated that in private aviation, this was more illusion than reality. The fatal accident rate in 1981 for light twins was actually higher per 100,000 hours flown than it was for single-engine planes. Even more surprising, at least 25 percent of the fatal accidents in light twins were related to engine failure, in sharp contrast to only 9 percent for those involving single-engine planes. Although this information seemed to fly in the face of common sense, it was nothing new: for instance, an earlier study on general aviation crashes that occurred between 1965 and 1969 revealed that while light twins were only half as likely to crash in the event of an engine failure as a single-engine airplane, if a light twin did crash, it was four times more likely to be fatal.[87]

One problem was that unlike multi-engine military or commercial aircraft, which were designed with enough power to remain aloft if an engine quit in flight, many light twins could not maintain a safe altitude on only one engine. For instance, with both engines operating, one popular light twin of the postwar era, the Piper Seminole, can fly well above Colorado's Pike Peak, which at 14,110 feet above sea level is the tallest mountain within the lower 48 states. If one engine fails, however, a fully loaded Seminole can maintain an altitude of only about 4,000 feet above sea level.[88] In low-lying parts of the continental United States, this single-engine ceiling will allow the plane to avoid any obstacles in its path as it limps to the nearest airport for an emergency landing. But if an engine were to quit somewhere over the Appalachian Mountains, then the pilot and passengers may be in serious trouble. Further west, there is no question about the outcome. Not only is the Seminole unable to clear Pike's Peak on one engine, but it cannot even remain airborne over the high, flat prairie west of the Kansas-Colorado state line. As a result, the pilot of a Piper Seminole who suffers an engine failure in this region ends up in a gradual but irreversible descent—even with the remaining motor running at full power—until the plane flies into the ground. The pilot of a larger, more powerful light twin such as the Cessna model 310R is not entirely immune to these problems either. With a single-engine ceiling of 7,400 feet—close to twice that of the Piper Seminole—a fully loaded Cessna 310R can barely clear the highest peak in the Appalachians with one engine shut down. Over the Rocky Mountains, it faces the same unhappy fate as the Seminole. Thus, the redundancy of two engines does not always translate into increased safety, leading to a grim joke among pilots that when one engine on a light twin quits, the primary purpose of the other engine is to "fly the airplane to the scene of the accident."[89]

Although limited single-engine performance is a very real concern, this is by no means the greatest challenge facing the pilots of light twins. When an engine fails in a standard light twin—that is, a plane with one engine mounted on each wing—the pilot must react quickly and correctly or risk losing control of the airplane. As one engine continues to provide full power, the dead engine on the opposite wing creates immense drag until the pilot can stop its wind-milling propeller by feathering the blades to minimize aerodynamic resistance. Failure to act promptly and properly to remedy this condition, known as asymmetric thrust, results in the plane turning steeply toward the dead engine. In a worst-case scenario, the plane may roll inverted and stall or spin. This is dangerous enough under ideal circumstances, but if it happens near the ground, or in the clouds during instrument flight conditions, the results are often deadly.[90] One pilot who witnessed an accident involving a light twin described how immediately after takeoff, "an engine quit, the airplane rolled upside down and scattered itself over the countryside."[91]

Pilots, instructors, safety experts, and manufacturers argued endlessly over whether to blame the plane or the pilot for the high rate of fatal accidents in light twins. The author of an editorial observed, "There are a lot of talented aviators flying twins, but there are a greater number who are ticking time bombs, waiting to go off at the first hiccup of an engine."[92] In response, one pilot bluntly declared: "My experience showed me over the years that 80 percent of all pilots flying their twin-engine aircraft would crash and burn after losing an engine at the most critical point on takeoff. Why? Bad procedures from bad instructors. . . . Throw rocks at the problems, not at twin-engine aircraft."[93] Another pilot blamed a combination of economics and ego but, again, not the technology itself: "The first cause of the dismal safety record of light twins is economics. Pilots who aspire to fly twins can barely afford the operating expense of the aircraft, let alone a competent instructor and satisfactory hours of instruction to ensure proficiency." He added, "The second problem with twins is pilot ego. The average individual who can afford a light twin is successful in business and isn't about to let some 'kid' [youthful instructor] tell him how to fly his new airplane."[94] These comments provide a glimpse into the attraction of light twins: aside from a slight increase in speed over a high-performance single (hardly economical considering that adding another engine instantly doubles the operating and maintenance costs), light twins had a reputation as demanding and unforgiving aircraft. This, in turn, implied that anyone who chose to fly them had better have the "right stuff" in terms of technical expertise, situational awareness, and good judgment.

In the early 1960s, Cessna, which already had a successful light twin on the market (the model 310), took a new approach to solving the asymmetric thrust problem. Tacitly acknowledging that conventional light twin design was to blame for high accident rates, Cessna engineers eliminated asymmetric thrust altogether by creating an airplane that had one engine pulling from the front and another one pushing from the rear. The result looked something like a standard, single-engine high-wing airplane with its tail chopped off (fig. 4.5). With one engine on the nose, a second engine installed facing backward immediately behind the passenger cabin where the tail would normally be, and two spindly booms extending back from the wings on either side of the rear propeller to hold twin tail fins in place, the Cessna Skymaster looked nothing like a conventional twin; one sympathetic writer described the plane's appearance as "decidedly bizarre." Because both engines were installed along the aircraft's centerline and not on the wings, this unconventional arrangement—officially known as "centerline thrust"—meant that if an engine failed, the pilot could cope with the issue of reduced power without also fighting to keep the airplane under control. The same author who labeled the plane's looks "decidedly bizarre" also noted, "This was an airplane a [single-engine Cessna 172] Skyhawk pilot could step into and, theoretically at least, fly safely. It was still necessary to obtain a centerline-thrust multiengine rating, but the Skymaster was an airplane that anybody should have been able to fly."[95]

Advertisements announced that the "radically new" Skymaster made flying a twin "as easy to fly as a single-engine plane," and like the Ercoupe two decades earlier, the plane initially received rave reviews in the aviation press.[96] An article announcing the new design ran under the title: "New 'Push-Pull' Twin, the Skymaster, Scores Hit at Cessna Sales Meet," and the AOPA's Max Karant lauded the centerline thrust concept as a "breakthrough" for aviation safety.[97] Cessna

Fig. 4.5. 1966 Cessna T337 Super Skymaster. In this in-flight view of a 1966 Cessna T337 Super Skymaster, the plane's unique push-pull twin engine configuration is clearly visible. Smithsonian National Air and Space Museum (NASM 00020201).

soon upgraded its first centerline thrust twin, the Cessna 336 Skymaster, with larger engines and retractable landing gear, and called this faster, more powerful plane the Cessna 337 Super Skymaster. During the Vietnam War, the U.S. Air Force purchased more than 500 militarized versions of the plane, designated the O-2, to conduct dangerous low-level Forward Air Control combat missions within shooting range of enemy ground troops. Eventually Cessna produced the plane with turbocharged engines and even built a pressurized version to allow pilots and their passengers to cruise in comfort high above bad weather. Although the company hoped to make flying safer by design, unlike the simple little Ercoupe that had once promised to make flying as easy as driving a car, the Skymaster was still a complex, high-performance airplane.[98]

The Skymaster developed a loyal following and enjoyed consistent though not particularly impressive sales for nearly two decades. However, it also collected a growing group of detractors. Some derisively dubbed it the "Mixmaster" after a popular brand of electric mixer, a clever parody of the plane's real name that suggested the Skymaster (and perhaps its pilots) belonged in the women's world of the kitchen instead of the manly domain of the sky.[99] Another derogatory nickname, "suck-and-blow," at once described the arrangement of the propellers (the front one "sucked" the plane through the sky, while the rear one "blew" it along) and suggested none too subtly the blowjob (fellatio).[100] In a postwar society that viewed heterosexual relationships as the norm and assumed that this sexual act involved an unequal power relationship between partners, women were cast in the role of submissive performers, men were the dominant recipients, and jokes about blowjobs (more specifically, who performed them on whom) provided an easy way for men to attack another man's masculinity.[101]

Even Skymaster supporters admitted that the plane was not widely respected within the aviation community. The author of a 1971 article about the plane began, "All Super Skymaster pilots are good sports. The fact is, when you fly a Cessna Super Skymaster you've got to be a good sport, because you're going to receive some good-natured ribbing about it."[102] Another article, published decades after the Skymaster made its debut, revealed in its subtitle that centerline thrust pilots still faced an image problem: "Thirty Years Later, Still Trying to Get Some Respect." Its author echoed comments from earlier pieces when he observed matter-of-factly: "The hardy soul arriving in one of Cessna's great multiengine experiments garners sneers from drivers of 'real twins'—you know, those airplanes with an engine on each wing. . . . The kind with the occasional foray into uncontrollability when its 'real' pilot reacts improperly to an engine-out emergency."[103] In a 1975 article exploring why the revolutionary twin failed to do well in the marketplace, *Flying* magazine's Richard Collins mused, "On two engines or one, the Skymaster minds its manners. Is it too nice to be a winner?" Obviously an advocate of Cessna's revolutionary concept, Collins attributed slow sales in part to what he called the airplane's " 'chicken' stigma (only cowards fly them)" among his fellow pilots.[104] And in a 1985 article about good aircraft designs that never quite caught on with pilots, Matt Thurber concluded, "One of the biggest problems in marketing the Skymaster was overcoming its image as an airplane for wimps."[105]

Not all comments from the fence crowd could be characterized as "good-natured ribbing." As the previous examples suggest, some aviators argued that because people who flew Skymasters never had to worry about facing a potential life-or-death situation involving asymmetric thrust if an engine failed, they were not "real" multi-engine pilots. From the start, the FAA formalized this sentiment into law, creating a new class of pilot certification—"Airplane Multiengine Land (or sea)—Limited to Center Thrust"—for individuals who were qualified to fly aircraft like the Skymaster but not traditional multi-engine planes that had their engines mounted on the wings. Since the Skymaster was the only centerline thrust personal plane on the market, private pilots who held this limited rating could only fly a Skymaster. By contrast, individuals who earned their multi-engine rating in a traditional light twin (with its more challenging engine-out flight characteristics) were automatically allowed to fly a Skymaster with no additional training or ratings.[106] Though based on real technical skills that pilots needed in order to safely fly different kinds of airplanes, this legal distinction further cemented cultural definitions of who possessed the right stuff to handle a complicated and potentially dangerous machine like a conventional light twin, and who did not. Given widespread views among pilots, anyone who valued masculine virtues such as technical competence or the ability to overcome danger with individual skill would likely think twice about choosing a Skymaster and risk being branded a "chicken" or a "wimp" by fellow fliers.

For a time, Cessna attempted to turn the Skymaster's mixed reputation and the perennial jokes about it into a marketing gimmick. In a none-too-subtle swipe at the dismal single-engine performance exhibited by most light twins, one advertisement featured a photograph of a Skymaster cruising effortlessly high above a jagged mountain range with its front engine shut down; underneath the caption read, "What's so funny about 18,000 feet on one engine?" The message was clear: pilots could joke about the Skymaster all they wanted, but the plane's superior single-engine performance—which beat other light twins like the Piper Seminole or even the Cessna 310 by a wide margin—was no laughing matter. Another ad tackled one of the Skymaster's negative nicknames head-on. Titled "Some people like to kid a winner," it shows a Skymaster taking off in the background as an air traffic controller in the foreground mimics the imaginary two-headed "push-me-pull-you" antelope from the children's story *Dr. Dolittle.* To counter potential accusations that centerline thrust was for sissies, in 1969 the company distributed an advertisement featuring two U.S. Air Force O-2 versions of the Skymaster with rocket launchers slung beneath their wings. The accompanying text asks readers: "'Flying Tiger' or 'pussy cat?'" and then immediately answers this rhetorical question: "Correct answer is 'both.'"[107] If Cessna's centerline twin was good enough to fly combat missions over Vietnam as a "flying tiger," then the fact that its docile handling rendered it a "pussy cat" in the air should not automatically relegate pilots of the so-called Mixmaster to the kitchen alongside their wives.

Later Cessna advertisements ignored the plane's negative reputation altogether and instead focused on how it delivered all of the best (and none of the worst) features of both twins and singles. One that was cleverly titled "Stay twin proficient and still have time to swing" shows a Skymaster pilot "swinging" his golf clubs on the fairway in the free time that he did not need to spend practicing single-engine emergency procedures in a conventional multi-engine plane. This approach implicitly shifted the market for Skymasters away from pilots who wanted to demonstrate their technical prowess to fellow fliers through their mastery of a potentially dangerous high-performance light twin, and toward those for whom flying was part of presenting themselves to the general public as affluent and accomplished members of the upper middle class. Another ad declared in bold, oversized print that pilots who bought a Skymaster for its "bigger useful load" and "single-engine flying ease" would have to "throttle back" if they wanted to "match speeds with its competition," which clearly referred to single-engine instead of multi-engine planes.[108]

Although the Cessna Skymaster was not plagued by asymmetric thrust when an engine failed, over the years it accrued a less than sterling safety record. NTSB accident statistics from the early 1970s indicated that the Skymaster had one of the highest fatal accident rates among light twins, hardly a glowing endorsement for a design that was supposed to make multi-engine flying *less* dangerous. Separate analyses of these and later accidents by *Aviation Consumer* and the AOPA Air Safety Foundation agreed that pilots, not the airplane itself, deserved most of the blame for this poor showing.[109] Indeed, accident reports and anecdotal evidence suggest that, like the Ercoupe before it, the Skymaster's inherently safe design may have encouraged complacency among pilots. For instance, Archie Trammell, an otherwise enthusiastic advocate of the centerline thrust concept, admitted in a 1971 article about the much-maligned design: "The interesting thing is, engine-out characteristics of the Super Skymaster are probably *too* good. A man I know flew his from San Antonio to Dallas one day, wondering all the way why his indicated airspeed was off seven knots [eight miles per hour]. It was because the rear engine had swallowed a valve."[110] With the front engine still running normally, the cockpit was noisy enough that the pilot failed to notice that his rear engine (with its "nice, muffled sound") had lost power. Still, a quick check of the exhaust gas temperature gauge would have instantly given cause for alarm.[111] Instead, the unnamed pilot continued with his flight, blissfully unaware that if his front engine also quit he would suddenly be flying a glider.

Not all pilots were so lucky. More than one Skymaster plowed off the end of the runway after its pilot tried to take off with only the front engine operating. A 1972 article by Richard Collins, who like Trammell was a steadfast fan of centerline thrust airplanes, but not necessarily impressed by everyone who flew them, observed: "This [type of accident] has happened a sad number of times, apparently after the rear engine died while the airplane was being taxied out for takeoff. One can only wonder if there aren't adequate clues early in the takeoff run . . . that would tell the pilot his airplane is accelerating under the influence of [only] 210 hp instead of 420 [total hp with both engines operating]." Collins was more critical of the person at the controls than Trammell, concluding: "Are pilots really determined to take off regardless of how soggy the airplane feels? The wise pilot should start his takeoff run with the idea of stopping if *anything* unusual appears as the airplane thunders down the runway."[112]

In 1973 Cessna announced that it would include an engine-out indicator light in the instrument panels of new Skymasters, much like a so-called idiot light on a car's dashboard meant to alert the driver that something was amiss, in an attempt to prevent this scenario from happening.[113] Apparently this wasn't enough, because in 1977, the FAA issued a mandatory Airworthiness Directive requiring that all Cessna 337 Skymasters be retrofitted with a placard in the cockpit to remind pilots, "DO NOT INITIATE SINGLE ENGINE TAKEOFF."[114] Long after the plane went out of production, commentators continued to voice the same complaints. As one author put it in 1993, "It seems freeing the pilot from asymmetric-thrust worries has merely shifted the accident causes elsewhere. Perhaps because the Skymaster is less susceptible to the conventional yaw-spin-boom twin accident, pilots do not take the [Cessna] 337 as seriously as other twins."[115] Mounting evidence that Skymaster pilots were less competent than their counterparts who flew conventional light twins simply revisited the long-running debate about whether pilot skill or technology held the key to safety in the sky. If designers built an airplane that was easier to use and therefore technically safer to fly, would that translate into fewer mishaps? Or would it instead, as accident statistics seemed to show, merely enable less-qualified individuals to get themselves into more serious trouble?

In the end, Cessna was unable to overcome the Skymaster's reputation as both the butt of easy jokes and a magnet for bad pilots. Faced with declining demand and rising liability costs, the company stopped building the Skymaster in 1980, but with used models still widely available, the controversy continued.[116] By this time, the revolutionary centerline thrust twin had become a symbol of what could go wrong when engineers tried to make flying *too* safe by design. In his 1985 editorial "Trouble with Twins," *Flying* magazine's Richard Collins (by now a senior editor at the magazine) asserted: "There are some who say that the airplane, not the pilot, should be fixed. This isn't a bad premise, but it is no cure-all. For practical purposes it isn't even a cure-a-little-bit." To prove his point, he invoked not one but two past attempts to design increased safety into personal planes. "Years ago, in an admirable effort to quell the stall-spin accident rate, the Ercoupe was developed by a talented engineer named Fred Weick. The airplane would neither stall nor spin, so it had an excellent safety record in that regard. But its overall record was worse than other airplanes in its class. Perhaps pilots just thought it wouldn't bite." Collins then turned to the issue of asymmetric thrust in light twins: "Later, Cessna pioneered twin-engine safety with the Skymaster centerline-thrust twin. No more loss of directional control because one engine failed. But [now] the Skymaster is out of production and didn't have an exceptional safety record."[117] He neglected to bring up the Skymaster's "'chicken' stigma" that he had described a decade earlier in his article that lauded the plane as "The Virtuous Twin." It is possible that by this time, with the weight of accident figures due to blatant pilot error added to the mix, his opinion about the virtues of centerline thrust had changed. But if so, this simply brought Collins a step closer to the ranks of so many other pilots who were convinced that people who chose to fly Skymasters lacked the self-discipline, skill, and judgment needed to fly it safely.

Cessna Skymasters, Beechcraft Bonanzas, and taildraggers represent a small portion of the postwar private aviation fleet. Even so, virtually every pilot recognized these aircraft, in part because they were so different from the majority of personal planes. In every case, what made these planes stand out visually was also directly tied to what made them more or less demanding to fly. As a result, pilots could hardly avoid overhearing the often contradictory reputations attached to these aircraft. Whether it was the Bonanza's distinctive V-shaped tail, the "little rear wheel" on a taildragger, or the odd-looking "push-me-pull-you" engine arrangement that defined the Skymaster, there was no hiding the fact when someone chose to fly a plane so different from the norm.

Bonanza pilots could reasonably expect that almost anyone who saw their plane would conclude that the owner was successful and enjoyed an upper-middle-class income. They also hoped that fellow fliers would label them as competent pilots who could handle a "hot," high-performance airplane, rather than an overconfident, underqualified pilot, an accident waiting to happen. Likewise, by the mid-1970s, those who flew planes equipped with conventional landing gear could also expect instant recognition and respect within the community of pilots.

The Cessna Skymaster also stood out visually both on the flight line and in the air. With its centerline thrust configuration, there was no mistaking a Skymaster for any other light twin. With its "'chicken' stigma," derogatory nicknames, and an accident record that called into question the competence and judgment of those who chose to fly this plane, pilots of the so-called Mixmaster had to put up with more than just a little "good-natured ribbing" from fellow fliers. In a competitive, hierarchical community of pilots that valued technical skill above all else, it could be a very fine line between admitting that you lacked the "right stuff" (or simply the time or money) to fly a conventional light twin and demonstrating to fellow fliers that you were made of the "wrong stuff" by deliberately choosing to fly a centerline thrust twin that was just as fast and expensive as most conventional light twins but required less skill to operate.

According to *Aviation Consumer*'s Richard Weeghman, flying V-tail Bonanzas, taildraggers, or anything else because it had a reputation for requiring above-average piloting skills represented a clear violation of common sense. By extension, deliberately shunning an airplane with designed-in safety features like the Skymaster was equally wrongheaded. In reality, Weeghman's broad indictment of the small yet significant minority of private fliers who preferred more demanding aircraft only partially explains the reasoning behind their actions. Many believed not only that these planes required better pilots but also that, by learning to fly these aircraft, and fly them well, they became better pilots. From this perspective, increased competency in the cockpit reduced overall risk and improved safety, no matter what kind of plane one flew.

Self-satisfaction was part of the reward for mastering a more demanding, and less forgiving, airplane. But recognition from one's peers was always a large part of the reward, regardless of whether pilots consciously acknowledged this or buried it behind arguments about the relative importance of skill versus technology in reducing accidents. The author of "Some Can, Some Can't: Tame a Taildragger" closed his 1970 article by observing: "There are few things in flying as satisfying as greasing on a PT-22 in a crosswind [making a smooth crosswind landing in a World War II–era military trainer] or winning a bet by making five consecutive uneventful wheel landings in a Cessna 140 [even more difficult than executing a standard three-point landing in a taildragger]. You can get out of the airplane thinking to yourself, 'I can fly that thing. It didn't do that. I did.' And you'll be right."[118] Nine years later, the owner of a two-seat Cessna 140 wrote proudly: "You can't imagine the boost one gets when a twin pilot walks over and says he wishes he could fly a taildragger because it looks like great fun. It certainly is!"[119] Thus, private fliers used technology and skill not only to set themselves apart from those who did not fly airplanes, but also to establish their place within the skill-based hierarchy in a community of pilots that valued technical competence and the willingness to take on—and overcome—risk.

Hog Wallow Airports, Hangar Flying, and Hundred-Dollar Hamburgers

Constructing Masculine Pilot Identity on the Ground

"The easiest way to get the sense, taste, and smell—and there are lots of good sensations, tastes and smells—of flying is to do a little airport bumming," Walter Boyne advised prospective pilots in his 1980 book, *Flying: An Introduction to Flight, Airplanes, and Aviation Careers*. "After you've checked in at the operations shack," the informal name for the main airport office, "just mosey around and watch the genial, contented groups fooling around with their airplanes." Boyne compared airports to marinas, "for owning an airplane is like owning a boat—always something to do." "But," he continued, drawing on a three-decade aviation career that began in the early 1950s, "there is another undercurrent at an airport, a low-keyed but unmistakable feeling that flying is still an adventure, worth doing for its own sake, and that the participants all enjoy a common fraternal bond." Boyne waxed eloquent about the type of airports that served private aviation, describing them as "those very special places whose flight operations shack is a strange combination of truck stop, fraternity house, and P.G. Wodehouse English club" where "you'll begin to learn the sometimes elaborate code of courtesy and conduct of the flying fraternity and of the informal but usually deadly accurate assessments made of pilots." Writing from the perspective of a seasoned insider, he assured neophytes: "You'll note that drop-in visitors and students are greeted warmly and get treated beautifully; after all, the airport is a business, and these are potential customers."[1]

That same year, another longtime pilot, Gordon "Bax" Baxter, went undercover for *Flying* magazine at airports around the country using the pseudonym "Hank Snee" to discover what would happen when a middle-aged businessman dropped in unannounced to inquire about flying lessons. At the larger, fancier, "big city" airports he visited, with their "nice pilot lounges and slick offices" and their "clean and perfect airplanes," he reported a universally cool reception. "With few exceptions, the general aviation industry talks to itself. Although

I sometimes got a very good sales pitch after I broke the ice, my overall impression as an outsider walking onto the flight line is much the same as walking uninvited into a private club." But Baxter's real surprise came when he visited several smaller fields that catered to private aviation. Writing mostly in the third person to convey his alter ego's experience, Bax described how "Hank wandered onto one placid airport. There being no office he could find, he went to the hangar. Ol' Junior was sitting in the hangar mouth, enjoying the breeze, watching the [aircraft] traffic come and go, two black cocker pups playing at his feet. Junior refused to shift his distant gaze or acknowledge me standing there, holding my leather briefcase." After an awkward wait, Bax/Hank finally broke the silence. " 'Where could I learn to fly an airplane?' 'Right here.' More silence, and the music of an old, old green Cessna doing touch-and-goes. 'Cheapest place on the field,' offered Junior. 'That's what they all tell us. We're the cheapest.' " Bax reported that "Hank looked all around at the gentle decay of airplanes, hangar, faded signs, flaking paint, rusting tools," and concluded, "The man was right, no doubt." Unimpressed by both the facilities and his reception, he took his business elsewhere.[2]

While searching for the office at another, "livelier down-home airport," Bax/Hank walked into a hangar crowded with aging airplanes in various stages of disrepair. He described the scene in detail: "a noseless Beech 18 [large twin-engine plane], congealing in its own tar. A silver [Piper] Cub stood tipped on its nose against one wall, to allow room for the bass boat and trailer. A desolate push-pull Cessna [Skymaster] carried its own messages written in its thick coating of gray fuzzy dust. 'Wash me.' 'Fly me.' 'You've waited too long.' " Looking further, he discovered that "against the other wall were the absolutely fascinating remains" of an aerobatic airplane "that had spun in under power." Of this grim scene, he recalled, "Only the tail feathers, stacked on top of the crushed, blood-stained cockpit, identified the mess." Reverting to the third person, he described how "Bax stood with the morbid fascination of a pilot looking at a really bad one [accident]. And Bax wondered how Mr. Snee felt seeing the insides of a crushed airplane."[3]

Baxter admitted to readers, "It was the kind of hangar and airport that all my best past flying tales have come out of." But what happened next came as an unpleasant surprise. "I let Mr. Snee find the office and just stand there until spoken too. The old pilots, my natural friends, just looked up, discounted me a 'townie' and went on with their family talk." After a bit, "Some students and young instructors came and went," again ignoring the well-dressed stranger in their midst. Finally, in a deliberate attempt to get someone's attention, "Snee put his briefcase purposefully up on the counter and studied the big wall sign advertising the flying lessons and rates." But even the receptionist, apparently in tune with the prevailing attitude of local fliers, studiously ignored him as she gossiped on the phone with a friend. In recounting the incident, Baxter lamented, "I was unaware of what snobs we are. I have always been a part of this rusty hangar and wildflower flying. . . . But now I saw it as Hank Snee would, and Hank was not a good ol' boy, not born in a leather jacket as part of the family."[4]

These two extremes—Walter Boyne's portrayal of airports as "special places" where everyone was welcome and a quiet sense of "adventure, worth doing for its own sake" pervaded the atmosphere, versus Gordon Baxter's depiction of these very same airports as austere, cluttered sites where wrecked planes stood mute witness to the dangers of the air and local "snobs" refused to acknowledge anyone "not born in a leather jacket"—represent opposite sides of the same coin. As both accounts suggest, the thousands of small and medium-sized airports that served private fliers throughout the postwar era were far more than just a collection of runways, hangars, gas pumps, and repair shops, and they served functions far beyond being places where aerial journeys began and ended. Instead, these intersections between earth and sky were physical stages upon which pilots gathered to perform "the sometimes elaborate code of courtesy and conduct of the flying fraternity" that simultaneously educated and entertained aviation insiders, served as a barrier to exclude those who didn't belong, and in the process perpetuated traditional associations between aviation and masculinity.

Airports as Masculine Spaces

Shortly after World War II, Lynn L. Bollinger and Arthur Houghton Tully Jr. visited 180 airports in 40 states while doing research for a detailed report that would be published in 1948 by the Harvard School of Business Administration. Although they focused on proposed solutions for problems with postwar fixed-base operators (aptly described by one pilot-journalist a few decades later as "aviation's all-in-one equivalent of the filling station proprietor, new and used car dealer, garageman, repairman, driving instructor, taxi and bus operator, accessory dealer, café keeper and rent-a-car entrepreneur"), like many business case studies, the authors also provided historical context for their topic.[5] Looking back on the interwar era, they noted that it was mainly a passionate, irrational "love of flying" that had kept many airports open through the long, lean years of the Great Depression. "Indeed," they observed, "many an operator subsisted on a chili and hamburger diet, slept on a cot in the back of his hangar, worked long hours, and stayed in business only by the mercy of his creditors." According to the authors, from a rational business perspective, "most would have been better off financially as employees in the filling station across the road. Their mechanical talents would have provided most of them more lucrative employment as automobile mechanics."[6]

This stark economic reality, combined with the "very love of flying which has led to neglect of the business office and its function," goes far to explain why so many prewar airports were little more than a converted pasture or farmer's field at the edge of town where a single hangar simultaneously served as a shelter for airplanes, repair shop, classroom, and airport office—and, in some cases, even the living quarters for the near-indigent individual who ran this whole ad hoc affair. Bollinger and Tully credited the "love of flying plus the tenacity and toughness of the old-time fixed-base operators" for keeping aviation alive in America during these bleak times and concluded that "without these qualities, 'unbusinesslike' as they may have been, the United States could scarcely have attained either its rapid aerial supremacy in World War II or its prewar airline development." However, they also noted that "the obvious shortcomings of the [present-day] aircraft sales-service business and the much maligned 'typical fixed-base operator' " could be traced to these prewar traditions regarding how to run an airport.[7]

As a result of their postwar survey, Bollinger and Tully observed that "the most outstanding characteristic of the fixed-base business today is its diversity." While they acknowledged that there were a few large, well-appointed facilities scattered across the country that served both private and business fliers in style, they also noted that "many operators today, especially at smaller towns and in rural regions are still at the first stages of the business's evolution. They are in spirit and physical facilities the counterpart of those first barnstormers who rented hayfields and converted old barns into makeshift hangars."[8] In other words, many airports of the immediate postwar era were little different from the muddy, makeshift fields that had served the intrepid birdmen of aviation's so-called golden age.

Bollinger and Tully apparently thought that "the obvious shortcomings" of postwar airports were so blatant that they required no further enumeration. Fortunately, numerous contemporary accounts describe both the nature of these "shortcomings" and how widespread they were (fig. 5.1). The title of one article published in *Flying* magazine in June 1945, "How Not to Run an Airport," neatly sums up the editor's scathing assessment of conditions at four of "the busiest airports" in the greater Chicago metropolitan area whose primary customer was "the private pilot and plane owner." Eight pages of photographs showed deplor-

Fig. 5.1. Austere Postwar Airport, circa 1946. This 1946 photograph of an Aeronca 7 Champion aircraft idling in the foreground illustrates the muddy taxiways, stark cinder block hangars littered with debris, and what appears to be a derelict World War II–era military trainer (minus its propeller) parked behind the last hangar, a scene typical of early postwar airports used by private pilots across the nation. Not shown is the airport office, which was often little more than a shack. Rudy Arnold Photo Collection, Smithsonian National Air and Space Museum (NASM XRA-3113).

able conditions, including dirty facilities and unkempt grounds, broken parts and trash strewn about hangars and repair shops, and leaky buildings that were little more than rundown shacks. At one airport, aircraft engines awaiting repair lay on the dirt floor of a hangar; at another, a single toilet shared by men and women was "dirtier than the worst backwoods privy" and had to be flushed by filling the bowl with water from the adjacent sink. According to the article, these photographs "illustrate the average airport, not the exception."[9] A related article, published in the next month's issue of *Flying*, described the "pine planked benches in dirty, cold 'lounge rooms' which bespeak an attitude of indifference to the customer by the average [airport] operator."[10]

The editor's opening remark in "How Not to Run an Airport" that "1,550 of the 1,600 fixed-based operators in the United States [97 percent] are private flying's—and their own—worst enemies" was almost certainly an unsubstantiated exaggeration meant to drive home the need for immediate action. However, *Flying* magazine was far from alone in sounding the alarm.[11] For instance, a 1946 article in *AOPA Pilot* reported that "hundreds of 'AOPA Pilot Report on Airport Conditions' cards" submitted by members revealed that "fewer than eight percent of all airports come up to the excellent rating."[12] An editorial in *Skylady* magazine that same year elaborated on the problem, declaring: "An Airport Cleanup Program is urgently needed NOW!" *Skylady*'s editor reported that although larger airports served by commercial airliners were "doing what they can to improve their facilities," the roughly 3,000 smaller airports "that cater to private, cross-country and student flying, are most of them in shocking condition." In fact, "many have no rest rooms at all. Very few have toilets which are clean or even acceptable. Practically none have sufficient attractive lounging space." She admitted that "those of us who have been flying for some time, put up with this since we want to fly. But a large share of the potential market doesn't—and won't."[13]

Even nonpilots were moved to comment. In another piece published by *Skylady* in 1946, Grace Jones, whose husband was a private pilot, complained, "We have had many wonderful trips, and I thrill over the clean sport of being in the air. But why—WHY does the ground part of it have to be so dirty? In all our trips I have found only one airport where the office and sanitary facilities are clean and neat. When it comes to a choice of mode of travel I vote for the car." She continued by invoking a damning comparison: "I would far rather visit the average filling station than the average airport. You don't have to walk a couple of blocks through stubble or mud, and when you reach your destination [the filling station's restroom] you find a place better than something resembling a pig sty. After all, aviation and I are both past the pioneer stage."[14] Such descriptions of airports by Jones and others in the years immediately following World War II, however, suggest that Bollinger and Tully were correct in concluding that this aspect of aviation was not so different from the prewar era after all.

Operating under the expectation that postwar private flying was poised to become big business with millions of middle-class Americans as its customer base, one man sought to change the nature of airports nationwide and turn a profit in the process. In April 1944, Henry J. Kaiser, who had made a fortune mass-producing Liberty cargo ships during the war, sent a report to the federal government proposing that his company build up to 5,000 new airfields across the United States. One of his stated goals was to "place every point [in the country] within 15 miles of our airport[s]" to support the coming air age. Kaiser envisioned four sizes of airports, each offering different levels of service. The smallest, "flight strip terminal," would feature a single unpaved sod or dirt runway, fuel pumps or a gas truck, a few hangars for locally owned airplanes, a lunchroom, an office equipped with a telephone and telegraph, and automobiles for rent to solve the perennial problem of local transportation once a pilot landed. Kaiser's plan called for these small airstrips to be built in conjunction with "regular highway projects" to take advantage of "single legislation, the equipment and personnel on the job, and the right-of-way and other complications ordinarily solved in projects of this nature." The next size up, known as the "'minimum' terminal," would be located adjacent to existing highways to serve "small towns, or sparsely settled locations" and act as a rest stop for pilots on long trips. In addition to everything that the bare bones flight strip terminal offered, amenities would include "a clubroom and lunch counter combination, rest rooms, and a manager's office" located in a small main building, along with "a repair shop and service station operating out of the hangar." The larger "'average' terminal" was designed for a city of 10,000 to 20,000 people in mind, and would include all the services of the smaller fields plus a cafeteria, lounge, and an on-site motel for transient travelers. The largest, "'extensive' terminal," reserved for major urban areas, was to be "a complete community center" unto itself. In addition to every imaginable service for pilots, Kaiser's plan called for multiple flying schools, "display rooms" for both auto and airplane dealerships, numerous dining establishments and snack bars, a wide variety of entertainment and shopping facilities—including a theater, sports fields, "curio and gift shops" and "stores of all types"—plus nurseries, playgrounds, and an infirmary. These airports would not merely be liminal places to begin and end a journey; they were to be destinations unto themselves.[15]

Drawing on his experience with wartime shipbuilding, Kaiser argued that standardization and uniformity were crucial to success. This philosophy is evident in every aspect of his plan, from a runway layout meant to make best use of limited land, to the use of prefabricated terminal buildings to be manufactured by a central factory—preferably his own—and then assembled on site by local general contractors. He also argued that while local franchise owners would actually operate the airports, "the terminals should be supervised by a large managing or operating company" (presumably also his own) to ensure, among other things, that "minimum and uniform standards of cleanliness, service, and accommodations will be maintained." This approach, he promised, would "make personal aircraft flying safe, practical, convenient and pleasurable."[16]

Kaiser's proposal came out just nine months before the CAA officially presented its National Airport Plan to Congress, and his staff openly acknowledged the CAA for providing information and advice.[17] This suggests that his company submitted this proposal for Kaiser Terminals in anticipation of the federal government's official call for a massive postwar expansion of the network of public airports. In 1940, on the eve of U.S. entry into World War II, there were only 2,231 airports, large and small, scattered across the country. Between 1941 and 1944 the CAA spent approximately $400 million to build or upgrade more than 1,600 facilities to meet the military's wartime needs, so that by July 1, 1944, there were 3,086 airports nationwide. Only 286 of the largest airports in 1944, around 9 percent of the total, offered scheduled airline service. At the other end of the spectrum, more than half of the nation's airports that year were categorized as Class 1 or Class 2 fields, which meant that they were suitable only for "private owner small type aircraft" or "private owner larger type aircraft." The CAA's plan for the postwar era called for spending up to $2 billion—half of this federal money, the other half in the form of matching funds provided by state or local governments—to double the number of airfields nationwide over the next 10 years. Of these 3,000 new airports, only 100 were to be "large air terminals" served by airlines, since the predicted surge in postwar private flying would mean, according to the CAA, that "the greatest demand will be for the smaller types of landing fields."[18] Thus, Kaiser had every reason to believe that airport construction represented a lucrative business opportunity.

It appears that the CAA never officially endorsed Kaiser's blueprint for centralized, standardized airports supervised by a single nationwide firm, but in the immediate postwar era it did promote his concept that airports should be more than just places where fliers took off and landed. For instance, in a 1945 publication intended to encourage small-town governments to build municipal airports, the CAA predicted that, in addition to local private fliers based at the field and transient pilots who needed somewhere to rest and refuel, it might be possible to entice the general population to utilize the airport as a public space. "Also, there will be those who do not fly and do not intend to fly, but are attracted to the airport by the aviation activities and other functions of public interest. The airport will, if properly developed, serve as a recreational facility. It may be possible to incorporate a swimming pool, tennis courts, sports area, or picnic ground in the layout, and thereby increase its value and appeal as a public utility."[19]

The famous shipbuilder's ambitious plan to blanket the nation with Kaiser Terminals might have improved airport conditions in the immediate postwar era, but it was never put to the test. When Congress finally passed the Federal

Airport Act in 1946, it authorized only $500 million over a seven-year period starting in 1947 (for a total of $1 billion including state and local matching funds), just half the budget suggested in the CAA's original proposal. And in reality, the federal government would spend even less. By 1954, less than $200 million in federal funds had actually been appropriated for airport construction and improvements at some 1,200 locations nationwide. Yet despite continued cuts in the program, including a yearlong fiscal "holiday" in 1954 during which no federal moneys whatsoever went to airport construction, the number of these facilities continued to grow. Fueled by the short-lived private pilot training boom as hundreds of thousands of GI Bill veterans flooded local flight schools, there were 6,484 landing fields of various sizes nationwide by 1949, actually exceeding the CAA's original 1944 plan to double the number from 3,000 to 6,000 over a 10-year period.[20] But instead of Kaiser's vision of an orderly, centrally planned and managed nationwide network of facilities for private fliers, most of these postwar airports were developed, owned, and operated by a hodgepodge of private and local concerns.[21] The numerous complaints and calls for reform suggest that most fell far short of the "minimum and uniform standards of cleanliness, service, and accommodations" that Kaiser had claimed would "make personal aircraft flying safe, practical, convenient and pleasurable."[22]

In keeping with the fierce independence that had characterized prewar fixed-based operators, and probably also in reaction to the growing threat of Soviet totalitarianism as the Cold War superseded the fight against fascism, most early calls to clean up the airport environment relied on local action instead of Kaiser's style of top-down, centralized planning and governance. At the same time, by proposing that the supposedly finer, domesticated sensibilities of women represented the best way to measure or even achieve success, many would-be reformers implied that troublesome airports suffered not from a lack of government oversight but rather from too much masculine influence. For instance, in 1946, Dudley M. Steele, manager of the Lockheed Air Terminal in Burbank, California, chided other airport operators: "The airport is *The Show Case of Aviation*. No one, your wife or mine, ever bought anything out of a dirty showcase. Clean restrooms, clean windows, lawns that are clean and well kept. Paint generously applied and frequently, inside and out, to all airport buildings will pay high dividends. If there are coffee shops, restaurants or cocktail lounges, let them be so located that the public can see them and not hidden in some dirty corner. Let those facilities be clean, neat and with excellent food and service."[23]

One year later in 1947, a *Flying* article titled "Two-Gal Field" described how Charlotte Niles and Margaret Lowell-Wallace, who had met during the war in the

WASP program, now leased and managed a small airport on Long Island. In addition to doing "their own flying, instructing, [and] servicing" (both women were pilots, and at least Lowell-Wallace was also a licensed mechanic), they had "built a small hangar and installed water, electricity, telephone." A large photograph on the first page features the small but attractive clapboard cottage-style building that served as the airport office, complete with flowers in the window boxes, a wooden rail and post fence, and an outdoor patio with chairs shaded by an overhead trellis. The caption reads, "Pleasant administration building attracts students and pilots for hangar flying." A smaller photo on the next page shows the two women gardening in front of the same building and notes, "Neat administration building is kept spick and span with fresh paint, plantings, and flowers."[24] Although *Flying* did not specifically call on other airport operators to install shaded patios and plant pretty flowers, coming on the heels of articles like "How Not to Run an Airport" (1945), "Here's How to Run an Airport" (1945), and "How Not to Run an Airport—One Year Later" (1946), it is clear that the editor featured this "Two-Gal Field" as a positive example for others to emulate.

Dr. Jack Rushing, a private pilot from Dallas, Texas, took a different approach when he addressed the Airport Management and Operations Conference in 1950 about a problem that went beyond mere dirt and disorder. "Here's something else important," he admonished his mostly male audience. "Mr. Doe comes out to your airport for some reason or other—not to learn to fly—and he sees old wrecks around. You will see him take a gander at a ball of metal, and he thinks 'somebody tried to fly that thing, now look.'" Rushing, a surgeon by profession, compared the negative publicity generated by littering an airport with wrecked planes to a hypothetical example from his own workplace, the hospital. What, the doctor asked airport managers, would happen if he left "a few stiffs lying around" in the lobby so that whenever a new patient walked in, he could "proudly point" to a corpse and say, "'I operated on this man yesterday'?"[25] In light of Gordon Baxter's / Hank Snee's reaction when he spotted the crumpled, blood-stained wreckage tucked away in the corner of a hangar exactly three decades later, Rushing's advice that wrecked planes might scare off potential students rings true from a business perspective. But then again, as Bollinger and Tully had pointed out in their 1948 Harvard business case study, airport operators, prewar and postwar alike, were famous for their passion for flying that led to "unbusinesslike" behavior. And given long-standing associations between skill, risk, and masculinity in aviation, it seems plausible that airport managers, as well as the local community of pilots who might have otherwise urged them to clean up their airports, may have seen nothing wrong with leaving the aerial equivalent of "a few stiffs lying around" to remind casual visitors that conquering the sky really *was* a demanding skill that deserved admiration and respect.

AOPA Pilot took this kind of "unbusinesslike" attitude to task in a 1953 editorial that proclaimed, "The leather-jacketed fly-boy, who felt he was doing the industry a favor by operating a field that was more a hog wallow than an airport, is rapidly becoming extinct." In his place, the editorial predicted, "bigger and better men—coldly calculating businessmen—are beginning to recognize civil aviation as a good long-range paying proposition."[26] However, despite this optimistic suggestion that poor conditions had merely been part of private flying's early postwar growing pains, "hog wallow" airports remained commonplace. Five years later in 1958, the same magazine's West Coast correspondent, Lois C. Philmus, described a typical long-distance trip by private plane. "You just landed after a tiresome cross-country. . . . It's almost over as you taxi off the runway, you think." But instead, the ordeal was just beginning. "Ground control (if you found any) wasn't much help in getting to the fixed-base operator. So you taxi around a bit and finally spot a hangar lean-to. This must be it. No signs, but it looks like one." Philmus continued her play-by-play narrative: "You taxi up front. Place is deserted but there's a light on in back. Tools laying around in the grease. Yep, this must be it. You cut your engines and sit for a minute looking out. Maybe someone heard you and will come out and tell you where to park. Well, apparently not. You climb out, and go looking. You slosh through the grease on the hangar floor. Eureka, signs of life—voices from the lean-to. You call out. Well, guess they didn't hear. You poke your head in—there are a couple of guys sitting in there. Must be the operators—they're covered with grease, too." They prove about as welcoming and helpful as the missing signs that failed to direct the visiting pilot to the airport office in the first place. For a total of five paragraphs, Philmus painted a gloomy (and greasy) picture of the typical small to medium-sized airport, ending with "No cabs, a telephone back in the darkest corner; no waiting area so you go outside and sit on your bags, hoping that a taxi will arrive within an hour."[27]

Complaints like these continued into the next decade and beyond. Author, pilot, and self-styled aerial adventurer Martin Caidin, who spent two months in 1960 flying around the United States with a photographer to gather material for a book project, grumbled: "Perhaps the greatest disappointment was the private airports of our country. My apologies to those really beautiful fields and airports which speak so handsomely for private aviation, but the majority of fields are eyesores, garbage dumps, rundown sewers, and the like. It is astonishing to encounter this as virtually a steady diet, and finally we simply quit flying into the small airports. . . . I did not want to follow this procedure, but I soon became sick and tired of grime, sloppiness, lack of places to eat, and so forth."[28] In 1964 Joseph Whitehill described how during a 2,600 mile cross-country trip he landed at an airport in Topeka, Kansas, that was "as rare as Oz, where we found the following: courtesy, cheerfulness, cleanliness, a free courtesy car, no landing fee, no parking fee, and no sour faces because we didn't buy gas."[29] But behind Whitehill's upbeat description lurked the not-so-subtle implication that such facilities were still "as rare as Oz." Two years later in 1966, C. Pfeiffer Trowbridge described taking "a dream flying trip" in which he spent a month flying around the country with his wife and two young daughters "on an itinerary that could only be described as 'anywhere, USA.' " Although he was more positive about the conditions at most airports than either Caidin or Whitehill, he saw fit to remark in a letter to the editor, "Since AOPA has done such a service with its book *Places to Fly*, I believe the membership needs the companion volume I am compiling, Places *Not* To Fly," suggesting that fellow fliers would appreciate help in avoiding subpar facilities (fig. 5.2).[30]

Throughout the 1960s, would-be reformers continued to maintain that airports could be improved by applying feminine standards of cleanliness and order, arguing that managers should domesticate these dirty, disorderly, masculine places. One 1961 editorial, written with the male pilot in mind, declared: "Small airports too often fail to recognize the requirements of today's growing general aviation. Has your wife reported to you recently on some of the rest room facilities she may have encountered?"[31] A story published the same year, humorously titled, "DANGER! Female Airport Operators at Work," described how one aviation parts salesman had initially balked in the mid-1950s at having to deal with a woman who managed an airport in his territory but had since come around to a very different opinion. "They operate the cleanest and friendliest airports in the country," he observed. "You don't find greasy instructors, two-year-old magazines, and dust an inch thick at the airports presided over by women." Apparently to assure readers that these changes in the airport environment posed no serious threat to the inherently masculine nature of flying (or, for that matter, his own masculinity), he added teasingly, "Of course, most of these ladies don't know push-rod housings [for an airplane engine] from FHA housing [referring to the Federal Housing Administration] . . . but that just makes us feel superior!"[32]

At least a few individuals tried to improve business during the 1960s by investing considerable time, effort, and capital in projects specifically intended to make the spaces that served pilots more appealing to women. Such was the case

Fig. 5.2. Austere airport humor, circa 1960. This cartoon reflects the austere and cluttered conditions private pilots continued to encounter at some small airports throughout the postwar era. "Five Bucks to Use the Runway?" [cartoon], *AOPA Pilot* 3, no. 7 (July 1960): 53. Courtesy of the Hagley Museum & Library.

when "grocery chain magnate" Walter Van Der Ahe and "former real estate–auto salesman" Finley Brinley joined forces to open Van Nuys Skyways pilot service center. Located on the outskirts of Los Angeles just a few miles from the heart of Hollywood, the bustling Van Nuys airport can hardly be considered a typical U.S. airport; still, Van Der Ahe's and Brinley's efforts, as well as the long-term fate of Van Nuys Skyways, are telling. In a 1961 article, pilot-correspondent Lois Philmus—who just three years earlier had penned her dismal depiction of sloshing "through the grease on the hangar floor" in search of the office at a typical general aviation airport—described how Van Nuys Skyways represented "a major revolution" in both how airports looked and how they treated their customers. Founded by two successful businessmen who were complete outsiders to aviation, she explained how the establishment had little in common with more traditional facilities run by those who had come up through the ranks of "the charmed—but inbred—general aviation circle." Van Nuys Skyways included a "glamorous Cessna [aircraft] showroom, with sharp overtones of a Cadillac sales agency," a full-service repair shop and parts department, a coffee shop complete with an outdoor patio and waterfall, and two small rooms and a shower for transient pilots. Philmus observed that "old timers—sadistically proud of their hardened austerity diet of minimum service, maximum inconvenience, and hard tack surroundings—literally cringed at the Hollywood-premier type grand opening and openly sneered at the guest list: executives, salesmen, civic leaders, movie stars—and their wives. Why, most of them didn't know a stabilizer from an aileron!" But, she noted approvingly, Van Nuys Skyways was not designed for these "old timers" who had already paid their dues. Instead, echoing Kaiser's 1944 proposal to improve airports, this approach targeted "the people who fly or would fly when flying is attractive, practical, convenient and comfortable."[33]

According to Philmus, Skyways cofounder and general manager Brinley believed that it took more than simply good service, clean facilities, and a fresh coat of paint to make flying "practical, convenient and comfortable" to a wider audience. Instead, he deliberately attempted to re-gender the airport environment by downplaying its traditional masculine atmosphere. For instance, to make sure that "every day is ladies' day at Van Nuys Skyways," Brinley introduced special daytime ground school and flight classes designed to draw affluent Hollywood housewives into private aviation. Implying that part of being a smart businessman meant that he recognized his own natural limitations when it came to attracting customers of the opposite sex, Brinley enlisted his "pretty, blonde vivacious wife Darlene" to lead "the growing group" of flying housewives. He also tasked her with "decorating a sitting room for the gals." As Brinley explained to Philmus, "The ladies told me they felt funny just hanging around outside when they are at the airport so I gave Darlene the room to fix up for them." Philmus declared the pair's efforts a success and described a scene that contrasted sharply with the norm for airports across the country: "The outside patio [at Van Nuys Skyways] sometimes looks like shopping day at Beverly Wilshire [Hotel on Rodeo Drive in Beverly Hills] when groups of pretty housewives gather, fresh and cool in sun dresses, and sip lemonades. The gossip's about the new styles—airplanes, though, not clothes."[34]

Around the same time but on the opposite side of the continent, the Flying W Ranch in New Jersey was hailed as "today's aviation showcase of the future" for airports that hoped to achieve business success by making private flying a family affair.[35] Founded in the early 1960s by former airline captain Bill Whitesell, the Flying W Ranch incorporated a western theme into every aspect of its operations. Whitesell made his rounds on horseback and insisted his employees wear western-style clothing: "That's an order." In addition to a full line of services for the private pilot—a Piper aircraft dealership, maintenance facilities, flight instruction, aircraft rental, along with plans for a motel and "a complete aviation supply store set up like an old-fashioned country general store"—the Flying W also offered a family-friendly restaurant, picnic area, playground, western-style riding stable and trails, fully stocked fishing pond, tennis courts, and "the only airplane-shaped swimming pool in the world." The author of one article on the Flying W explicitly noted that Whitesell had gone out of his way to make his airport a more welcoming place for women. "He has a fine sense of customer relations, so badly needed in general aviation. His major effort is to attract and please the women of the family, knowing full well that the Flying W already has plenty of attractions for the men." Whitesell himself declared, "I've done everything I can to let the gals see general aviation under the most favorable circumstances. . . . If they don't want to fly, there's plenty for them to do while their husbands are taking instruction." He added, almost as if he were comparing his airport to Van Nuys Skyways on the other side of the country, "Our facilities aren't multi-million-dollar, but they're neat, clean and attractive." So impressed were the editors of *AOPA Pilot* that they featured the Flying W on the cover of the July 1962 issue, and American Mercury Insurance Company published a full-page advertisement saluting Whitesell and his new airport in the same issue.

Despite optimism that "a major revolution" was afoot, Van Nuys Skyways and the Flying W would remain, like the Topeka area airport that had so surprised a transient pilot in 1964, "as rare as Oz." The Flying W closed in the early 1970s after Bill Whitesell went bankrupt in another failed business venture. It would remain closed until 1984 and went through several owners—and another bankruptcy—until it was eventually purchased by new owners in 1996 who hoped to restore the facility to its 1960s "glory days."[36] The upscale pilot center at Van Nuys met with hard times, as well. A 1965 article indicated that "after some growing pains," which included hiring a new general manager, the business was "booming" and future prospects looked bright. However, one year later an investment group purchased Van Nuys Skyways, and by the early 1970s it had completely disappeared from the pages of aviation periodicals.[37] Competing fixed-base operators at Van Nuys continued to offer upscale service to the area's well-heeled private and business fliers, but there is no indication that they made a concerted attempt to attract women or re-gender the airport environment in the tradition of Finley Brinley's Van Nuys Skyways. Apparently pouring large amounts of time, effort, and capital into sprucing up airports, improving service, and going out of the way to accommodate "the gals" was not the path to success.

Van Nuys Skyways and the Flying W represent failed attempts to remake individual airports into less overtly masculine spaces. Meanwhile, by the mid-1960s there was a new, top-down effort at the national level to accomplish the same general goal. In the spring of 1967, *Flying* magazine columnist Sally Buegeleisen complained, "The flowers that bloom in the spring . . . rarely do so at an airport." She proceeded to list the now familiar catalog of shortcomings: "Too often, airport landscaping seems to be fashioned haphazardly from weeds and neglect, and outdoor decoration on many airports consists almost entirely of old cans and barrels, greasy containers and debris, and the bits and pieces left over from defunct airplanes." But after jokingly assuring her mostly male readership that she was "not suggesting that we plant roses to climb up the hangar walls, or arrange potted geraniums tastefully around the [runway] approach lights," Buegeleisen continued on a more serious note: "It is not so much the lack of effort at beautification that disturbs me as the apparent attitude [by men] that appearances don't matter." She then described how women were finally taking matters into their own hands.[38] One year earlier, the FAA's Women's Advisory Committee on Aviation (WACOA) had cosponsored a program with the Ninety-Nines (the nation's largest organization of women pilots) to encourage fixed-base operators and municipal governments to clean up their airports. Although the airport program lacked any real teeth for enforcement, it was inspired by the legally binding Highway Beautification Act passed by Congress in 1965, which among other things mandated the removal of automobile junkyards that could be seen from public highways. President Lyndon B. Johnson's wife, Claudia Alta "Lady Bird" Johnson, had spearheaded the effort to clean up America's highways; in 1967, to show her support for the aviation version of this program, the First Lady personally presented the first ever Airport Beautification Award to Sky Harbor Municipal Airport in Phoenix, Arizona. The fact that in 1966 *Flying* magazine had reported on WACOA's early efforts at airport beautification under the tongue-in-cheek subtitle, "Willful Women," is just one more indicator of how pilots tended to view this "feminine" intrusion into how they maintained their manly domain.[39]

By the early 1970s, the Johnsons were out of office, and, possibly in deference to the feminist movement of the previous decade, fewer and fewer observers suggested that a woman's touch would improve airport conditions. However, the fight to rid the nation's small to medium-sized airports of derelict aircraft continued. "Most of the big airports used by the airlines are no problem," the FAA explained in 1970. "The airlines aren't going to allow any damaged or dismantled aircraft to sit around where their customers might see them."[40] That year, echoing the warning Dr. Rushing had delivered to airport managers exactly two decades earlier, FAA administrator John H. Shaffer announced: "These junk aircraft not only degrade the appearance of airports but tend to convey the erroneous impression, to both the flying and nonflying public, that aviation is inherently unsafe." But even top-down pressure failed to alleviate the matter. By the end of 1972, more than five years after the First Lady delivered the first Airport Beautification Award in person, the FAA claimed that it had identified 1,262 derelict planes still cluttering 581 airports, and Shaffer announced yet another aggressive "two-year campaign to get rid of the junk aircraft that litter airports and give general aviation a tarnished image."[41] The persistence of this problem suggests that there was more behind this grass-roots resistance to cleaning up airports than, as Sally Buegeleisen had suggested, men's "apparent attitude that appearances don't matter." Viewed in the context of the relationship between masculine identity and aviation, it seems instead that appearances *did* matter, and that the real issue was that many private pilots did not want the airports that they used to conform to a feminized domestic ideal.

The desire to carve out a masculine space separate from everyday work and home life was not unique to postwar private pilots. *Life* magazine opened 1954 with a feature on "The New American Domesticated Male." Mixing statistics with amusing anecdotal observations, the article depicted "the average U.S. man" shopping for furniture with his wife, participating in PTA meetings, fixing supper, even babysitting so that his wife could go "have her hair done, shop, [and] go to club meetings." But the accompanying illustrations also showed him constructing an outdoor brick barbeque with his son in the backyard and "Building a boat for junior" in his basement workshop. According to one caption, the introduction of complex machinery meant that even yard work had become a distinctly manly endeavor: "Lawn mowing, now that it can be accomplished behind a noisy engine, gives a man a sense of power and a gadget to tinker with. He spends almost as much time fussing over his mower as he does mowing his suburban plot."[42] Thus, the so-called domestication of the middle-class American male did not mean that he had given up certain activities—and spaces—that were specifically gendered as masculine.

Even with their yards and basements as masculine retreats, more than a decade later men were still apparently trying to find places where they could feel and act like men. Acknowledging that women managed and decorated most parts of the postwar American household, a 1967 *McCall's* article suggested women should give their husbands "A Room of His Own" so that he might feel more comfortable within his own home. According to *McCall's*, "Nothing is more surely calculated to delight a husband than a room or area decorated to reflect his very special tastes and interests." Ironically, the popular women's magazine seemed to miss its own point; rather than actually provide the husband with a room that was truly "his own" to do with as he pleased, it instead directed his wife to choose the space for him and then decorate it with a manly theme on his behalf. This approach could all too easily result in a tastefully furnished den that was better suited for sitting and reading (or for the wife to show it off to visitors) than for actually *doing* something messy. Instead of seeking out a male equivalent to the parlor or living room, amateur "ham" radio operators, model makers, woodworkers, and other hobbyists who used tools to tinker with technology preferred (or were forced) to sequester themselves in utilitarian spaces such as basements, attics, or garages that fell outside of their wives' immediate control. These spaces typically had minimal furnishings and decorations and were dominated by the tools and technology associated with their hobby, thus visually marking them as "masculine" to visitors.[43]

Airports shared a similar appearance, and served a similar function, as the cluttered ham operator's "radio shack" tucked out of sight in the basement or attic, or the home craftsman's basement or garage workshop, which was also hidden safely from the sight (and influence) of the feminized, domesticated household. In twentieth-century America, dirt—or lack thereof—served as a highly visible marker of both gender and social class. Since women were considered responsible for maintaining standards of cleanliness, both within the domestic domain of the household and when it came to sending their freshly scrubbed and neatly groomed families out in public, there was a gendered component to cleanliness.[44] Thus, rundown airports with their "greasy instructors, two-year-old magazines, and dust an inch thick" conveyed a clear message: these were places created by and for men.[45] The rough-around-the-edges airport represented, in the words of Walter Boyne, "adventure, worth doing for its own sake," its dirt and disorder a haven from the feminized domesticity of the suburban middle-class home and an escape from the ordered, regimented, and (some argued, then and now) emasculating white-collar workplace.[46] The same conditions that marked airports as masculine spaces also kept many women away, or at least made them feel uncomfortable during their visits there. This is aptly illustrated by the pilot's wife who in 1946 declared that she preferred traveling by car because the conditions at roadside service stations were less appalling than what she routinely encountered at airports. Despite her assertion that "aviation and I are both past the pioneer stage," not only did many postwar airports continue to resemble the rough and inhospitable fields of aviation's pioneering prewar era, but the persistence of these conditions for decades after she penned this complaint suggests that more than a few pilots wanted to keep them that way, too.[47]

During the 1970s, even as the FAA continued to wage war against "junk aircraft," the number of airports continued to grow. By 1970, the federal government had spent $1.2 billion on building or improving airports during the quarter century since World War II ended. However, unlike the CAA's original vision for postwar airports, which had called for focusing on small and medium-sized fields to support an expected (but ultimately unrealized) surge in private flying, most of these funds—roughly 88 percent between 1947 and 1960, around 94 percent in 1970–71 alone—instead went toward building longer runways, upgrading passenger terminals, and making other improvements at the nation's largest airports to accommodate the airlines' transition from propeller-driven transports to bigger, heavier jets.[48] But even though most of the money went to a handful of large facilities, most airports still existed to serve general aviation, including private flying. For instance, of the 10,040 airports operating nationwide at the end of 1970, only 786 were served by scheduled airlines. At 8 percent of the total, this was almost identical to the 9 percent ratio back in 1944, despite the fact that the total number of airports had tripled during that time. Although many medium-sized airports could support the business jets and larger piston-engine twins increasingly used by corporate aviation, these remained in the minority. Only about a third of all airports in the country had paved runways in 1970, a similar number (probably mostly the same facilities) had runway lights to allow nighttime use, and less than a third of all airports had a runway longer 3,000 feet. For reasons of runway length alone, this meant that more than two-thirds of all airports in the United States were suitable only for small personal planes, crop dusters, and helicopters.[49] As the total number of private pilots climbed toward its all-time high in 1980 and the number of airports increased to more than 12,000, nearly 60 percent lacked any runways longer than 3,000 feet.[50] In other words, as late as 1980, most airports still existed mainly to serve private fliers.

Yet despite this continued growth, Gordon Baxter's 1980 account of "the kind of hangar and airport that all my best past flying tales have come out of," seen anew through the eyes of aviation outsider Hank Snee, provides a poignant example of how little conditions had changed at some of these "down home" airports. And Baxter was not alone. Diane Ackerman's autobiographical account of learning to fly in the early 1980s described the fixed-base operator out of which her flight school was run as having Styrofoam walls "grayed by the dense smoke frequently hanging like a weather system over the chairs and desk" and implied that women and men shared a single lavatory down a dark corridor that was equipped with "two stalls and a sink with a mirror too high for most women to look into."[51] Other pilots recorded their impressions from the perspective of the disgruntled consumer. In his 1985 *Flying* magazine article, "Happier Landings: How to Improve Airports," Matt Thurber neglected to comment on appearances and décor, but he did provide a detailed description of the more practical problems he faced on a routine basis. These ranged from muddy taxiways, potholed runways, and unreliable runway lights to a general lack of good service from airport personnel and a dearth of ground transportation to and from town. After landing one evening in French Lick, Indiana, to check on weather conditions before continuing his flight, Thurber discovered to his dismay that the only telephone was inside the airport office, which was closed and locked for the night. In this era before cell phones, he found himself stranded far from a hotel. "I didn't want to scud run [fly beneath a low overcast] in darkness, yet I couldn't call a taxi to get into town; the only option left was to sleep in the airplane." He also complained that, "Finding a convenient airport isn't as hard as finding good food on the airport," and noted that on-site restaurants, even bad ones, were so rare that "I usually cross my fingers and hope to find at least a [vending] machine or two serving pilot victuals." With the exception of complaints about restrooms, which had dropped significantly in letters and articles by the 1970s (possibly as a result of gradual change within the industry, but just as likely a result of increased enforcement of building codes), Thurber's description of airport conditions in 1985 reads almost word for word like those submitted 40 years earlier by the first generation of postwar private pilots.[52]

A Phillips 66 Aviation Service advertisement published the same year as Thurber's article suggests that conditions at many medium-sized municipal airports that served both business jets and private fliers also fell short of expectations. A half-page photograph depicts a dilapidated pilot's lounge with a corporate pilot holding his head in shame as his passenger, an immaculately attired senior executive, waits in obvious discomfort for a car to take him to an important business meeting. Cheap plastic chairs, an overflowing wastebasket, a battered naugahyde sofa conspicuously mended with duct tape, and an empty self-serve coffee carafe dominate the room. In the corner, a seedy-looking, cigar-chomping manager leers from behind a battered counter decorated with a lopsided, handwritten sign announcing, "No Checks." The advertisement declares:

"The good news is that you've arrived twenty minutes ahead of schedule. The bad news is you've got to spend them here." Phillips 66 promised far better conditions if pilots stopped at one of the company's Aviation Performance Centers.[53] But the fact that there were only 70 such officially designated Centers nationwide begs the question of how many hundreds, or perhaps thousands, of other airport fixed-base operator lounges and "operations shacks" resembled the one in the photograph.

Although most pilots who bothered to write about airport conditions did so in order to voice complaints or urge reform, others agreed with Walter Boyne's glowing assessment that these places were fine just as they were. For instance, a 1985 *Los Angeles Times* article about the threat of urban encroachment on Meadowlark Airport in Orange County, California, described the tiny airfield, "hemmed in by the tract homes, liquor stores and office buildings that have sprouted up around it," as a destination for area pilots in search of local color and camaraderie in spite of its generally shabby condition, including "potholes" in the aircraft parking area and "runway lights that haven't worked in six months." After speaking with fliers who were based at the field, the reporter described it as "a place where, on a Saturday afternoon, you can drive out to the café, have a hamburger at one of the picnic tables near the runway and talk airplanes with pilots from all over Southern California who have flown in for the afternoon." Attorney–private pilot William Gramble compared Meadowlark's run-down yet easygoing setting to Orange County's far-larger, better-maintained metropolitan airport. According to Gramble, at the larger airport, "you fly your plane in, you get out and you park it, and that's the end of the line. There's no atmosphere, no congeniality." In contrast to this sterile and impersonal environment, he described Meadowlark as "a hub for people that are more involved in aviation itself, people who don't necessarily fly just to get somewhere, but more for the sake of just flying the plane."[54] Gramble's opinions about small airports—like those of thousands and perhaps hundreds of thousands of postwar pilots who accepted as the norm or even actively sought out the "atmosphere" and "congeniality" that places like Meadowlark offered—would have likely remained invisible to history had he not been interviewed by a reporter writing a story about the uncertain future of a tiny, tenuous, and, in the words of Bollinger and Tully's case study published nearly four decades earlier, extremely "unbusinesslike" airport.

Tall Tales and Lessons Learned

The 1985 *Los Angeles Times* article's portrayal of Meadowlark airport as a weekend destination for pilots who wanted to enjoy a hamburger and "talk airplanes,"

Walter Boyne's 1980 description of the typical airport office as a "combination of truck stop, fraternity house, and P.G. Wodehouse English club," and Gordon Baxter's lament, also from 1980, that local pilots looked up from their "family talk" barely long enough to decide to ignore a stranger who was obviously "not born in a leather jacket" all suggest that pilots constructed airports as masculine spaces not only through their demonstrated preference for clutter over cleanliness, but also in the way that they behaved toward each other—and outsiders—while they were on the ground. Nowhere is this more evident than in the long-standing tradition of hangar flying. Simply put, *hangar flying* means talking about aviation. Although the term was commonplace in aviation periodicals by the late 1920s, the practice itself dates from the earliest days of powered flight when pilots, grounded by inclement weather, gathered in a handy hangar to discuss their favorite subjects: airplanes, flying, and women, not always in that order.[55] Like the "unbusinesslike" practices of fixed-based operators that spurred Bollinger and Tully to publish their Harvard School of Business Administration case study in 1948, the prewar tradition of hangar flying was passed down from one generation of pilots to the next, becoming a regular feature at postwar airports. And, like the physical appearance and cleanliness of airport facilities, hangar flying also came under fire by critics who warned that it was bad for business.

In 1949, as aircraft sales and the number of new student pilots plummeted, Cessna Aircraft Company's R. E. Dutton beseeched a convention of airport managers: "Let's stop hangar flying at the airport." He went on to blame tall tales, spun by pilots at local flying fields across the country, for perpetuating private aviation's reputation as a risky pastime, and asked his audience, "Is it any wonder that 80 percent of the people in these United States think that an airplane taking off is still an accident just hunting for some place [*sic*] to happen?"[56] At the next year's meeting of the same group—the very gathering where Dr. Rushing warned airport managers to clean up junk airplanes lest they scare away potential students—Frank Trumbauer, an assistant regional manager with the CAA, echoed a similar opinion about how pilots talked about their exploits. "Sampson killed 1,000 men with the jawbone of an ass," he complained. "We kill that many sales and ideas in aviation every day, using the same weapon." Trumbauer suggested that when airport operators and instructors came in contact with potential students, "instead of explaining or elaborating upon our close calls and hazardous flights, why can't we talk to these important people about how to fly [safely in inclement] weather . . . and how to use the radio; how to fly cross-country, and how they can make use of the existing facilities that are available to them."[57]

Although Dutton and Trumbauer focused solely on hangar flying's negative impact on selling private aviation to the general public, their complaints implied that pilots used hangar flying to boost their personal reputations among fellow pilots and nonfliers alike, telling stories—real or embellished—intended to highlight the high degree of technical prowess and ability to keep calm under pressure that was required to overcome the dangers of the sky.

Just as early calls to clean up airports went unheeded, the same was true regarding efforts to rein in hangar flying. In an anonymous letter to the editor of *AOPA Pilot* published in 1957, "the Bookkeeper" at Crystal Skyways in Minnesota complained, "One of our 'thoughtful travelers' just left, and I must say it is a relief. I am sorry I could not say: 'Thanks come again.' He came in, bought no gas, tied down for free. I dialed his telephone numbers for him, and did everything but dust off his shoes. He returned the courtesies by taking up nearly two hours of my time with a constant barrage of unimportant questions, interspersed with amazing, heroic episodes of his life as a 'pile-it.'" According to the Bookkeeper, this boastful and inconsiderate "pile-it" (obviously a nickname for someone with a propensity to "pile it on" a little too thick) was not just a boorish exception to a courteous norm. Instead, the Bookkeeper groused, "There are so many transients, as well as dropper-inners and hangar flyers, who . . . feel that a flight office is a place of entertainment, minus highballs and foot-rail." Although the disgruntled Bookkeeper viewed hangar flying as a wasteful form of "entertainment," this individual, like Dutton and Trumbauer, seemed convinced that the underlying reason fliers engaged in this practice was to prove their heroic manliness as "pile-its." The editor refused to condemn hangar flying outright, noting that "most of us 'eat and sleep' flying" as a simple matter of fact. Instead, the editor simply reminded readers to exercise discretion: "We may be hypnotized by our own fascinating yarns . . . chances are that the operator has heard then all before."[58]

George P. Haviland expressed more ambivalence than the Bookkeeper in a 1958 article titled "YOU Can Sell Flying," in which he provided fliers with a laundry list of common practices that were sure to drive off potential converts to the cause of private aviation. "Visit your local airport," he advised readers. "You'll find pilots hangar flying almost any hour of the day. Listen, and I'll wager you'll hear flying stories that would make your head swim if you weren't in aviation yourself. Strangest of all, when some inexperienced fledgling joins the group, the stories become more distorted than ever." Haviland, a career air force officer, was quick to point out that he wasn't completely against hangar flying, noting that it both had educational value and, when performed before the proper audience, could be entertaining as well. But, he explained, "there is a difference between exaggeration for the gang and padding the story to impress some neophyte."[59]

Like most who commented on hangar flying during this era, Haviland made no attempt to discern any cultural functions beyond suggesting that it was a friendly game of one-upmanship among aviation insiders that tended to get out of hand whenever there was an "inexperienced fledgling" to impress. But in addition to impressing outsiders, hangar flying also served to define who belonged—and who did not—to the close-knit community of pilots. Gordon Baxter became fully aware of "what snobs we are" only after his fellow fliers shunned him "as a 'townie' and went on with their family talk" when he pretended to be a neophyte in 1980. And despite Walter Boyne's assurances to readers that same year that "you'll note that drop-in visitors and students are greeted warmly and get treated beautifully; after all, the airport is a business, and these are potential customers," this sort of treatment was apparently not uncommon throughout the postwar era.[60]

Men who were shut out of hangar flying conversations were unlikely to record their experiences in writing; either they persevered, learned to fly, and thus earned a place within the circle of insiders, or else they gave up and abandoned aviation. George Moloney, the self-described "middle-aged student" introduced in chapter 2, was a rare exception in that he persevered but then complained about it after the fact. When he "wandered into the flight office at a rather raunchy field" in the late 1960s to ask about learning to fly, he recalled, "several seedy characters were deeply involved in telling what we used to call 'sea stories' and I suddenly felt that I should be back in the office or someplace other than this place." When the former World War II naval officer-turned-successful businessman dared to interrupt, the storyteller barely acknowledged him before returning to his " 'one engine out and a load of ice' story with this other character who was waiting impatiently for him to finish so he could tell about *his* feat—real or imagined—of derring-do."[61] Moloney was by this point in life accustomed to being treated with respect rather than ignored. However, in the world of aviation he was a rank beginner, and his advanced age, business acumen, and accumulated wealth counted for little within a community that that sorted individuals into a hierarchy based on flying skill and their willingness to follow "the sometimes elaborate code of courtesy and conduct of the flying fraternity" that Walter Boyne would describe two decades later. As such, Moloney was expected listen and learn, just like anyone else who showed up to take lessons. By barging in and rudely interrupting a hangar flying story midflight, he had immediately violated one general criterion for judging newcomers: "Have you been courteous and thoughtful?"[62] Furthermore, given the fact that he was a successful middle-aged businessman, his actions likely led the pilots sitting around the airport office to write him off as an arrogant outsider who thought he could throw his weight around simply because he earned more than they did. In this case, as with Gordon Baxter / Hank Snee two decades later, sometimes locals used hangar flying to exclude rather than impress the nonflier.

Other nonfliers, especially the wives and girlfriends of pilots, also complained that aviation shop talk excluded them from the conversation, even when they were standing right there. A 1958 advertisement for Embry Riddle Aeronautical Institute titled "What's Linda Lacking?" capitalized on this phenomenon. The full-page ad featured two young men at a soda fountain demonstrating aerial maneuvers with their hands as they pore over a brochure about aviation careers. In the foreground stands Linda, young, attractive, but totally ignored as she forlornly sips her soda through a straw, obviously unable to comprehend (or compete with) whatever these two future aviators are discussing. Rather than passively pout like the fictional Linda, at least a few real-life women who associated with pilots tried to pry open the linguistic gateway to the flying fraternity. In a 1947 article titled "The Pilot's Missus," Ruth Downie advised pilots' wives to avoid becoming an "airport widow" by trying to "at least learn some of his flying lingo." As she explained, "When hangar flying rears its head on a social occasion—as it always will when two or more pilots get together—don't groan and start talking about clothes. Listen in and try to follow the flow of conversation. When a word is used that isn't clearly understood, ask what it's all about. It not only adds to your accumulation of terms, but sometimes flatters your beloved's ego." A 1961 article described how two women who ran an airport café completed ground school (with apparently no intention of actually becoming pilots) because they "wanted to learn enough from the course to understand the 'hangar-flying' of their customers." And nearly two decades after Ruth Downie penned her advice, Grace Ellis told aviation writer Don Downie (Ruth's husband) in 1966 that she had started taking flying lessons in part because "it makes it a lot easier around the house when you can at least understand a little of what these pilots [her husband and his friends] are talking about."[63]

Hangar flying served other functions beyond impressing or excluding outsiders. For instance, in 1956 Gill Robb Wilson, editor and publisher of *Flying* magazine, encouraged pilots to create local flying clubs in order to promote more hangar flying. "Just owning and flying your own plane can be a lone wolf business," he explained. "Most people don't want to be lone wolves. The club idea solves that situation. And it's a great meeting ground for old and new pilots, a medium for hangar flying."[64] As a forum for socializing, sharing information, and educating new pilots, Wilson—who had remained actively involved in aviation ever since he learned to fly during World War I—viewed hangar flying as a valuable tradition that could make private aviation both safer and more enjoyable to participants. When most pilots described this activity, they echoed Wilson's assertion that these ad hoc get-togethers provided an excellent forum for sharing lessons learned. For instance, in 1958 Robert W. Helber Jr. recalled that after he and his wife flew into bad weather and lost sight of the horizon, "strangely, my mind flashed back to a conversation Audrey and I had with Charley and Dede Moncure of Seattle during our stay at Acapulco. Once, during our drinks and hangar flying on the veranda of the hotel there, Charley told me he was caught without instruments while flying a Cessna 140, and had kept the plane on an even keel for 50 minutes by using the gimbals of the compass as a guide" [author's note: do not attempt this technique]. Helber reported that he tried this emergency substitute for an artificial horizon, and he and Audrey survived to describe their own close call in a whole new round of hangar flying tales.[65]

Aviation periodicals even formalized this version of hangar flying in print. Starting in 1939, more than a decade before Wilson took the helm of *Flying*, the magazine (then titled *Popular Aviation*) began publishing short cautionary accounts submitted by readers under the title "I Learned about Flying from That!" In 1958, when the AOPA launched its own full-length membership journal titled *AOPA Pilot*, it introduced a competing monthly feature titled "Never Again!"[66] In fact, Helber's story about how hangar flying had saved his and his wife's lives was featured in the second issue of this new publication. To stave off occasional complaints that tales like these glorified risky and irresponsible behavior, editors of both periodicals noted that they presented these accounts so that pilots could share their misadventures, in the spirit of face-to-face hangar flying sessions, with a broader audience so that others might learn from and avoid their mistakes.[67]

Although hangar flying was, by definition, an earthbound activity, it also provided an excuse for group excursions by air. John Purner, a Chicago-based businessman and private pilot who published a series of popular books on fly-in eateries, told a reporter in 1999, "Down in Texas, where I grew up and learned to fly, you'd go to the airport every weekend, especially in rural or suburban areas, and there are people standing around, swapping lies about flying, until about noon when everyone's gotta go fly someplace for lunch. They'd all say they were going out for their 'hundred-dollar hamburger' because the burger cost only about $3, but it cost another $97 to get there and back."[68] Purner's 1998 book, titled *The $100 Hamburger*, made the term ubiquitous in aviation circles, but the practice long predates the name.[69] In 1969 Keith Connes humorously described the "$12 hamburger" as follows: "The chopped meat comes to 45 cents and the direct operating costs of the plane total $11.55 for the round trip—at a conservative figure." He went on to explain in parentheses, "A conservative figure is one in which you don't add in all your costs; otherwise you'd quit flying." Four years later, apparently adhering to a less "conservative" formula, Marion C. McDonald mentioned "$25 hamburgers" in a 1973 article about private flying. By the mid-1980s, according to another article, the price had doubled again to $50.[70]

In more than a few "Never Again!" or "I Learned about Flying from That!" accounts, a trip to a nearby airport for lunch or dinner served as the backdrop against which some cautionary tale played out. In the mid-1960s, Betty Denné Haesloop, author of one such article, became lost over unfamiliar territory on her way to meet a group of fellow fliers for lunch. When she and her passengers finally arrived, the meal was nearly over, and as she recalled: "The silence was deadly as all eyes turned questioningly in our direction. 'Betty, where *have* you been?' 27 male voices asked in unison." However, food plus fliers resulted in the inevitable lunchtime hangar flying session, and, as she recounted, "When the friendly kidding finally simmered down, then came the reassurances and amazing stories of each one's 'lost' experiences. No group of friends could have been more helpful in bolstering my ulcerated pride." Haesloop's male friends demonstrated that they were a good-natured and forgiving lot, but they also managed to get in a paternalistic jab at the female flier later that day. During the flight back to her home airport, she reported, "the Unicom [radio frequency] fairly buzzed all the way. The boys were making sure there was no repeat performance. 'Five-Four Delta [the call sign for Haesloop's airplane] . . . WHERE are you now? Can you see that brush fire dead ahead? What is your heading? Correct eight degrees right.'"[71]

"I Learned about Flying from That" and "Never Again!" remain regular features in *Flying* and *AOPA Pilot* to this day, and *Flying* published three "best of" compilations of stories between 1976 and 1993. *The $100 Hamburger* entered its third edition in 2006, and by early 2015 Pruner's website by the same name, which offered paying subscribers up-to-date information on airport eateries, claimed that it represented a "community of more than 50,000 Recreational Flyers!" This combined longevity and marketability attests to the continued popularity of the hangar flying tradition among pilots throughout the postwar era. But a closer reading of the *Flying* and *AOPA Pilot* stories also reveals hints of what Walter Boyne described as "the sometimes elaborate code of courtesy and conduct of the flying fraternity" that pilots used to form their "informal but usually deadly accurate assessments" of other fliers.[72]

For instance, Robert T. Shaw began his "Never Again!" account with this introspective note: "Funny thing about flying—when a pilot has had a close call with the Dark Angel, he doesn't like to talk about it unless he can put the blame squarely on someone else. Yet, as time passes, this reticence usually diminishes. Some rainy afternoon, when the local flying is being done within the confines of the hangar, he will tip his chair forward, join in the discussion and spill out the details of his own brush with death." Shaw wrote this in 1964, some six years after a "close call with the Dark Angel" in the late 1950s.[73] And he was not the only flier who was reluctant to "spill out the details" until long after the ink had dried in the logbook entry for an almost-fateful day. In the 1978 article "To Thine Own Self Be True," flying physician W. G. Bradley introduced his own story by noting: "Hangar-flying being a reasonable pastime for an aging birdman, I've been thinking of a very foolish thing that I did some 15 years ago and decided at last, to tell about it."[74] Many pilots were less than willing to come rushing back to share such tales with others until enough time had elapsed that they could attribute their mistakes to past inexperience or the hubris of youth, rather than risk having the incident reflect negatively on their current flying ability. Confession might be good for the soul, but if improperly performed, it could also harm one's reputation for the good judgment and technical prowess expected of pilots.

Such was the case when Doris Purdy penned a brief "Never Again!" article in 1966 admitting that she had carelessly allowed herself to be "swallowed" by clouds after a flight to watch the sunset over Oakland, California. This mistake, though potentially deadly, was hardly unusual: accident reports and other "Never Again!" articles of the era indicate that getting caught in the clouds happened to other non-instrument-rated pilots all too often (indeed, this is the very error that both Shaw and Bradley gravely confessed to years after the fact). However, as part of her story, Purdy admitted that although she had completed many hours of blind-flying instruction in a ground-based simulator—exactly the type of training that could have kept her out of trouble that night—she had never taken it seriously. "Often, when I was in the middle of a very difficult problem, I would throw open the hood and say, 'I quit.'" To make things worse, she also disclosed her hope to use her own aerial misadventures as the basis for a television comedy about a woman pilot. In response, V. Alan Mode fired off an outraged letter to the editor. He wrote, "If the Never Again article by Doris Purdy . . . is an example of her 'blunders' through the air which are to be portrayed in comedy, we differ greatly in our views on flying. I fail to see how any television program can portray flying or a pilot in comedy and serve the interest of any portion of general aviation." After railing on for another paragraph, in which he lectured "Miss Purdy" for failing to realize that she had been "only a few minutes from death" as a result of her poor choices, he concluded, "Blunders of a pilot seem poor material for any type of comedy."[75] To him, any attempt to deal with one's mistakes through humor represented the wrong approach to flying by the wrong kind of pilot. Mode's tirade that aviation was no laughing matter echoed two decades later when Diane Ackerman's male instructor angrily admonished her, "Flying isn't supposed to be fun. It's *dangerous*."[76] Ultimately, Purdy's most damning mistake was not that she flew into the clouds, but how she related the story afterward.

Although women made easy targets because of gendered assumptions about their abilities as pilots, they were not alone in overstepping the bounds of hangar flying. For instance, in a "Never Again!" article published in 1960, two male pilots who claimed that they were "always cautious about the weather" described how they took off one night with two teenager members of the local 4-H Club aboard, in the rain, with the temperature just above freezing, into marginal weather conditions. Before takeoff, they consciously decided against checking to see whether the screws holding a wing panel in place had worked loose, even though they were dissatisfied with the results after tightening these screws for an earlier flight. Their reason for not rechecking the screws was simple: it was cold, dark, and wet outside. Shortly after takeoff, the screws came out and the panel blew off, ripping a large hole in the wing in the process. Robert Caswell, the pilot who later related the incident, noted: "To make matters worse, we knew that we were over the roughest terrain in central Iowa. Below us was the Des Moines River and Ledges State Park, nothing but steep, heavily-wooded hills." Caswell and his copilot were able to nurse their crippled craft back to the airport and land without further incident. To Caswell, there were two lessons to be learned: they should have delayed the trip long enough to fix the problem in the first place, or—failing that—at the very least they should have gotten out to check whether the screws had worked loose again before taking off into the night.[77]

At least one fellow pilot, Chester A. Adams, was so unimpressed by Caswell's actions and subsequent self-analysis that he felt compelled to write a letter to the editor. "What does he [Caswell] mean 'always cautious about weather?' Single-engine, night flight, light rain, temperature a few degrees above freezing, and admission of mechanically defective aircraft. I can barely imagine a set of circumstances more conductive to suicide and murder. Mr. Caswell mentions only two loose screws. He counted wrong. There were four when he and his copilot friend took over the controls."[78] According to an outraged Adams, Robert Caswell and his copilot were guilty of far more than simply taking off in a faulty aircraft without thoroughly checking out a known problem. In addition to this mistake (which was bad enough), they had, by their own admission, violated a laundry list of other commonsense rules for safe flying. Caswell's final mistake was in how he told his tale: by asserting that he was "always cautious about the weather," but then proving by his own admission that he was not, he opened the way for other fliers to question his overall competence and judgment as a pilot.

A 1978 piece by *AOPA Pilot* columnist K. O. Eckland put a humorous spin on a different kind of hangar flying pitfall from those that had ensnared Doris Purdy and Robert Caswell. "Boomer Bullhorn is one of our favorites around here," Eckland began as he lampooned a fictional version of someone he assumed his readers had met, or were bound to someday meet, at their local airport (fig. 5.3). "There isn't anything he hasn't done and there isn't a plane he doesn't have time in or an airport he hasn't been into at least once. The fact that he only has a hundred hours total has nothing to do with it. He's managed to cram a lot of experience into those brief hours. He's very good at cramming. He has more tales to tell than Mother Goose, only hers are believable." Eckland continued, "Despite having the personality of an IRS agent and the appeal of an oil leak, he manages always to have an audience of impressionable student pilots who are open-mouthed at his stories. Luckily, most are yawning." Eckland's caricature contrasted Boomer's impressive hangar flying feats—for instance, "He's the only pilot to make it across [the] huge mountains and desolate swampland" in the hundred or so miles between Gary and Fort Wayne, Indiana—with a scathing assessment of Boomer's *actual* flying abilities, delivered in a rapid-fire staccato of one-liners. "In truth, Boom has about as much aeronautical expertise as a cannon ball. He still brings a map to taxi to the fuel [pumps]. . . . He spends so much time in the [airport traffic] pattern that the [control] tower uses him for a check point. . . . The only FBOs [fixed-base operators] who will rent him a plane are those who are going into bankruptcy anyway."[79]

Eckland clearly intended his column to elicit a chuckle from readers, but he also sought to impart some not-so-subtle advice regarding hangar flying in the process. His depiction of Boomer Bullhorn simultaneously confirmed complaints that some hangar flyers exaggerated or fabricated stories solely for the sake of enhancing their reputation and cautioned readers that trying to talk the talk if they couldn't walk the walk would leave them looking foolish in the end. However, unlike those who decried the practice of hangar flying altogether,

Fig. 5.3. Hangar flying humor. K. O. Eckland's fictional character "Boomer Bullhorn" inadvertently demonstrates some of the pitfalls of breaking the unspoken rules regarding hangar flying (even the airplane looks bored). K. O. Eckland, "The Old Aerodrome," *AOPA Pilot* 21, no. 5 (May 1978): 82. Courtesy of K. O. Eckland's family and the Hagley Museum & Library.

Eckland's parable implied that some degree of honesty and credibility were crucial components of this time-honored tradition. Pilot identity was based on technical skill, and stretching the truth about one's abilities beyond belief, as Boomer Bullhorn routinely did, threatened the very foundations of the aviation community.

Just as airport conditions improved somewhat starting in the early 1970s, at least one aspect of hangar flying began to evolve around the same time. This may in part reflect a generational changing of the guard within the aviation community, but it was also almost certainly influenced by the way members of the nonflying public viewed private pilots. In the summer of 1969, *Flying*'s Richard Collins advised fellow fliers in an editorial, "We are often pictured by the general press as a reckless bunch of fatcats, and if you wear wings to a cocktail party, some folks can become almost hostile when they find out you fly for personal or business reasons and not for an airline." He blamed this reputation in part on the pilots themselves. "Maybe the way we talk on occasion has affected our image: 'There I was, on my back at 800 feet, with six inches of ice on the wings, the radios and instruments out, ceiling zero, thunderstorms all around. . . . ' Or, 'That LaGuardia is a real madhouse. Why I got lost taxiing and wound up on the runway, and you should have heard the tower.' " Collins acknowledged that "pilots do like to tell war stories, and this is one of the oldest and most cherished privileges of belonging to the fraternity." "But," he continued, "the tales should be told only to equally old and bold pilots."[80]

This was not just another admonition to cut down on hangar flying; indeed, Collins was clearly a fan of this pastime. Instead, he was reacting to a recent public relations crisis that had rocked private aviation. The previous summer, airline travelers had experienced unprecedented delays, due not to weather but to an overloaded air traffic control system. In August 1968, *Time* magazine compared "America's aerial arteriosclerosis" to a dangerously clogged circulatory system and described how "the traffic jam in the skies" had "shifted from acute to chronic." The story continued, "The glut that has all but congealed the New York City metropolitan area's 'Bird Cage'—Kennedy, La Guardia and Newark airports—now spreads confusion across the country and abroad, shredding connecting schedules in Los Angeles and squeezing service in Miami. [A] fortnight ago, 'Black Friday' choked the Golden Triangle between New York City, Chicago and Washington with 2,079 delays. Black Friday now is every day."[81]

Most private fliers were not directly affected by this "traffic jam in the skies," since it actually involved only a handful of major hubs through which most of the nation's commercial air traffic passed. The problem was that the airline industry claimed "that little airplanes were using too much runway time" and that private pilots who insisted on equal access to large airports on a "first come, first served" basis were the primary culprits for the delays.[82] In 1969, a year after the crisis grounded travelers and made national headlines, Donald Bain published *The Case against Private Aviation.* Bain, a former airline public relations executive who would later go on to write more than one hundred books (including the popular *Murder, She Wrote* mystery series), came down squarely on the side of the airlines in his damning description of private pilots as "mediocre" amateurs who failed to maintain minimum flying proficiency, routinely flew while intoxicated, and could only afford their expensive hobby thanks to considerable subsidies paid by the (nonpilot) American taxpayer.[83] In the fall of 1969, one of Bain's most dire predictions came true when a small Piper Cherokee flown by a student pilot collided with an Allegany Airlines DC-9 airliner near the Indianapolis airport, killing 83 people.[84] Although this tragic accident occurred several months after Collins published his editorial, it was in this highly charged climate that he had urged pilots to be more circumspect with regard to their hangar flying audience and save their best (or worst) tales for each other.

This helps explain why, when Gordon Baxter made his appearance at a small airport masquerading as a nonflier in 1980, local pilots turned their backs on this obvious neophyte and "went on with their family talk" among themselves rather than try to impress him with their aeronautical prowess. At the time, Baxter expressed surprise at their standoffish manners. But by the time he published a new book for potential students one year later, he had come up with an explanation for this behavior. "In flying there is the 'official face,' as in textbooks, talking to the press, or to any outsiders, and there is the 'other face,' the way pilots really talk to each other. The two never mix." He went on to devote an entire chapter to the joys, pitfalls, and etiquette of "the two faces of flying." After tempting readers with examples of hangar flying stories—including how one friend enjoyed flying a Stearman biplane "high over the sunny hills of California with his lady sunbathing [apparently nude] in the rear cockpit," and explaining how pilots could join the legendary "Mile High Club" by engaging in sexual intercourse in an airplane flying at or above 5,280 feet ("Remaining overnight in a Denver hotel will not qualify you," he advised laconically)—Baxter concluded, "So when, you may rightfully ask, do you get past your new-boy-at-this-club status and become part of the other face of flying?" The answer, of course, was that fledglings had to earn this privilege, just as they had to earn their wings. Except for the way pilots talked about flying in front of outsiders, very little had changed about the norms, content, or cultural motives of hangar flying since the earliest days of postwar private aviation. And, according to Baxter's chapter-length description, the practice remained a foundation of the masculine fraternity of pilots.[85]

Hangar flying and airport conditions worked in tandem to shape the experiences of anyone—pilot or nonpilot—who visited almost anywhere that private flying took place. A 1985 *AOPA Pilot* aircraft review, ostensibly about a vintage Cessna 140, illustrates this as well or better than any article or essay written specifically about the "common fraternal bond" that pilots strove to maintain throughout the postwar era. Author Mark Twombly opened by describing how Bernard A. (Bernie) Funk flew his lovingly restored 1946 Cessna 140 only "about 70 hours a year, or 80 minutes a week." "This isn't much," Twombly conceded, "but . . . time does not measure the satisfaction he gets from aviation. The [Cessna] 140 is only a part of an airport community . . . in which Funk is a leading citizen." He went on to report how Funk and his flying friends, who were based in Frederick,

Maryland, had "the best of explanations for spending so much time at the airport for so little time in the air: They are having fun."[86]

Twombly reported that for these "Sunday pilots" who flew mainly for recreation, "hangar socializing" and after-hours get-togethers were "as much a part of their flying as three-point landings." For instance, Funk and "about two dozen other regulars" met every Friday night in the dining room of the airport's aptly named restaurant, the Final Approach. Anyone who came through the door—stranger and friend alike—was greeted with a rousing variation of the *Mickey Mouse Club* theme song, followed by "an invitation to hoist a few [drinks] and swap flying tales" with the group. "Membership in the club," Twombly added with a touch of humor that likely had some basis in truth, was "open to anyone who doesn't turn and run."[87]

Similar gatherings took place at Funk's hangar, informally known as "the clubhouse" by local pilots. Conveniently located with an unobstructed view of the runway and within easy walking distance of the Final Approach restaurant, its contents and décor suggest a decidedly masculine space that served both practical and symbolic functions beyond simply sheltering an airplane from the elements. "The walls are papered with aviation posters slowly taking the texture of parchment and fading photographs of friends and airplanes. Furnishings include a school bus bench seat, a well-stocked refrigerator that is operated on the honor system, concert hall stereo speakers, two full-size rolling tool boxes, work bench and parts closet, a handful of lawn chairs and a collection of airport transportation conveyances, including golf cart, go-kart and motorcycle." In addition to this comprehensive inventory, Twombly described the social scene that centered on the hangar. "It's guaranteed that on weekends and warm evenings Funk and a half-dozen friends will be parked out front in lawn chairs, logging a few more hours of hangar flying. Between tales, runway arrivals and departures are critiqued. Colorless tricycle-gear aircraft rate silent jeers, while taildraggers and biplanes are treated with elder-statesmen-like respect."[88]

In one of the nine photographs accompanying the article—every one of them featuring Funk's beloved Cessna 140—an unidentified young woman lounges in a lawn chair alongside the plane's owner and two other men in front of his hangar clubhouse.[89] This single visual clue suggests that these gatherings enjoyed a certain degree of mixed company. But aside from pointing out that Funk was unmarried, an observation meant to explain why he had the freedom to spend "almost every afternoon and most weekends at the airport," the article contained no other references to wives or girlfriends, much less women pilots. Although this informal airport community was technically "open to anyone who doesn't turn and run," the conspicuous absence of women in Twombly's description, combined with the masculine décor in Funk's hangar and the group's boisterous behavior whenever they took over the restaurant, suggests that it was essentially a club created by, for, and about men who loved to fly. Indeed, what is perhaps most remarkable about this account is how little had changed about airport culture in the 40-odd years since Funk's plane rolled out of the factory.

Thus, the dual stories of what postwar airports looked like and how pilots interacted with each other at these places are more complicated than the litany of complaints about dirty restrooms, junk airplanes, wildly exaggerated tales, and boisterous or rude behavior might first suggest. The aviation community's long-term resistance to cleaning up airports suggests that for every flier who openly complained about how messy and inhospitable these places could be, there were legions of others who were at least willing to accept—and perhaps even openly welcomed—such inconveniences. Just as a few wrecked planes served as a reminder that flying still required considerable skill, dirt and disorder signified airports as masculine places, a setting for action, freedom, and potential adventure free from the constraints of having someone constantly scold private pilots to pick up after themselves. For much the same reason, pilots also persisted in hangar flying for reasons beyond the entertainment and educational value that this time-honored tradition provided. By recounting past experiences and misadventures, they could demonstrate their aeronautical prowess to others who had not personally witnessed the flight in question and sort each other into a skill-based hierarchy. In a community that valued technical prowess and personal character, how one recounted an event counted as much, if not more, than the actual details. Although Bernie Funk and his friends may have truly believed that they were "just having fun" whenever they got together at his cluttered airport clubhouse to talk about flying, they—like so many other private fliers throughout the postwar era—were also engaged in a carefully orchestrated performance that relied on long-standing associations between aviation and masculinity to define what it meant to be a pilot.

Gendered Communities

Negotiating a Place for Women in Private Aviation

In 1965, in the midst of the burgeoning feminist movement sparked in part by Betty Friedan's best seller, *The Feminine Mystique*, Dr. Milton Horowitz published a scathing article titled, "For Men Only?" in *Flying* magazine. Horowitz, a professor of psychology and contributing editor for *Flying*, argued that women should stay out of the pilot's seat, period. While conceding that there were a few competent female fliers, he insisted that, "*in my view, aviation is, by and large, a masculine activity, and the woman who enters this arena does so at the risk of becoming a second-rate aviator or a less feminine female*." Horowitz provided a detailed, scholarly sounding rationalization, laced with terminology from his profession, to support his argument that women were naturally unsuited to the demands of operating a complex machine in the unforgiving environment of the sky. He also complained that the very presence of women in aviation threatened to compromise the joys associated with this manly endeavor: "I feel the same about women in flying as I do about women in golf or tennis. Who wants a mixed foursome or mixed doubles? It takes your mind off your game and dulls your competitive instincts. I am convinced that you can't have it both ways: you cannot enjoy the spirit and competition of the game and your feminine companionship at the same time." Although his article spanned several pages, Horowitz summed up his views in a single sentence: "What I do assert is that flying and airplanes are suited to a way of life and a set of activities that are as masculine as they can be."[1]

In the months that followed, *Flying* published an unusually large number of reader responses to the article. Of 17 letters to the editor, only 3 agreed with the professor's tirade against women in the cockpit. Bart E. May wrote, "I don't dislike women in their place. I heartily agree with Dr. Milton W. Horowitz. I get so sick of equal rights for women that I could vomit! How in hell can a man treat a woman as a lady and still as an equal?" In May's view, a woman's "place" clearly did not include the pilot's seat. Bruce M. Fonoroff also agreed with Horowitz, writing, "'For Men Only?' so completely expresses my own views on women in aviation that I would replace the question mark with an exclamation point. My heartiest congratulations." The third letter, submitted by Dolly Kearn, reveals that men were not alone in believing that women made poor pilots. She wrote, "My pilot husband and I were flying the other day and I tried to land our [Beechcraft] Bonanza. I quote him verbatim: 'You haven't got it! You gotta man-handle a plane! You've gotta think like a man!'" Her closing remarks suggest that just as her husband had constructed a masculine identity around his ability to "think like a man" and "man-handle a plane," she defined her own femininity in direct opposition to these supposedly gender-specific traits: "My handsome roommate landed the plane. But somewhere in his shouting did I detect a ring of pleasure at my un-manliness?"[2]

"For Men Only?" was one of five full-length articles—and the only overtly negative voice with regard to female fliers—in an issue of *Flying* devoted specifically to the theme of "Women and Aviation." In the same issue, Robert B. Parke addressed the topic in his editorial titled "The Feminine Case." After humorously restating some of the many reasons that women were supposedly unfit to fly, he observed, "the astonishing thing to most of us [men] is that in spite of their shortcomings, the several thousand woman pilots who are now flying are pretty good. Their skills are high. Their safety record is excellent and even their judgment is often worth respecting." Regarding a potential future "influx of women to aviation," he warned, obviously tongue in cheek: "It's scarcely a tidal wave yet, but it's enough to sound the alarm. Of course, the presence of hordes of women in an activity does revolutionize the activity, and this we should be braced for. You can almost certainly expect potted geraniums outside the hangars, curtains and rugs in the pilots' lounge and clean rest rooms." Beyond regendering the traditionally masculine airport environment, he predicted on a more serious note that "airplane sales would truly zoom—not if prices were cut, not if airplanes were prettier, not if the National Safety Council endorsed the airplane—but if women took to them." Along with these positive developments for private aviation, he acknowledged that the mostly male community of pilots might experience some changes to its time-honored traditions as well. In a not-so-subtle jab at Dr. Horowitz, Parke noted, "To some, the feminization of flying will be akin to holding a women's bridge party in a monastery. And it's true the day of the rowdiness of the Quiet Birdmen and the all-night hangar session may pass." But, he concluded, leaving no doubt where the editorial leadership of *Flying* stood on the question of women pilots, "the beauty of the sky is that it is big enough for both of us."[3]

Articles, editorials, and letters to the editor of this nature indicate that individuals held vastly different views regarding the relationship between gender and aviation. Yet because these sources represent the viewpoints of their authors, and in many cases were filtered by an editorial staff that sought to advance a particular agenda, it is difficult to determine whether they reflect widely held attitudes within the broader community of pilots. For this reason, national-level aviation organizations that provided their members with opportunities to enjoy each other's company, sharpen their skills and increase knowledge, or use aviation to perform community service can serve as a more reliable window into the role gender played in postwar private flying. Four such groups—the Quiet Birdmen, the Ninety-Nines, the Aircraft Owners and Pilots Association, and the Experimental Aircraft Association—represent different missions, values, and constituencies within private flying. Examining the specific types of activities that these groups sponsored for their members, as well as the venues in which these took place, not only illustrates how deeply masculine identity was entwined in private flying but also reveals that several competing versions of masculinity existed within the postwar aviation community.

Quiet Birdmen and Ninety-Nines

The Quiet Birdmen, whom Parke mentioned in his 1965 editorial with no further explanation beyond an aside about the "rowdiness" of their "all night hangar session[s]," is one of the oldest organizations for fliers in the country. Also known as the QBs, this invitation-only club was founded in New York City in 1921 by a handful of former military pilots to rekindle the camaraderie they had enjoyed in France during the Great War. In the years that followed, members continued to establish new local chapters, known as "hangars," across the country, a practice that continued after private flying grew exponentially following World War II. As the group's full name suggests—"Ye Ancient and Secret Order of Quiet Birdmen"—there is little reliable information available to outsiders about past or present activities at local and national meetings. Indeed, an undated, "confidential" handbook for members of the New York Hangar, published sometime after 1939 (and most likely after World War II), advised that one of the few formal rules was: "What goes on in QB meetings is confidential and is not to be discussed outside."[4] Despite this penchant for secrecy, rumors circulated about the group's raucous behavior. For instance, in 1982 *Flying* magazine columnist Gordon Baxter joked that no women "that I know of ever lasted through a Quiet Birdman's meeting. Not in the role of pilot, that is."[5] Nearly two decades later in 2001, Tracey L. Potter, president of Hagerstown Aircraft Services in Hagerstown, Maryland, was more explicit when he responded to a fellow pilot's query about the Quiet Birdmen: "Years ago I think that the association was used as a good excuse for a bunch of men to get away from their wives, get drunk, and watch partially clothed women."[6]

Although what happened during local and national QB meetings remains largely cloaked in mystery, there is no doubt that women were not allowed to join their ranks throughout the postwar era. For instance, the New York Hangar's booklet explained that among other qualifications, "QB membership is limited to male aircraft pilots with acceptable personal qualifications who soloed prior to 1919 or who have a minimum of 250 certified solo hours [later apparently doubled to 500 flight hours], and who are over 21 years of age."[7] The rule barring women was still in effect in the early 1990s when a state Supreme Court justice from West Hartford, Connecticut, was informed that his membership in this flying fraternity was an obstacle to becoming a federal judge. In his letter of resignation to the head of the local QB hangar, he noted: "Because of the QB's policy limiting membership to males only, the Chairman of my confirmation hearing before the U.S. Senate Judiciary Committee asked me to resign. I agreed to do so."[8]

Ironically, one of the most detailed accounts of what went on at a QB meeting comes from a woman. In 2010 Martha Lunken—a longtime pilot, occasional contributing editor to *Flying* magazine, and self-avowed traditionalist ("I'm anything but a feminist and my panties don't get wadded up about guys clinging to their male-only traditions")—wrote an article describing how she had spied on an elaborate QB "Wing Ding" that took place in the mid-1970s. From her perch in a distant tree, Lunken watched through binoculars as the men milled about drinking and joking on the grounds of a private estate in Cincinnati, Ohio. At around 6:30 p.m., a bus pulled up close to the pool, but since no one disembarked, she concluded that it was "probably a band" for entertainment later that evening. The QB's leader eventually called the meeting to order, and at his command the men turned to the west in unison and toasted their departed brethren. After that, they sat down to dinner. As they "carved into big steaks and slathered ears of corn with butter," Lunken despaired of anything truly interesting happening. But then, just as she was about to give up and slip away into the darkness, " 'things' finally got under way."[9]

Lunken admitted that she had imagined that "spying on a storied Wing Ding" would be "like watching forbidden native rites on [the island of] Bali Hai in [the movie] *South Pacific* . . . ," and now at last she caught a glimpse of what she'd come to see, despite "dire warnings" from fellow pilots that she would "end up bare-ass in the pool" if she were caught in the act. "If memory serves," she recalled with obvious humor some 35 years after the event, "people emerged from the bus, but they didn't have musical instruments; they were like mermaids who'd forgotten their bathing suits. They all jumped into the pool and started performing a kind of (but not exactly) Esther Williams swimming routine. I couldn't really see the choreography but it must have been great, because it sure had everybody's attention. Then when they got cold or tired of swimming and climbed out of the pool, they headed back into the bus followed closely by some of the Birds [Quiet Birdmen], which was odd because I'm pretty sure none of them had been in the pool." From this and her other efforts over the years to learn more about the group (including sneaking into a local hangar meeting disguised as an elderly man who belonged to a distant QB hangar), Lunken concluded that these get-togethers were mainly "about drinking, topless bartenders, big steaks, male bonding, raunchy jokes and hugely exaggerated tales of flying and sexual prowess."[10]

While the organization's views on female pilots—aside from its refusal to admit them to its ranks—are cloaked in secrecy, other shreds of evidence suggest the Quiet Birdmen agreed with Milton Horowitz's 1965 assertion that flying was a masculine endeavor. As recently as 2004, the publicly accessible homepage of the Somerville [New Jersey] Hangar of Quiet Birdmen featured an emblem drawn in the tradition of a squadron patch like those worn on military flight suits. The top half of the emblem includes the organization's name, "Somerville Birdmen," immediately above the image of a fully armed World War II–era fighter plane in flight, linking the QBs (many of whom had likely once flown in the service) to the kind of physical and technological power traditionally associated with military masculinity. The bottom right corner displays two brown jugs labeled "XXX." In keeping with a long-standing tradition of men demonstrating their masculinity by maintaining a certain degree of self-control even when inebriated, virtually all accounts of QB gatherings, no matter how vague, suggest that alcohol played a key role in the organization's rituals. The bottom left quadrant of the emblem features a drawing of a shapely woman sunning herself topless on a beach, reflecting a hypermasculine culture that viewed women primarily as sex objects. Combined, these three images convey the idea that the QBs of the Somerville Hangar were men who enjoyed fast planes, hard drinking, and hot women. In their get-togethers, these same men replicated the storied traditions of the interwar barnstormers and airmail pilots who boozed and womanized their way across the country, the wartime military fliers who partied to relieve stress between missions in their makeshift overseas officers' clubs, and the larger-than-life postwar test pilots whom Tom Wolfe immortalized in his 1979 book *The Right Stuff.* In this culture that demonstrated power and control through the mastery of technology, drinking, and sexual conquest, there was definitely a place for women in and around aviation, but it wasn't in the pilot's seat.[11]

Then again, membership in the QBs did not necessarily mean that a man fit the classic profile of a hardcore male chauvinist who refused to accept women in roles other than as sexual objects. Fred Weick, the aeronautical engineer whose many accomplishments include inventing the supposedly foolproof Ercoupe described in chapter 3, not only mentioned in his autobiography that he was a member of the Quiet Birdmen but also freely admitted that he derived "a good deal of pleasure 'hangar-flying' with pilots from all around" at the group's meetings. Very much a product of his times, he accepted common stereotypes of the era that women were generally less able to master complex technology than men, as evidenced by the way he frequently used stories of women who learned to fly in an Ercoupe as proof of his design's inherent safety and simplicity. Yet Weick also seemed to have no problem with the idea of more women taking to the sky, and indeed he was proud that the Ercoupe had to some extent made this possible.[12]

To some fliers, regardless of whether they actually belonged to the QBs, the organization represented the "good old days" when flying was still an all-male club and men could talk and act as they pleased without worrying about offending women. In March 1982, pilot Gordon Baxter titled his monthly column for *Flying* magazine, "The Airperson's World," a gender neutral takeoff on "The Airman's World" (a popular monthly feature written by *Flying*'s editor Gill Robb Wilson during the 1950s and early 1960s).[13] Although Baxter's column was laced with humor, he noted with genuine ambivalence, "In truth, I welcome the woman pilot and cry out for the passage of the ERA [equal rights amendment], but will regret the passing from our pilot's language of such colorful terms as the one used by old-time pilots to describe an application of maximum thrust when both the throttle and mixture controls are advanced to their full forward limit of travel" (by which he clearly meant "balls to the wall," though he refrained from actually using this common aviation term in his piece).[14] He also gleefully admitted that despite his supposedly enlightened views, he was in many ways still a throwback to an earlier era. For instance, Baxter described how one night when he told a gathering of female fliers that "some of my best friends are women pilots," in response "they arose as one man, er, body, laid hands upon me, carried me outside the Holiday Inn banquet room, and flung me into the pool." And after describing the famous aviatrixes of aviation's early years—"Without exception, all the pioneer women [fliers] were beautiful"—he playfully flirted with what some might have labeled sexual harassment, even in 1982, by concluding, "Had they flown with me, the only hazard might have been an improper scan" (a takeoff on the term for a pilot's highly focused and repetitive "scan" of the instrument panel while flying without any outside references during inclement weather). Although Baxter did not reveal whether or not he was a QB himself, he slyly hinted that he knew (and perhaps approved) of the nature of their gatherings when he observed that no women pilot he knew had ever made it to, or through, a meeting of this men-only club.[15]

Individual women pilots reacted to aviation's pervasive culture of masculinity in different ways. A few embraced the cult of manly behavior that had long surrounded aviation. Florence Lowe "Pancho" Barnes represents the best-known example of this type. A former socialite who made a name for herself during the interwar years as a stunt pilot and air racer, Barnes ran a ramshackle bar in the postwar era called the Happy Bottom Riding Club in the desert near Muroc, California. Her favorite patrons were the wild young military test pilots stationed at nearby Edwards Air Force Base, and by all accounts she could drink and swear with the best of them. According to Tom Wolfe's description of Barnes in his best-selling nonfiction book *The Right Stuff*, she routinely "shocked the pants off" even the brazen young test pilots "with her vulcanized tongue." For instance, "Everybody she didn't like was an old bastard or a sonofabitch. People she liked were old bastards and sonsabitches, too." And it was not unusual to overhear her say something like, "I tol' 'at ol' bastard to get 'is ass on over here and I'd g'im a drink."[16] Although Barnes by no means represented the typical postwar female flier, other women pilots acted like "one of the boys" to varying degrees. Some, like Pancho Barnes, obviously enjoyed doing so, but others adopted this approach as a method to negotiate the mostly male world of flying. By acting like one of the boys, these women set themselves apart—at least at the individual level—from the negative stereotypes that associated female fliers with incompetence and lack of aggressiveness.[17]

At the other extreme, some women consciously played up their femininity in order to disarm resentful men and gain acceptance within the community of pilots. Such was the case with Nancy Batson Crews, who claimed that she had "majored in Southern belle" at the University of Alabama in the late 1930s. As a young college student, she successfully sweet-talked her adoring father into letting her participate in the CPTP, then into buying her an airplane so that she could continue to build flight time. This set the stage for her to fly military aircraft with the WASP program during World War II. According to her biographer, after a stint as "a typical 1950s stay-at-home mom," she opened her own business instructing students and towing gliders. On more than one occasion throughout her long flying career, Crews—who adamantly refused to identify herself as a feminist—deliberately deployed her feminine charms to win over male pilots who were generally opposed to sharing the sky with a woman.[18]

The challenges of finding a place within the highly masculine culture of prewar aviation inspired some early female fliers to form their own organization in 1929. Named the Ninety-Nines at the suggestion of Amelia Earhart, the organization's first president, the group took its name from the total number of women who signed on as charter members. Following World War II, this all-women pilots' group continued to flourish. Local chapters of the Ninety-Nines sponsored events that simultaneously brought members together socially and served the greater good, be it on behalf of the community of pilots or the local communities in which they lived. Projects included aerial marking (making "signs" visible from the air, often painted on the rooftops of large buildings, to guide pilots along their route), sponsoring air shows, and providing educational programs to boost community support for airports and to encourage everyone—girls and boys, women and men—to consider learning to fly for pleasure or to pursue a career in aviation. Members of the Ninety-Nines also volunteered their services and their planes to transport blood for the Red Cross and to airlift patients and medical supplies long distances.[19] Like many all-male organizations, the Ninety-Nines created their own auxiliary to help out with events, only in this case it was made up of men, mostly the spouses of the group's members. As Amelia Earhart's husband George Putnam humorously explained in a 1932 article, these auxiliaries to Ninety-Nines called themselves the "Forty-Nine Point Five Club" (which later evolved into the term "Forty-Nine and a Half" for individual men) because they "reckoned arithmetically as fifty per cent of our better halves."[20]

One of the greatest benefits derived from belonging to the Ninety-Nines was mutual support in the predominantly male world of aviation. The group's Amelia Earhart Memorial Scholarship Fund helped women to pursue expensive advanced aeronautical ratings that they might not otherwise be able to afford. But most of this support took place at the local, informal level. Barbara Goodwin, a Michigan schoolteacher whose husband was already a private pilot when she decided to learn to fly in the 1980s, later recalled seeking solace on many occasions from members of the local Ninety-Nines chapter as she struggled under a particularly demanding male instructor. In the end, she not only earned her license but went on to become a flight instructor herself and mentor other women who were learning to fly. Fran Bera reflected that participation in events created by and for women pilots held many attractions, "such as making friendships with other women pilots that have lasted a lifetime. These were women from all parts of our country and the world. They understood the pure joy of flight and the beauty of seeing our country unfold from coast to coast under our wings, while we tested our skills to the utmost. . . . It was wonderful to communicate all of this to other women and know they understood."[21] Bera's thoughts illustrate the support and camaraderie that many members of the Ninety-Nines valued about the all-female organization of pilots and suggest that they had difficulty finding similar support within the broader and mostly male community of pilots.

Even semantics posed an obstacle for women fliers intent on establishing their identity within a community—as well as the larger world outside the airport fence—that expected pilots to be men. As one sympathetic article noted in 1958, "Troublesome too, is this problem: whether to call women who fly airplanes 'aviatrixes,' 'aviatrices' or 'aviators.' The answer is so very simple: they are *women*. Specifically referred to, when flying airplanes, they are *pilots*. And for more than 30 years they have been trying their best to convince the world at large of this simple answer to that nagging problem."[22] Perhaps it should come as no surprise that it was a group of Ninety-Nines who threw aviation writer Gordon Baxter into the Holiday Inn swimming pool sometime in the late 1970s or early 1980s after he tastelessly joked, "Some of my best friends are women pilots."[23] By insisting that there is no contradiction between being a woman and being a pilot, the mere existence of the Ninety-Nines challenged the gendering of aviation as masculine.

Yet in their activities and public statements, the Ninety-Nines were careful to temper this implicit challenge to the status quo. The All-Women Transcontinental Air Race, better known as the Powder Puff Derby (a whimsical label originally coined by humorist Will Rogers to describe the 1929 National Women's Air Derby), was unquestionably the most publicized event sponsored by the Ninety-Nines.[24] From its inception in 1947 until the final commemorative race in 1977, the AWTAR served both as a serious competitive event for participants and as a venue to showcase women's participation in aviation.[25] Recognizing that an all-women air race might come across as unfeminine and thus create negative publicity for women pilots, organizers established an official dress code in the early 1950s that banned slacks and shorts and required all contestants to wear dresses or suits. One observer wrote in 1952, "This regulation stems from the same desire for serious recognition that has led . . . participants to shun the title of 'Powder Puff Derby' which was originally applied to their big race."[26] That statement suggests that the Ninety-Nines believed that regardless of their accomplishments in the sky, women could achieve "serious recognition" as pilots only if they conformed to gender norms of the era—including feminine dress and appearance—when they were on the ground. From this viewpoint, if members of the Ninety-Nines looked and acted too much like men, they would suffer the same fate as aviatrixes of the first half of the twentieth century, who were widely viewed as exceptional anomalies within the all-male world of flying rather than respectable role models for other women to emulate.

Likewise, the Ninety-Nines' nonflying activities mirrored those of many other respectable middle-class women's groups: fundraising, community service, and boosting their particular cause through education and outreach without creating too much of a stir. A leading historian of women in postwar aviation notes, "In spite of the seemingly unfeminine activity of its membership, the Ninety-Nines chose to follow the social dictates of American society in determining its organizational activities and priorities." Instead of a radical feminist organization, "It was a women's club that politely broke with convention."[27] Contemporary accounts by outside observers and comments by members of the Ninety-Nines support this conclusion. For instance, in the same 1965 issue of *Flying* that carried "For Men Only?" by Dr. Horowitz, another piece—this one by aviation writer and former fighter pilot Richard Bach—addressed the conflicting desire by some women pilots to advance their cause without overstepping mainstream gender roles of the era. Descriptively titled "The Invisible 99s," Bach's article opened with the line, "Struggling 'manfully' under a strange cloak of anonymity, an aeronautical coffeeklatch of women aviators strives to be recognized—or taken for granted." Bach clarified that many members of the organization hoped to make flying acceptable for women without directly challenging men's place in the air and quoted one who explained almost apologetically: "I don't feel that we are competing with the men in aviation. I think when they invented the airplane they had this love of flight, and I don't think they should be selfish with the thrill they get out of it."[28]

In fact, at least a handful of Ninety-Nines clearly did long for the day when women could move into military and airline cockpits alongside men. Jacqueline Cochran, wartime president of the Ninety-Nines and one of the organization's most famous members after Amelia Earhart, had used her position as head of the WASP program to lobby for the full, formal acceptance of women pilots into the U.S. Army Air Forces during World War II.[29] Ultimately she was unsuccessful; the program was disbanded in 1944, and 20 years later when Bach wrote his article about the Ninety-Nines, women were still barred from flying for major airlines or the military. Geraldyn "Jerrie" M. Cobb, also a member of the Ninety-Nines and a full generation younger than Cochran, was a highly experienced civilian test pilot with thousands of flight hours and several aviation world records to her name by the time she volunteered for astronaut selection in 1960. Although Cobb and 12 other women passed the same rigorous physical and psychological tests as the famed all-male "Mercury Seven" astronauts, NASA ultimately declined to send women into space for another two decades.[30] But if any of the Ninety-Nines whom Bach personally interviewed for his article harbored similar desires as Cochran or Cobb, they kept this to themselves as part of a collective strategy to carve out a nonthreatening place for women pilots within the masculine status quo.

As subcultures that existed within the broader community of pilots, the Quiet Birdmen and the Ninety-Nines illustrate the influence of masculinity on private flying throughout the postwar era. However, both of these organizations were relatively small. The Ninety-Nines reported a membership of 2,500 in 1965, 14 percent of all active women pilots but only about 0.5 percent of all civilian fliers at the time.[31] Membership numbers for the QBs, like all other details about this group, remain a closely guarded secret, but anecdotal evidence suggests that it too was never particularly large. For this reason, examining what went on within larger aviation organizations that admitted both men and women, and that recognized the role of the nonpilot wives of private fliers as well, helps to complete the picture of how gender roles within the aviation community changed or stayed the same in the face of broader social upheaval within postwar American society.

From Round-Ups for AOPA's Little Guy to Plantation Parties for Mr. General Aviation

The oldest and largest national organization open from the start to all pilots, regardless of gender or race, is the AOPA, created in the spring of 1939 by five flying businessman who sought to protect aviation's "little guy," the private pilot, from government attempts to restrict access to America's skies. As the organization grew throughout the postwar era—roughly 81,000 members in 1960; 141,000 members in 1970; and nearly a quarter million members, or one in three active civilian fliers, by 1980—it remained true to the broad goals expressed by one of its founding fathers in a July 1939 letter: "To make flying more useful, less expensive, safer, and more fun." It did so by promoting general aviation to the public and by vigorously lobbying in Washington, D.C., against new government regulations, user fees for pilots, and other issues that the AOPA leadership believed posed a threat to private and business aviation. However, as private flying evolved after World War II, the organization increasingly viewed its primary constituents as middle- to upper-middle-class private pilots who were also affluent, educated consumers interested in making informed decisions about the equipment they purchased and used. This in turn shaped the types of membership events that it held. From their beginnings as good-times-on-a-budget affairs for the "little guy," the AOPA's annual gatherings increasingly came to resemble the upscale professional conventions that its members likely attended as part of whatever white-collar career paid for their private flying.[32]

Early AOPA gatherings were local fly-ins in the truest sense; they typically took place at or near an airport during the weekend, and many participants arrived by plane. For example, in what was billed as "one of the first post-war breakfast flights," in late 1945 or early 1946 the Colorado chapter of the AOPA held a fly-in at the Sterling Municipal Airport, roughly 100 miles northeast of Denver. Despite less than ideal weather conditions, 78 people arrived in 46 airplanes. Pilots tied the tails of their planes to the bumpers of cars parked on the airfield as a precaution against high winds. The Sterling Chamber of Commerce sponsored the breakfast, which was served on makeshift tables inside the city's municipal hangar, and 10 young women of Beta Sigma Phi (a nonacademic service and social sorority) served as waitresses for the visiting pilots and passengers.[33]

Within a few years, the national headquarters for the AOPA began holding an annual "Round-up" at Rehoboth Beach, Delaware. Organizers of the inaugural 1947 Round-up promised: "Contests, get-togethers and surprises will crowd the two-day stay, and some of aviation's outstanding personalities will be present. Highlight of the flight is a beach party, clam bake and shore dinner Saturday night. . . . The flight's date promises perfect bathing and uncrowded beaches. Surf fishing and golf will be featured on the recreation program." The AOPA deemed its 1947 Round-up a success, and an article describing the event enthused, "Perfect weather brought nearly 300 planes and 550 pilots" and noted that "pilots from 12 states flew as many as 400 and 500 miles to attend the weekend festivities." The Rehoboth Chamber of Commerce sponsored the event and booked overnight lodging for attendees (who received special rates of $1.50 to $3.00 per night at area hotels), and Delaware State Police troopers and National Guard detachments guarded the planes while the pilots and their companions ate dinner, attended a beach party, and then danced late into the night at the Rehoboth Country Club.[34]

The 1948 AOPA Round-up at Rehoboth was an even bigger success, with 412 airplanes and some 1,000 AOPA members and guests flying in. A photograph shows scores of men and women sitting in the sand around bonfires, and one keen observer noted that the "giant shore dinner on the beach" consumed "250 chickens, 1,000 pounds of potatoes, 10 bushels of oysters, 1,440 eggs, 150 pounds of ham, 55 pounds of cheese, 1,500 clams, 35 cases of beer, 25 cases of Coke, and 10 cases of milk."[35] A party, for sure, but not exactly posh. An editorial about the third annual Round-up, held in the next year in 1949, declared: "If you had any doubts about the type of person who's flying these days, you soon got over them around the campfires at the beach party that night. . . . The vast majority were housewives and their husbands—men who owned small-town businesses, men who worked for large corporations, men who sold shoes, men who farmed."[36] This description of who the organization thought its members were helps explain the relatively down-to-earth, beach party setting for these early gatherings.

The AOPA continued to hold annual fly-ins at the same seaside airport for several more years, but bad weather constantly threatened the event. For instance, in 1952 a decision of whether to cancel the get-together reportedly hung in the balance to the very last moment in "a nerve-wracking game of nip and tuck with what was officially known as Hurricane Baker, the season's second big blow, which was headed directly for the flight's general area as late as Friday [the day before the fly-in]." As it turned out, the organization deemed that year's Round-up "a success," and stated that "weather on the flight's big day—Saturday, September 6—was perfect."[37] In 1953, "the general area west and northwest of Delaware" experienced "quite a bit of thunderstorm activity," grounding most pilots except those arriving from the south. As a result, only 525 planes made it to Rehoboth for the weekend, roughly half what organizers had hoped for.[38] Then in 1954, after organizers decided to move the event north up the coast to Nantucket Island, Massachusetts, the turnout was even worse. "Weather did it again," complained an *AOPA Pilot* article afterward, noting that the original gathering was postponed owing to Hurricane Edna, and marginal conditions plagued the alternate "rain date" as well. As a result, "three-quarters of the expected 1,500 people had to cancel their reservations at the last minute when they found they couldn't make it." Those hardy souls who made their way to Nantucket anyway found that the weather affected more than just flying conditions: "Because of the cold wind, and wet beaches, the planned beach party became an indoor affair at the Nantucket Yacht Club." As a consolation for the low turnout,

"AOPA'ers were served 'seconds' and 'thirds' of fried chicken dinners, depending upon their desires and capacities," and there were far fewer attendees to compete for the door prizes.[39]

In the aftermath of the disastrous turnout on Nantucket in 1954, the AOPA announced that its next Round-up would not be a fly-in at all, but rather a six-day cruise to Bermuda aboard a "600-feet-long, seven-deck-high luxury liner" that could accommodate "an expected 731 AOPAers." When the AOPA proposed this new format, members reportedly "reacted heavily in favor of the planned jaunt that could not be postponed, cancelled, nor spoiled by bad weather." No doubt the AOPA's leadership was also ready to take a break from its yearly guessing game with meteorological conditions. A May 1955 announcement advised: "Voyagers may arrive in New York City by private plane or other transportation anytime before the ship sails."[40] However, a later announcement omitted any mention of traveling by private plane, stating simply: "Remember that you get to and from New York by your own transportation."[41] In just a few short months, the 1955 Round-up had quietly evolved from a genuine "flight-cruise" into a floating vacation that was a "fly-in" only in spirit, with no personal plane required (or even expected) in order to attend.

Although a preview of the cruise itinerary mentioned meetings in passing, the focus of this seagoing Round-up was almost entirely social. "The voyage will feature parties, a president's reception, dancing, AOPA meetings, deck games, swimming, and normal shipboard activities." Following a brief stay in Bermuda, "the return trip will highlight a Wednesday midnight deck buffet supper, and the annual AOPA banquet on Thursday night." The organization counted on its members being enthusiastic enough about hangar flying that they would be willing to give up the actual flying aspect of their annual fly-in. But the AOPA was somewhat optimistic when it chartered an entire cruise ship for this event. In May 1955 it announced: "Reservations are going fast—100 in the first 10 days."[42] A June update noted: "Over 25% of the *Queen of Bermuda's* more than 600 available spaces have already been reserved at this writing." By August, just a month before the first-ever shipboard Round-up was to set sail, *AOPA Pilot* warned readers to "Get Reservations Now." Reading further, however, reveals that organizers were in fact having trouble filling the ship; the same notice quietly disclosed that the AOPA had opened this previously members-only cruise to "all interested flying enthusiasts and related aviation groups." And the September issue of *AOPA Pilot*, which must have arrived in mailboxes mere weeks, if not days, prior to the September 3 sailing date, included a frantic "Last Call for Bermuda Flight-Cruise."[43]

According to an AOPA account published afterward, the cruise was a success, "with not one small gripe or misadventure to mar the week of fun and fellowship." Coming from the organization that sponsored the event, one can assume a bias toward favorable reporting. However, given the long-standing and widespread tradition of hangar flying, it is easy to believe the following description of how most members on the cruise spent their time: "Things got off to a bang with the President's pre-dinner reception the first night out. Introductions were hardly made before members launched into comparisons of flying conditions, of weather, of airport facilities in all parts of the country, exchanging data and experience. With such an unusual opportunity for hangar flying it's not surprising that this became the undertone for the whole six days on sea and shore." But these rosy descriptions of shipboard camaraderie and nonstop hangar flying masked one fact that AOPA Round-up organizers must have found disappointing: in the end, only 310 people (and not necessarily all of them AOPA members) signed up for the cruise, a far cry from the original estimate of more than 700.[44] This low turnout may reflect a miscalculation by the organization's leadership. A full decade after the end of World War II, AOPA members in the mid-1950s were definitely older and almost certainly more affluent than they had been just a few years earlier. But when it came time to actually make reservations, few, it seems, were ready, willing, or able to commit both the money and time needed to take a weeklong vacation at sea.

Over the next several years, the AOPA continued to search for the right formula for its annual get-togethers. In 1956 it abandoned the cruise format and hosted three regional fly-ins around the country. Of these, only one—held at a beachside hotel in Biloxi, Mississippi, and nicknamed the "Plantation Party" after the Saturday evening banquet and musical review that was the highpoint of the weekend—warranted a review in the pages of *AOPA Pilot*. This is especially telling since only 36 people, including two AOPA staff members, attended.[45] In 1957, the organization's leadership tried a combination of previous events, hosting both a Caribbean cruise and a landlocked national convention based on the precedent set by the 1956 Plantation Party. Not including transportation to and from home, the seven-day cruise cost $175 to $335 per person, whereas the three-day Plantation Party—including accommodations plus the banquet and ball—ran $41 to $48 per person (with even cheaper rates for late arrivals). Members voted with their pocketbooks. With only 107 AOPA members and their guests occupying a ship with 400 berths, "the number of participants was disappointing." Nearly three times that number showed up in 92 airplanes for the hotel-based Plantation Party that year.[46] Despite the poor showings in previous years, the

AOPA tried a cruise once more in 1960. Building upon daytime forums for pilots held during the 1957 shipboard Round-up, this time the itinerary noted: "Of special interest to pilots will be the series of aviation clinics held on board which will highlight what pilots want to know concerning safety, regulations and new developments in aviation."[47]

Meanwhile, the fly-in Plantation Parties of the late 1950s simultaneously set the pattern for future conferences and picked up where these cruises left off. Starting in 1961, what had previously been a strictly recreational affair combined serious educational seminars for pilots with social events for everyone, and the annual Plantation Party soon came to resemble the professional conventions that many doctors, lawyers, and businessmen regularly attended to sharpen their professional skills and network with others in their field. The event, though hardly secretive like a QB meeting (and far less raucous), was open only to AOPA members, their families, and invited guests. Seminars ranged from classes on instrument flight procedures and how to get the most from a Flight Service Station weather briefing, to town hall style meetings with senior government officials (which some years included the chief of the FAA).[48] Pilots could also sign up for flight training with instructors brought in especially for the event in order to sharpen existing skills or start working toward a higher rating.

The 1970 Plantation Party in Hollywood, Florida, was typical of AOPA conferences held from the mid-1960s through the mid-1980s (the meeting was renamed "AOPA Convention and Industry Exhibit" to sound more businesslike in 1979, but the format remained largely unchanged). According to an account published in *AOPA Pilot*, for attendees "the serious business of 'partying'" took many shapes. "It could take the form of lolling on the beach or around the swimming pool and absorbing sunshine, or it could be the starting of a heavy flight-training schedule, or examining the vast array of general aviation products in the Exhibit Hall, or attending the afternoon AOPA Clinic sessions or the AOPA Aviation Law and AOPA Aviation Medicine Seminars." In addition, "for the outdoor types, there were fishing and golf tournaments. There were special events for the ladies, too: a fashion show . . . [a] wig show and a bingo game. Then there always were the floating hangar-flying sessions, which started spontaneously when two or more pilots got together." A travelogue film titled *Wings to the Virgin Islands*, narrated by its producer, drew an audience of 1,500. Other attractions included a lunchtime talk by noted trial lawyer and social commentator F. Lee Bailey. In addition to the beach, pool, and the "special events for the ladies," those not interested in aviation could also enjoy golf, tennis, water-skiing, or sailing. In the evenings, there were poolside cocktail parties followed by lavish dinners. Entertainment included performances by Metropolitan Opera soprano Jean Fenn and by pop singer and former Miss America contestant Anita Bryant. The top door prize at the Plantation Party banquet was a brand-new four-seat Piper Cherokee airplane. Other prizes included a two-week trip for two to Africa, a mink coat, and an RCA color television.[49]

Like other organizations for white-collar professionals, the AOPA moved its meeting over the years from city to city. Locations were chosen in equal parts for both tourist attractions and climate and included Palm Springs, California; Las Vegas, Nevada; and Hollywood, Florida. Good flying weather was a must, not only because many participants arrived by private plane but also because of the flight training programs associated with the event. Instead of airports, AOPA members met at fancy hotels complete with fully staffed banquet halls and conference facilities. However, even though the AOPA chose not to use airports as the focal point for its annual convention, the organization was still unquestionably run by—and for—fliers, and no conference was considered complete without airplanes. This meant either shuttling participants out to the local airfield to look at airplanes or bringing the airplanes directly to the convention site.[50]

Starting in 1960, organizers routinely chose the latter.[51] Since the conference hotel was often located some distance from the nearest airport, AOPA officials arranged to have city streets closed to traffic so that the planes could taxi en masse from the airfield to the hotel grounds. A description of the 1964 convention noted: "The static display of new aircraft on the hotel's parking lot drew its share of attention. The planes were flown from Fort Lauderdale–Hollywood International Airport to the [Diplomat Hotel's] golf course and landed on the 1,506-foot No. 2 fairway, then taxied about a mile to the hotel's parking lot. The taxiing of the planes to the exhibit area was one of the spectacular events of the Hollywood meeting."[52] Things did not always go as planned. The next year's convention in Las Vegas involved "a lot of milling around . . . early in the morning on the opening day" with "a two-hour delay in getting 24 airplanes onto the boulevard for a trip of two miles from McCarran Field to the parking lot of the Dunes [Hotel] for a static display." After the Nevada Highway Patrol blocked their way, organizers belatedly discovered that "no proper permit had been obtained" in advance. However, speedy "teletype action from Carson City [the state capital] got the taxiing planes underway," hinting at the level of political clout that AOPA members enjoyed.[53] Five years later in 1970, this practice of taxiing from the airport to the hotel parking lot resulted in the unlikely spectacle of nearly 20 airplanes lined up on a road in Hollywood, Florida, as they waited for a sailboat to pass through a raised drawbridge.[54]

This annual "parade of planes," as the author of one article described it, garnered significant media attention as the AOPA took its pro-aviation message to—as well as literally through—the city streets.[55] In aviation's rough-and-tumble golden age between the world wars, itinerant barnstormers had exhibited a type of masculinity based on daring and risk-taking when they buzzed Main Street and put on an impromptu air show overhead to drum up local interest before setting down in a farmer's field on the outskirts of town to sell airplane rides. Through the public spectacle of this annual parade of planes, AOPA members demonstrated a different, arguably more mature, form of masculinity based on flexing their economic and political power as they temporarily took over the city and unleashed private flying from its usual confines behind the airport fence.

Like the parade of planes, at least a few pilots became regular fixtures at these conventions. For instance, *AOPA Pilot* considered it newsworthy at its 1980 meeting in San Diego, California, when Jack Nunemaker of Cherry Hill, New Jersey, and Jim Tilford of Palm Beach, Florida, both "arrived for their twentieth consecutive convention, adding the twentieth medallions to their bolo ties." The accompanying photograph shows two grinning men, their ties conspicuously weighted down with the inch-wide metal and enamel pendants. The photo includes a third person as well, but it was not the runner-up who had attended only 19 conventions and was reportedly "waiting for a chance to catch up." Instead, the caption acknowledged the smiling woman standing between the two 20-year Plantation Party veterans with five simple words: "That's Mrs. Nunemaker in center."[56] The article says nothing else about this pilot's wife, yet her smiling but otherwise silent presence at his side in many ways symbolizes the expected place of women in postwar private aviation: the enthusiastic and willing partner in what sociologists describe as *companionate marriage*.

One scholar sums up this concept—first popularized in the 1945 book, *The Family: From Institution to Companionship*—as the ideal of "single-earner, breadwinner-homemaker marriage that flourished in the 1950s. Although husbands and wives in the companionate marriage usually adhered to a sharp division of labor, they were supposed to be each other's companions—friends, lovers—to an extent not imagined by the spouses in the institutional marriages of the previous era." As a result, "much more so than in the 19th century, the emotional satisfaction of the spouses became an important criterion for marital success."[57] Many, or indeed most, American families did not actually achieve all aspects of the companionate marriage ideal, yet it pervaded postwar American society and culture. Ward and June Cleaver, the fictional parents in the television series *Leave It to Beaver* (1957–63), are probably the best-known personification of these ideals in popular culture.[58]

Had Ward Cleaver been a private pilot, then June would have no doubt been portrayed as his supportive, willing, and able helpmate on flights that involved the entire family. This image of pilots' wives as the homemakers of the air appeared throughout the pages of aviation magazines of the era. For instance, in 1954 Holly K. Robinson wrote a letter to the editor extolling the virtues of a new postwar technology—the Tupperware container—as spill-proof drinking cups for in-flight use. Nearly 15 years later, Mrs. Clint Allen provided a detailed description of the duties and responsibilities of the pilot's wife. "You can learn to serve coffee, tea or milk at the appropriate times, or reach to the back seat for necessary items for the pilot during flight. . . . You can keep an alert eye open for other planes in the vicinity, and also help by spotting small airports and airstrips when necessary." To Allen, however, helping out did not actually extend to assisting her husband with flying the airplane: "If you can smile at a rough landing, keep your hands and feet off the controls while in flight, keep chewing gum and lifesavers in your purse, refrain from any back seat driving or questioning flight procedures, you too can join the . . . [other wives of pilots] that fly the skies these days."[59]

Similar articles and letters submitted by both women and men throughout the postwar era reinforced the message that a woman's job in the cockpit was an extension of her duties as housewife and mother at home. Key responsibilities included keeping children fed, comfortable, and occupied. She also had to ensure that her solutions to these problems in no way interfered with her husband's deadly serious business of flying the family plane. After advising airborne moms to "Keep them full," Anne S. Moore warned in a 1978 article, "Thin-spread peanut butter and jelly sandwiches and fruit are better snacks than crackers and chips. Dad doesn't appreciate crumbs in the cabin, and a few Baggies and napkins on hand help control debris." Moore also observed that proper planning by mom could have a positive influence on the family's ability to complete the flight as scheduled, noting that "a midmorning snack can extend your [flying] range another hour or two" between rest stops.[60]

Even before the AOPA's 1969 Pilot Poll suggested that nearly 85 percent of civilian fliers were married, its leadership had long assumed that most of its members were responsible, middle-class family men.[61] As a result, from the very start, the organization's annual meetings—in all of their incarnations—reflected an underlying commitment to reconcile private flying with companionate marriage. The fly-in Round-ups of the late 1940s and early 1950s featured dinners on the beach and dancing late into the night. The shipboard excursions of the late 1950s were also designed with couples in mind, although daytime meetings for pilots and hangar flying sessions that "became the undertone for the whole six days on sea and shore" may have intruded upon marital togetherness.[62] Like conferences for white-collar professional groups, the resort-based Plantation Parties of the 1960s and 1970s provided continuing education and networking opportunities for male pilots while their wives went sightseeing, sunbathed by the pool, attended a fashion show, or enjoyed other "special events for the ladies." After presumably spending much of their day apart, the couple was expected to come together each evening to enjoy a varied schedule of receptions, cocktail parties, dinners, and first-class entertainment arranged specifically with middle-aged, upscale couples in mind.[63]

Yet all was not well within some postwar flying families. According to accounts published in aviation magazines and journals throughout the 1960s, the "antiflying wife" was the bane of many male pilots and a significant drag on the private aviation industry as a whole.[64] For instance, in a 1960 article titled "Beware the Skirt Barrier," flight school owner Paul Anderson bluntly summed up his thoughts on the matter: "Women are a minus factor. We can trace about a third of the drop-outs in our flying program to home pressure." Anderson's wife, who described herself as an enthusiastic passenger despite that fact that she wasn't a pilot herself, explained her husband's sentiments more fully: "A lot of women are uneasy about flying. Not only the danger, but they look on it as anti-togetherness, rather like golf, only more expensive and risky." She pointed out that these women would rather spend the family's disposable income on wall-to-wall carpet, new appliances, "or, better still . . . something the whole family can use together like a boat." The same piece quoted the wife of another flight school owner who described how "many a man who tries to get his wife to agree to his learning to fly faces an uncomfortable night on the davenport."[65] Three years later, the author of a 1963 article mused, "It's a standard axiom among old-time aircraft salesmen: 'If the owner's wife isn't interested in flying, the pilot is going to keep his airplane or his wife. One or the other will go.'"[66] Complaints of "anti-togetherness," nighttime exile to the living room sofa, and even divorce were hardly the makings of a companionate marriage.

Those whose livelihood depended on selling private aviation were not alone in bemoaning the lack of support for private flying on the home front. For instance, James R. Old, M.D., of Beaumont, Texas, reported that although his wife agreed to join him at the 1964 meeting of the Flying Physicians Association in Palm Springs, California, she flatly refused to fly. Instead, she took the train while he flew himself and another couple there and home again by private plane. In 1965 Hank Strickland wrote the editor of *AOPA Pilot* to complain, "Most, or should I say many, [private pilots] who fly can barely afford to do so. I know many like myself whom I meet around the airfields speak of the same troubles and have wives who want to talk us out of flying."[67] Whether a woman feared going aloft, objected to the cost, or felt some combination of the two, ultimately the outcome was the same: an antiflying wife who threatened her husband's continued participation in private aviation.

Firsthand testimonials from women reinforced widespread assumptions that fear of the unknown lay behind many wives' objections to their husbands' involvement in private aviation. When Norval Moore decided to take flying lessons in the late 1950s, his wife Emma declared, "If you want to learn to fly at *your* age, go ahead. But don't think you will ever get *me* up in a plane with you." She also admitted (afterward) that as the day of his written exam drew closer, "Scaredy me, I actually prayed that he would not pass, for I feared all the implications of a pilot's license."[68] In 1963, Doreen Sherwood, a housewife and mother of three who had been married to a pilot for 12 years, admitted that she had avoided flying with her husband whenever possible. "I was terribly, terribly frightened. I was afraid that if something happened to Bill with us up there, what would I do? You see, I was firmly convinced that if I put just a finger on the wheel I would cause a fatal accident."[69] In 1965 a pilot's wife identified only as Hilda told an interviewer, "For 17 years I've been flying with him, and every flight has been one of nerves, nausea, nuisance and sheer terror!"[70] And in 1969, in response to a "how-to" article that described ways men could win over the antiflying wife, Katy Dance of Cleveland, Ohio, described how her own deep-rooted fears of flying not only turned "a lovely day . . . into a nightmare" from her own perspective but also "took the joy out of my husband's flying."[71] Dance's confession reveals the crux of the problem: in the ideal companionate marriage, both partners were supposed to spend their leisure time together, not go their separate ways. Thus, if a wife did not like flying, there was considerable pressure within the relationship for her to ride along anyway in misery, or for him to give up flying for an activity that both could enjoy, a classic no-win situation since neither solution brought happiness to both parties.

As fliers and general aviation industry boosters alike became increasingly concerned about unhappy women restricting men's access to the sky, the AOPA took action by introducing the Pinch-Hitter Course. Although chapter 2 describes the "nuts and bolts" of how this program sought to re-gender basic flight instruction to make it more appealing to women, revisiting the origins and evolution of the course sheds light on pilots' views of a woman's proper place in their traditionally masculine community.

When the AOPA hosted the first Pinch-Hitter Course in the fall of 1963, the program was an instant success. *Time* magazine, which provided national coverage of the inaugural class, reported that "many a woman became almost enthusiastic" about aviation as training progressed, a considerable achievement since most had reportedly started out as reluctant fliers at best.[72] Firsthand accounts backed this up. Larry Hunt, president of a multistate Cessna aircraft distributorship, enthused after the first class, "This is the greatest thing that's happened to aviation in years. I just had a husband and wife come in after she completed her Pinch-Hitter flying. I've been trying to sell them a new twin [engine plane] for years, and now *she's* the one who's eager to buy." Another graduate admitted in an interview that she had previously objected when her husband proposed buying a new radio for their plane. After becoming a Pinch-Hitter, however, she remarked that now she understood why they needed it, then added: "It'd make a nice Christmas present, don't you think?" While a few women did back out after discovering that their husbands had enrolled them without first asking their permission, this seems a reasonable response given that their husbands' actions clearly violated the fundamental ideals of companionate marriage that the course was intended to promote.[73]

Over the next several years, participants and their grateful husbands continued to write letters to the editor praising the course. Flying physician Fred F. Brown Jr., M.D., reported in 1969 that the Pinch-Hitter Course was "the best investment I've made in a long time" and expressed amazement that just two days of training had converted his wife from being "scared to death of an airplane" into someone who could cheerfully comment, "Touch-and-goes [practice landings] in a [Cessna] 182 are really a lot of fun, don't you think?"[74] In 1971 *AOPA Pilot* reprinted an unusually long and detailed 700-word letter to the editor from one Sara Cowart that opened, only partly in jest: "Thank you for keeping my marriage intact. Your Pinch-Hitter Course did what many others couldn't do. It kept an airplane from coming between my husband and me."[75]

In 1973, a full decade after the program's debut, Marion C. McDonald, an editor for the *New Haven (CT) Register* who admitted that she "didn't even know where the starter button was" on her husband's airplane prior to taking the course, interviewed several of her 41 fellow Pinch-Hitters months after graduation for a full-length article about the experience. When it came to flying, most fit the profile of the reluctant or downright fearful wife; however, McDonald painted a more nuanced portrait as she sought to uncover "what motivates a

Pinch-Hitter." In the process, she discovered an apparent contradiction: "I found out these gals are truly liberated women, married to adventurous guys who appreciate them. They work at hobbies and jobs, take courses, take care of families, or run businesses, yet most of them took the course because their pilot husbands wanted them to." (Aside from acknowledging that four men also completed the course, she gave them no other mention whatsoever.) After providing readers with numerous minibiographies to show her classmates' varied backgrounds and motives, McDonald closed her article with figures, presumably provided by AOPA, indicating that "statistically, 32 percent of those who have taken the course over the years have gone on to solo, and 25 percent have earned their pilot's licenses." At least a few of McDonald's classmates reportedly took additional flight training, and she interviewed one fellow Pinch-Hitter who had later soloed (it was unclear from the article whether she intended to pursue her private license); however, the numbers for her peers clearly fell far below those reported by the AOPA.[76]

Right from the start, the organization responsible for creating the course predicted that Pinch-Hitters would revolutionize private aviation. In his December 1963 editorial titled "Welcome Aboard, Honey!" AOPA's vice president Max Karant proudly predicted that the new program would convince formerly reluctant wives to become full-fledged private pilots. And in a magazine advertisement released that same month, the AOPA boasted that the Pinch-Hitter Course had set the stage for "a new air age . . . of women!"[77] Coming on the heels of Betty Friedan's bestselling book *The Feminine Mystique*, published earlier that year, it might be tempting to take the organization's leaders at their word and believe that they really did intend to open the skies to legions of liberated women. However, the details surrounding the origins of the Pinch-Hitter program indicate that they were instead mainly concerned with securing freedom of the skies for men by converting the antiflying wife into an ardent supporter of her husband's hobby.

Looking back years later, course organizer Ralph Nelson tacitly admitted as much when he recalled, "We didn't really mean it to be a tool for recruiting new pilots, but that's what happened."[78] Nelson had every reason to be proud of the Pinch-Hitter Course, but his and others' claims that it significantly influenced the number of female fliers is open to question. Although the proportion of women in the pilot population doubled from 3.1 to 6.2 percent during the 15 years following the first class, other figures suggest that this program probably played little direct role in this long-term increase.[79] Between 1963 and 1966, some 4,000 individuals, mostly women, graduated from the course, and up to one-third of these Pinch-Hitters reportedly went on to earn their private pilot's license. Thus, in those first three years alone, when the total number of women pilots (including students) increased from 11,757 to 20,265, the roughly 1,300 new private pilots who began as Pinch-Hitters represent a small but still measurable contribution (15 percent). But in the three decades that followed (1967–97), only about 7,000 completed the hands-on flight portion of the Pinch-Hitter Course, an annual average of fewer than 250 individuals. This figure represents a statistical drop in the bucket compared to the 11,000 to 27,000 women who enrolled in traditional flight instruction each and every year during the same period.[80] Then again, more than 100,000 women did complete just the classroom portion of the course (with no accompanying flight instruction). How many of these were wives of pilots with no aspirations for learning to fly, and how many went on to enroll as traditional student pilots without bothering to complete the dumbed-down Pinch-Hitter flight training, is impossible to determine. As a result, the Pinch-Hitter Course should be viewed for what it was originally intended: a means to reconcile the masculine hobby of flying with the ideals of companionate marriage.

By the mid-1970s, as a new ideal supplanted the companionate marriage model of the immediate postwar baby boom years, the Pinch-Hitter program faded from prominence in aviation journals. Labeled "the individualized marriage" by some scholars, this new version of family life held that "each person should develop a fulfilling, independent self instead of merely sacrificing oneself to one's partner."[81] Nonetheless, the AOPA continued to offer the Pinch-Hitter Course, although it would eventually drop the flight training portion. And in the early 1980s, the organization further committed to catering to traditional middle-class values by trying to turn its annual meetings into a genuine family affair.

For instance, the 1981 conference, by then known as the "Annual AOPA Convention and Industry Exhibit," took place in Orlando, Florida, where organizers attempted to provide "Something for Everyone." In addition to the usual speeches, seminars, flight training programs, and aviation exhibits (including "the latest in avionics and other goodies" in the hotel exposition hall and "50-odd new airplanes on display at the airport"), there was a "beat the pro" golf tournament for grownups who were "less interested in academic pursuits," as well as day trips "for kids young and old" to nearby Disney World and Sea World. One evening, "16 buses transported hundreds of AOPA members and their families to Disney World's Contemporary Hotel. There, steak dinners were on hand, along with a musical presentation by Disney World's 'Kids of the Kingdom' that entertained the crowd." Later that week, convention-goers were granted up-close access to antique airplanes and automobiles during a barbeque dinner held at the Wings and Wheels Museum. A NASA official presented a film on the Space Shuttle program and brought samples of the shuttle's ceramic thermal protection tiles for the audience to examine. Star-struck attendees could meet Academy Award–winning actor Cliff Robinson, who was on hand to be recognized for his longtime support of general aviation; those more interested in hearing from the halls of power in Washington, D.C., could attend the "Meet the FAA" program featuring J. Lynn Helms, chief administrator of the FAA. The final banquet harkened back to the organization's long-running "Plantation Party" theme. "Dancing and dinner were to the tunes of the Vaughn Monroe Orchestra. Hostesses dressed like ante-bellum southern belles saw to it that each of the ladies received flowers."[82] Although male pilots were still the primary audience for the convention, and they still had their chance to hangar fly with buddies over drinks, these family-friendly activities reflect a very different kind of masculinity from that performed at get-togethers of the Quiet Birdmen.

EAA Fly-Ins: A Mecca for Grass-Roots Aviation

Compared to the annual AOPA Plantation Parties, there was nothing posh or white-collar about the EAA's yearly fly-ins. After founding the Experimental Aircraft Association in the basement of his home in 1953, Paul Poberezny served as its president for 37 years before handing over the reins to his son Tom in 1989. The son of a Ukrainian immigrant, Poberezny grew up during the Great Depression in an era of widespread enthusiasm for all things aeronautical. As a child, he built makeshift planes out of wooden crates. As he grew older, he divided his time between reading everything he could about aviation and working at the local airport in Milwaukee, Wisconsin, where he polished planes and learned all he could from pilots and mechanics. In the mid-1930s, a sympathetic high school history teacher gave him a damaged Waco training glider. In short order, the intrepid teenager restored it in the family garage, arranged to have it towed aloft behind an automobile (a common method of launching manned gliders at the time), and at the age of 16 made his first solo flight without the benefit of formal flight instruction.[83]

An early biography described Paul Poberezny as a "self-made man," and his hands-on, self-taught, blue-collar mentality permeated the EAA from the very beginning. At first, Poberezny dedicated the organization to serving the interests of pilots who chose to build their own aircraft, which they then certified and flew under a special "experimental" category permitted by the federal government. However, during his nearly four decades at the helm, he steadily expanded the

EAA's mission to encompass the interests of nearly all private pilots, regardless of whether they had any interest in building their own planes. This attitude significantly broadened the organization's membership base and prompted loyal followers to bestow nicknames on Poberezny such as "Mr. Grass Roots Aviation."[84] In many ways the AOPA and EAA complemented rather than competed with each other. Whereas the former lobbied for all of general aviation, ranging from private fliers up to corporate flight departments and air charter operations (essentially anything short of airline or military aviation), the latter focused only on pleasure flying in the broadest sense, with a niche market of do-it-yourselfer aircraft builders at its core. These two organizations—and their annual meetings—reflect different versions of masculinity that coexisted within the broader culture of postwar private flying. The AOPA represented the powerful upscale consumer-pilot who was used to first-class accommodations and getting his own way, while the EAA catered to the fiercely independent "grass-roots" flier who, regardless of background and profession, relished the thought of getting grease under his fingernails as he painstakingly assembled his own airplane.

Compared to the somewhat uneven trajectory of the AOPA's early annual meetings, the history of the EAA's yearly get-togethers seems relatively straightforward. The first fly-in, which Poberezny scheduled to coincide with the Wisconsin Air Pageant in order to provide public exposure for the EAA, took place at Milwaukee's Curtiss-Wright Airport on a September weekend in 1953. Some 140 pilots and 21 airplanes participated. The agenda was loosely organized, with members taking each other up for flights in their homebuilt aircraft, swapping technical tips, and talking about aviation, as was the norm for any gathering of pilots. The Miller Brewing Company sponsored the event, providing a tour of its brewery and hosting the Saturday night dinner. Over the next several years, the convention evolved into a combination of fly-in, air show, and annual convention, and the EAA moved it to other locations in order to accommodate the ever-growing attendance. In 1959, some 7,000 people (1,400 of them EAA members) attended the fly-in convention in Rockford, Illinois. By 1970, the year the EAA moved the fly-in to its current location at Wittman Field in Oshkosh, Wisconsin, the organization reported that the gate count for the weeklong event reached 300,000. Throughout the 1960s, while the AOPA event settled into a format resembling a conference for white-collar professionals, the EAA's annual meeting maintained a grass-roots, do-it-yourself, working-class flavor as it expanded its offerings. Additions included professional air show performances, hands-on educational forums about how to build your own airplane, aviation industry exhibits, and on-airport camping for visitors who chose not to stay in local hotels.

Starting in the early 1960s, the EAA fly-in also garnered attention by causing its host, the normally sedate Rockford Airport, to become the world's busiest airport at the height of the convention. In 1963, the first year this happened, the fly-in generated a world record of 4,766 takeoffs and landings within a 15-hour period. EAA fly-ins were so busy that the FAA had to assemble a handpicked team of air traffic controllers each year and publish special arrival and departure procedures just for the event.[85]

If there was ever a total immersion experience for aviation enthusiasts, the EAA convention was it; over the years, numerous articles aimed at both pilots and the general public described Oshkosh as the "Mecca" for private fliers.[86] Visitors spent the day wandering the flight line examining hundreds of carefully constructed homebuilt planes and painstakingly restored vintage aircraft and warbirds (former military aircraft) on display.[87] There were industry exhibits to visit and plenty of new and used parts and accessories for sale. An article recounting the 1970 fly-in described the educational aspect of the gathering: "[Homebuilt aircraft] designers, and individuals specializing in certain phases of aircraft construction, held forums on such topics as wood/metal construction, blowing canopies, Fiberglas [*sic*] application, retractable landing gears, propellers, and rotary wing aircraft [helicopters and gyrocopters]. Additional forums covered such subjects as maintenance and preservation of World War II aircraft, building replica World War I types, and powerplants [aircraft engines]. In fact, you name the subject—it was discussed."[88] Every evening around dinnertime, convention-goers spread out along the runway or watched from their campsites as world-class professional air show pilots performed low-level aerobatic demonstrations. After dark, cartoons and movies entertained the children while their parents attended talks and other aviation-related programs. And, as at any gathering of pilots, there was plenty of hangar flying, whether it was on the flight line, in the forums, or over meals in the campground.[89] The thundering air show acts, tent-camping under the wing, and classes on how to build an airplane that occurred during EAA fly-ins stand in sharp contrast to AOPA conventions held on cruise ships or at upscale resorts. These aspects of EAA's annual gatherings reflect a strain of working-class masculinity within private aviation that placed a greater value on understanding and working directly with the technology of aviation than on professional success or simply operating the aircraft.

However, the masculine cultures of the EAA and AOPA shared common views on men's roles in the family and on the place of women in aviation. Like the AOPA Plantation Parties and their predecessor events, the EAA's annual fly-ins provide a window into how members transposed prevalent gender roles into the culture of private aviation, thus retaining and reinforcing its associations with men and masculinity. As early as 1963, the EAA convention program included a brief section titled "For the Ladies . . . " which promised "a visit to downtown Rockford for sightseeing, shopping, etc.," along with "a noon luncheon and fashion show ($1.75 per person), followed by bridge for those so inclined." This "Ladies" portion of the program ended with a gender-specific request: "Some volunteer help in watching the smaller children is asked of the mothers." Apparently the entire family was expected to attend the fly-in with dad, thus making it a true family vacation, but EAA anticipated that mom and "the smaller children" might not have much interest in the aviation-centered world of the fly-in itself.[90]

The EAA's 1966 convention did not describe any formal program directed specifically at women (although it did offer special daytime and evening programs for children and free bus service from the airport to the downtown shopping center).[91] By 1974, however, "Women's Activities" were again a standard fixture in convention programs, which by then had grown to magazine-length publications. These activities included daily seminars, social gatherings, arts and crafts, and day-trips away from the airport, a format that remained unchanged for decades. In 1975, the EAA added the ground school portion of the AOPA's popular Pinch-Hitter Course to its lineup of activities aimed at women; as the "world's busiest airport" for a week, however, Oshkosh was no place to conduct the flight training component of the course. By 1978, the EAA was even offering technical workshops on welding, wooden aircraft structures, aircraft upholstery, and working with fiberglass as part of its "Women's Activities" program.[92] Like the AOPA's Pinch-Hitter program, this change was probably less about recruiting women to build their own planes than it was about getting women interested in their husbands' hobby. Also like the Pinch-Hitter program, the EAA believed that the key to success was creating an all-female environment that seemed less threatening than the regular classes that taught exactly the same skills to a mostly male audience of hard-core homebuilders. And this strategy seemed to work. As a result of this growing emphasis on the wives of pilots, Beth Blagaich enthused in a 1979 article: "Last August, EAA women were enjoying the Convention as much or more than their husbands and children. Only a few years ago, the first week of August was another time for many EAA wives to decide, 'Should I stay home . . . or tag along?' " She went on to claim, "Since 1974, and the beginnings of Women's Activities at the Convention, Oshkosh has become a word meaning a vacation that women make plans for and get excited about as much as their husbands."[93]

The introduction to the "EAA Women's Activities" section in the 1981 convention program provides a preview of the many ways the organization attempted to reach out to pilots' wives: "If you want to know something more about aviation, COME TO THE WOMEN'S ACTIVITY CENTER. If you are tired of hearing about nothing but airplanes, COME TO THE WOMEN'S ACTIVITY CENTER. Whatever YOUR interest . . . there's something which should interest YOU at the WOMEN'S ACTIVITY CENTER." Having hopefully grabbed the reader's attention with this liberal use of capitalized letters, the program devoted the next four pages to describing what the reader could find at the Women's Activity Center. She could "Get a lift and a new look with [a] free facial offered under the auspices of Mary Kay Cosmetics," participate in daily aerobic dance classes led by Sharon Poberezny (wife of Fly-In Convention chair Tom Poberezny), or attend a three-day emergency lifesaving and CPR course. The center also offered activities resembling traditionally masculine, aviation-related topics such as introductions to aircraft building techniques (including learning to weld at special workshops for women taught by Dorothy "Carrot Top" Aiknoras, an airline pilot, aircraft mechanic, and experienced builder of aerobatic airplanes whom the program described as "our 'Jill of all trades' ").[94]

Like the AOPA, the EAA had good reason to provide programs aimed specifically at the wives of pilots: failure to do so might lead to women deciding to stay at home from the convention and pressure their husbands to do the same.[95] But this tactic also reaffirmed existing assumptions about gender and aviation. EAA conference organizers did not need to set up a special "Men's Activities" program because the entire fly-in was, in effect, already one immense program for men. Furthermore, like the AOPA, they did not need to worry that hiring women instructors to entice other women to try aviation would come across as being too "progressive" to socially conservative men—or even women—within their ranks. Because women in aviation remained scarce throughout the postwar era, the presence of a few highly accomplished individuals like Mrs. Velta S. Benn, who instructed the first class of Pinch-Hitter students for the AOPA in 1963, and "Carrot Top" Aiknoras, the aeronautical "Jill of all trades" who taught welding and other aircraft construction techniques for the EAA in the 1970s and 1980s, posed no more real threat to the masculine status quo than the aviatrixes of the prewar era.

Because the massive EAA fly-ins relied so heavily on volunteers, these events provided more opportunities than AOPA conventions did for nonpilot wives to become active participants. There is no indication that the EAA ever prohibited—or even discouraged—women from directing aircraft on the flight line, working crowd control duties, or doing other jobs generally associated with male volunteers. However, the "Women's Activities" section focused on recruiting for one particular function: "OPERATION THIRST [To Help Immediate Recovery of Starving Troops]," a program started in 1976. The 1978 convention brochure described what this entailed: "Each morning at 8:00, a group gathers at the beautiful large trailer in the Women's Activity Area to help prepare sandwiches and drinks. These are then driven onto the flight-line and remote parking areas and given 'for free' to all of the volunteers who work so many hours in the hot sun, directing traffic, parking airplanes, and doing many of the other jobs which help to make the EAA Convention work." The description ended with a plea: "Volunteers are needed for both the preparation and delivery of OPERATION THIRST. Sign up in the Women's Activity Office." A photograph of "Thirst Quencher" volunteer Mindy Criswell serving a cup of coffee to EAA Director Steve Wittman was accompanied by a caption that read in part: "No flagman or security patrol volunteer is so remote on the vast fly-in grounds that Operation THIRST can't find him (or her)—ready with sandwiches, cold and hot beverages and a beaming smile."[96] By the mid-1980s, the EAA had expanded this service so that it provided three meals a day for the small army of unpaid volunteer labor that made the entire fly-in possible. A 1985 article described the kitchen and dining hall, which were run by a mostly female volunteer corps: "The meals aren't what you would consider fancy, but they are what every roadside restaurant boasting 'home cooking' claims to be—delicious. The kitchen crew is headed by Nancy Lichtenberg who, together with her crew . . . perform palate-pleasing feats of magic, morning, noon and night. Entrees are served right along with friendly 'hello's,' 'how are ya's' and 'how ya' doin's,' creating an intimate atmosphere where conversation is king and the fixin's are fine, thank you."[97]

Thus, the women who volunteered for OPERATION THIRST and worked in the dining hall were no mere tag-along spectators to their husbands' hobby. Instead, they played a vital role in making the annual fly-in happen, regardless of whether they had any personal interest in aviation. Just as their husbands took pleasure in admiring airplanes, watching air show performances, learning new skills, and hangar flying with fellow fliers, these women seemed to enjoy the camaraderie of their own mini-community of pilots' wives as they worked assembly-line style to prepare hundreds of meals each day to feed the rest of the volunteers. Describing her experiences at the 1977 convention, Jayne Schiek wrote, "Operation THIRST proved to be one of our most satisfying activities and everyone enjoyed their volunteer hours spent working on it."[98] Photographs of the Operation THIRST volunteers from the late 1970s, as well as a photo of the kitchen staff for the dining hall for volunteers at the 1985 fly-in, show smiling, apron-clad, silver-haired women; not only were they old enough to be grandmothers, but they could have easily been the same young women from Beta Sigma Phi who four decades earlier had served breakfast at the AOPA's first Colorado fly-in immediately following World War II.[99] These volunteers who made sandwiches and delivered food in the 1970s and 1980s had found a place in the community of pilots not by breaking with traditional postwar gender roles but by working within them. As a result, with the exception of the strictest adherents to the "Men Only" approach to flying voiced by individuals like Milton Horowitz and practiced at meetings of the Quiet Birdmen, the presence of these women did little, if anything, to dilute the sense that flying was still a masculine adventure.

Throughout the postwar era, whether women learned just enough to take over the controls of the family plane in an emergency, golfed and shopped and shepherded children to amusement parks with other wives while their husbands talked flying with other men, served as a ladies auxiliary to support a mostly male event, or even tried their hand at welding in a classroom full of other women, none of these activities challenged traditional associations between masculinity and flying. Indeed, they reinforced middle-class masculinity by placing the woman in a supporting role and leaving the man ultimately in charge. But whether it was Ralph Nelson playfully tossing his wife Jane into a swimming pool as "punishment" for learning to fly behind his back at the 1963 AOPA Plantation Party (an event described in the opening paragraphs of this book's introduction), Professor Milton Horowitz's less amusing 1969 rant that women could not fly without either compromising the inherently masculine nature of the activity or sullying their own feminine identity, or the Quiet Birdmen's continued ban on women at their gatherings (with the exception of scantily clad entertainers), female fliers were a different matter.

Taken together, the views and activities of the Quiet Birdmen, Ninety-Nines, AOPA, and EAA provide a window into varied views that members of the aviation community held about the gendering of aviation. In their own very different ways, each group reflected—and through its actions reinforced—established norms that prescribed the proper roles and activities for women and men in postwar American society. Members of the QBs created a hypermasculine environment where they could drink to excess, talk about and possibly ogle beautiful women, and swap stories with fellow pilots; in short, they used the community they created as a temporary escape from the domesticity of middle-class society. The Ninety-Nines, on the other hand, took the opposite approach by co-opting middle-class gender norms in order to incorporate femininity into private aviation without ruffling the feathers of their male counterparts, creating a subculture where members could simultaneously be women and pilots at the same time.

The AOPA and EAA, with a combined membership totaling several hundred thousand by the mid-1970s, represented a much broader cross section of American pilots than the smaller, sex-segregated QBs and Ninety-Nines. These two larger groups took a middle path as they simultaneously catered to their predominately male base and consciously carved out a place for the wives of private pilots. Men could (and did) regularly gather for extended bouts of hangar flying at national conventions. However, much of this male bonding happened during the day, and the evening's entertainment, designed with respectable, middle-class, middle-aged couples in mind, in no way resembled what reportedly took place when the QBs got together. Although the AOPA and EAA promoted the concept that couples should fly together, they gently reinforced rather than challenged existing gender roles within aviation and in the broader society. Companionate marriage dictated that the wife fly along with her husband and, ideally, enjoy the experience as well. At the same time, just as this postwar ideal for marriage championed a strict division of labor between breadwinner and housewife, the same type of division of labor—and power—extended to the cockpit. Indeed, it was because of the importance of carrying that domestic division of labor into the cockpit that the audience was amused when Ralph Nelson pushed his wife Janie into the pool after she bucked convention and learned to fly on her own. And it was also for this reason that several decades after Jane Nelson's impromptu swim, wives of pilots still gathered behind the scenes to feed the volunteers who ran the EAA's massive fly-ins.

Although all of these organizations reflected, in one way or another, societal norms of the time, the nature of their meetings also reveals that masculinity in aviation was far from monolithic. The secretive QBs, with their reputation for "drinking, topless bartenders, big steaks, male bonding, raunchy jokes and hugely exaggerated tales of flying and sexual prowess," represent the archetypal stereotype of the ultramasculine, "right stuff" fighter pilot. Although many, if not most male pilots of the postwar era did not (and did not even wish to) live up to this image, at least a few of them embraced it willingly. And as a result, the reputations of the rest—as well as the culture of private flying writ large—benefited from what might be called "manliness by association," in which the actions of a few reinforced widespread impressions that flying was a masculine endeavor.[100] Meanwhile, the more staid members of the AOPA proved, through their upscale annual conventions and their ability to take over a town for a week, that they were respectable men who demonstrated their masculinity through their wealth, professional success, and political power. The EAA showed yet another side of private flying, the gritty, grease-under-the-fingernails do-it-yourselfers who gathered in legions to be educated, entertained, and awed during their yearly pilgrimage to Oshkosh, Wisconsin, the Mecca of all things aviation. Taken together, the activities of the Ninety-Nines, the Quiet Birdmen, the AOPA, and the EAA provide a comprehensive portrait of the depth and diversity of masculinity's influence in the culture of postwar private flying.

Conclusion

In the earliest years of powered flight, pioneering aviators like Claude Grahame-White adamantly insisted that "the air is no place for a woman." In the century that followed, the sky remained an almost exclusively white male frontier, even in the realm of private flying, where, unlike military and commercial aviation, there were never any laws formally prohibiting participation by women. Even so, it is wrong to assume that the highly masculine nature of postwar private aviation simply reflects a continuation of this pervasive attitude cultivated by prewar pilots.[1] Instead, widespread social mores and geopolitical events far beyond the airport fence shaped who took up flying in the postwar era, as well as their expectations and experiences of this pastime. Thanks to U.S. involvement in World War II, the previously thin ranks of civilian fliers swelled by more than 1,000 percent in just a few short years. At the same time, America's cultural norms meant that white women and African American men had to fight for access to the cockpit of military aircraft, and women were forbidden altogether from serving in combat. Combined, these factors meant that most military fliers trained during the war, as well as most postwar GI Bill pilots, would be young, white, and male. This initial cadre of World War II–era fliers formed the cornerstone of postwar private aviation, and those pilots—both as flight instructors and as members of the fence crowd at thousands of local airports across the country—became the formal and informal gatekeepers who screened future generations of aspiring aviators in the decades that followed.

The community of pilots used a variety of methods to vet new members, and acculturation began at the moment a newcomer stepped through the airport gate. For one thing, wartime considerations virtually guaranteed the homogeneity of the first generation of postwar private pilots. This in turn contributed to an undeniable demographic fact: prospective students were invariably greeted by a mostly male, almost completely white fraternity of fliers when they dropped by the local airport, sending a silent but inescapable message about who did and did not fly. And this was just the beginning. More than a few newcomers reported that they had to run a gauntlet of elitist insiders just to sign up for their first lesson, and untold numbers turned their backs on private flying forever as a result. Acculturation continued throughout flight instruction, a process in which *what* students learned from their instructors, as well as *how* they were taught and treated both in and out of the cockpit, simultaneously imparted practical information about flying and drove off individuals who couldn't or wouldn't measure up. The first generation of postwar pilots had been trained in the pressure-cooker environment of wartime military flight schools (or by former military instructors who inflicted the same traditions on their GI Bill students). As a result, they took for granted that learning to fly involved drill-sergeant-style yelling and verbal abuse intended to weed out those who weren't up to dealing with the dangers of aerial combat. When they in turn became flight instructors, they passed this tradition on to new generations of fledgling fliers, purposely or inadvertently eliminating those who couldn't or wouldn't stand for this treatment. The broader aviation community had a vote, as well, and if the local fence crowd collectively decided that a fledgling pilot did not belong, he or she received a cold shoulder from the tight-knit flying fraternity.

Members of this community also determined who did—and did not—deserve the title of pilot by defining what constituted a "pilot's airplane." When manufacturers offered aircraft that promised to allow almost anyone to take off and "drive" through the sky, paying consumers, most of whom were already licensed pilots, complained that replacing skill with technology would actually diminish overall safety. As a result, supposedly foolproof designs ultimately failed in the marketplace. This had long-term consequences for future generations of students, who had no choice but to learn to fly in less user-friendly trainers that had won prior approval from experienced pilots and consumers who celebrated individual skill over built-in safety and ease of use. Even after students made the cut and were admitted into the ranks of pilots, skill mattered to the point that some based their identity on what kinds of airplanes they were capable of flying. Although many were content to fly models with relatively modest performance and docile handling, since this alone set them apart from the larger nonflying public, others sought personal satisfaction and recognition from fellow fliers by deliberately mastering planes that had reputations for being difficult or even dangerous in the hands of the average operator.

Fliers also used the places where they gathered, the ethos that guided how they treated each other, and the organizations that they joined to further define private aviation as a masculine pursuit. For instance, airports served as stages where pilots performed the time-honored tradition of hangar flying. At these gatherings, men shared tips and tales about airplanes and women, taught newcomers how pilots were supposed to behave, and passed judgment on fellow fliers who violated the "elaborate code of courtesy and conduct of the flying fraternity."[2] Despite repeated calls for reform throughout the postwar era, many of these airports remained dirty and rundown, a highly visible advertisement that these were manly spaces free of a woman's supposedly civilizing touch. Seeking the company of like-minded individuals, pilots flocked together in formal clubs or associations. Whether these were segregated by sex or open to all, each organization reacted to or reflected the postwar ideals of companionate marriage in which both husband's and wife's happiness mattered, but ultimately the husband still ruled the roost.

These practices worked symbiotically to create a sum that was greater than its parts. In informal hangar flying sessions, the fence crowd coined value-laden insider terms like "Mixmaster," "doctor killer," and "taming the taildragger" and taught newcomers that the Ercoupe wasn't a "real airplane." These and other cultural traditions came together at fly-in conventions, which served as national-level venues for acting out, disseminating, and reinforcing the masculine values that underpinned the postwar community of pilots. Thus, when a middle-aged Ralph Nelson and Vic Kayne tossed Ralph's wife Jane into the hotel swimming pool in 1963, they were playfully reassuring the audience (mostly made up of middle-aged, white, male pilots like themselves, many with their nonflying wives sitting dutifully beside them) that although women were welcome in the cockpit, there was no real danger of them taking over the controls of the family plane.

By the mid-1980s, the generation of World War II–trained pilots who originally established postwar private flying, and then dominated it for so long, saw their beloved community starting to crumble. A government study published in 1971 forecasted steady growth in civilian pilot numbers over the next decade, with a projected total of more than 1.5 million by 1982. The same report predicted that 26,500 new general aviation aircraft would be built in 1982 alone. Instead, the pilot population peaked at an all-time high of 827,000 in 1980, far short of projections from a decade earlier, and then began to decline, plummeting by nearly 15 percent in just five years. Manufacturers fared even worse: the 4,300 general aviation aircraft produced in the United States in 1982 represented only one-sixth of those predicted in 1971, and starting in 1987 annual production hovered around 1,000 for most of the next decade.[3] Worst of all, almost none of these new airplanes were built with the private pilot in mind. For many, it seemed like the beginning of the end for private aviation.

In 1897, Samuel Langhorne Clemens, aka Mark Twain, famously responded to rumors about his failing health by declaring that "the report of my death was an exaggeration."[4] Twain's humorous statement proved prophetic, and he would live on for more than a decade. The same near-term outcome proved true of dire predictions about the imminent demise of private flying. After general aviation's "big three" manufacturers—Cessna, Piper, and Beech—ceased production of small single-engine personal planes in the mid-1980s, a handful of upstart companies, most notably Cirrus Aircraft, stepped into the void and introduced new models to take their place. Then, following a tumultuous decade that culminated in national-level tort reform and the General Aviation Revitalization Act (GARA) of 1994 to reduce the burden of product liability on manufacturers, Cessna resumed production of its old standby, the entry-level model 172.[5]

In 2004, 10 years after Congress passed GARA, the FAA introduced a new category of Light Sport Aircraft (LSA) and a new Sport Pilot certificate that was easier to obtain than the standard private pilot's license.[6] Computerized cockpit displays with a global positioning system (GPS) now provide amateur pilots with far more information than was available to professional airline and military pilots flying top-of-the-line aircraft just a few decades earlier. Rocket-deployed airframe parachutes, which lower an aircraft in distress to the earth with the pilot and passengers still strapped safely in their seats, not only are available as optional equipment for many new aircraft (with retrofit kits available for many existing models) but even come standard along with airbags in several personal planes currently on the market.[7]

The homebuilt-experimental aircraft movement, already in existence for decades, also gained momentum during the 1980s. Dozens of companies emerged to serve burgeoning demand, offering everything from blueprints for "scratch built" airplanes (where homebuilders purchase all needed materials from suppliers and then follow detailed plans to complete every aspect of construction on their own, including welding the metal framework, forming sheet metal into complex shapes, and driving thousands of rivets), to "quick build" kits that come from the factory with the most complicated sections preassembled by professionals. Dick VanGrunsven, or "Van" as he is universally known by homebuilders, started the most successful of these companies in 1973 when he began selling plans for the single-seat RV-3. Since then he has sold more than 18,000 sets of plans and kits for various one-, two-, and four-seat RV models, of which more than 7,500 were reportedly finished and flying by the end of 2012. Even more impressive is the way business has taken off in recent years, with approximately 500 new aircraft completed by homebuilders annually. In a recent interview, Van explained: "One thing I've learned along the way is that there's a bit of Walter Mitty to [pilots' dreams of flying]. You envision what it's going to be like to fly something like a World War II fighter. But you have no idea how heavy those controls really are, or that they're not well-harmonized controls. . . . [The RV] is a good Walter Mitty airplane, because it flies like you think a fighter flies but doesn't."[8]

Despite these signs that earlier reports of private aviation's imminent demise were, in Twain's words, "an exaggeration," there are several sides to this continuing saga. The unprecedented 9/11 terrorist attacks on U.S. soil in 2001 involving four hijacked airliners and the deaths of nearly 3,000 people resulted in a new wave of concerns over the future of private flying. For a time, Congress threatened to severely curtail private pilots' access to America's skies. Although most initial restrictions were eventually lifted, largely through the lobbying efforts of the AOPA and EAA, new threats of a different sort loomed on the horizon. By 2002, the number of active fliers old enough to have served in World War II (by then aged 75 or older) had declined to just 1.7 percent of the total pilot population. Meanwhile, general aviation experts noted that the "graying of the general aviation pilot population" extended well beyond this first cohort of postwar fliers.[9] In 1969, when Mr. General Aviation was in his prime, the average age for civilian pilots in the United States was just over 35. By 1990 the average age was 40.5 years old, and two decades later it had risen to the mid-40s. For holders of a private pilot's license (thus excluding students, who still tended to sign up for flying lessons in their late teens or twenties), the average age was even higher, reaching 48 by 2007 before leveling off again. Meanwhile, since reaching an all-time high of 827,000 in 1980, the number of active civilian pilots in the United States fell by roughly one-quarter over the next three decades.[10] Not only was the total number of pilots declining, but the age of those still actively flying was going up. In short, there weren't enough young new pilots in the pipeline to sustain private flying over the long term.[11]

The EAA responded by developing the "Young Eagles" program, which relies on members to "Give a kid a flight in a small plane and plant the seed for future participation in aviation." Twenty years after the program debuted in Oshkosh in 1992, the EAA could boast that volunteers had provided more than 1.7 million Young Eagles flights. Meanwhile, a consortium of general aviation companies created "Be-A-Pilot" to educate and attract a new crop of student pilots, and the AOPA devised "Project Pilot"—with Erik Lindbergh (Charles Lindbergh's grandson) as its first spokesperson—to match new students with experienced pilots to mentor them through the long and often daunting process of learning to fly. Despite these and other efforts to raise public interest in flying, including promoting June as "National Learn to Fly Month," the long-term future of private aviation remains uncertain.[12]

Even as the last members of the World War II generation retired from the cockpit in the first decades of the new millennia, decades-old themes continued to echo through the private aviation community. In 1975 Don Downie's article, "Taming the Taildragger: Toil, Tears, and Sweat—But Worth the Trouble," informed *AOPA Pilot* readers that mastering a tailwheel-equipped airplane would simultaneously make them better aviators and confer skill-based status within the flying fraternity. Nearly three decades later in a 2003 article titled "You're Not a Real Pilot," AOPA president Phil Boyer explained why he'd recently taken a week's vacation to earn his tailwheel endorsement. "Throughout my 35 years of private flying I have consistently been admonished by fellow pilots that 'you're not a real pilot until you have flown a taildragger.' I hate to admit it, but during my entire flying career, accumulating more than 6,500 hours, I have never logged any time in a tailwheel-equipped airplane." Boyer was careful to quote fellow fliers who claimed that "you're not a real pilot" rather than to come right out and make this claim himself. To have done so would have risked offending many AOPA members, the majority of whom likely had no tailwheel experience. Indeed, he addressed this fact directly in his introduction: "Don't be too quick to criticize me, since I am probably the typical private pilot who learned to fly in the past three or four decades. My introduction to flying was in the training airplane of the time, a Cessna 150. The tricycle-gear airplane allowed a fresh young pilot to land in crosswinds without incident. Then, like most pilots, I transitioned through the single-engine piston Piper and Cessna models so prevalent at the time. All had a nose gear, and to many of us that was the norm—why fly anything else?" Still, Downie's message had made its mark on the generation of taildragger pilots who teased Boyer for not joining their elite ranks. And it clearly worked on Boyer, too. For despite his impressive level of flight experience in other aircraft, the AOPA president admitted that until he too had finally completed the requisite training, "I never thought of myself as a complete pilot when those around me were telling their stories of Piper Cubs, ground loops, and flying taildraggers."[13]

Ten years later, in February 2013, two editors of *AOPA Pilot* went head-to-head in print about whether it was appropriate for private pilots to wear military-style flight suits. Senior editor Dave Hirschman argued that "flight suits are just plain useful" in terms of protecting not only clothing against the inevitable grease and grime involved in flying small planes but also the pilots themselves against the possibility of an in-flight or post-crash fire. "Yet some fellow pilots—our own self-appointed airport fashion police—insist that we should never wear them because the garments are uncool, we are unworthy, or both." Hirschman's response to accusations that wearing a flight suit exposed a private pilot's underlying Walter Mitty fantasies—"If I wanted current apparel advice, I'd ask my 16-year-old daughter"—suggests that he, just like AOPA's Mr. General Aviation of 1969, had reached middle age. Meanwhile, relative youngster Ian J. Twombly, the grown son of longtime *AOPA Pilot* writer and editor Mark Twombly, asserted: "By wearing a flight suit, you are making a statement. It's not enough that your wallet holds a pilot certificate. You must announce to the world that you are a modern-day Maverick [the recklessly macho lead character in the 1986 action movie *Top Gun*], ready to buzz the [airport control] tower at a moment's notice." Twombly's four-word title for his contribution, "Don't Be 'That Guy,' " suggests that there were at least a few examples of "that guy" running around local airports wearing their macho-fighter-pilot fantasies on the sleeves of their military-style flight suits.[14]

Deliberately courting controversy, a few months later *AOPA Pilot* revived a debate that began more than a half-century earlier regarding whether light twins were better or worse than single-engine aircraft in terms of safety, performance, and economy. Editor Dave Hirschman sided with the long-standing consensus that favored singles, while Peter A. Bedell, a professional airline pilot and co-owner of a light twin that he flew for personal use, argued that "in the hands of a well-trained pilot" twins beat singles in all categories. Still, after nine paragraphs of carefully marshaled data and analysis, Bedell revealed that there was more to the appeal of light twins than facts or figures could ever convey. "Finally, there's the cool factor. Twins look cool, sound awesome, and convey that you've achieved a sort of pinnacle of aircraft ownership and advanced pilot status." And, he concluded, "They are an absolute hoot to fly with all of that power on tap in your fist."[15]

These three recent examples—the continued allure of "taming the taildragger," warnings against dressing up to look like a fighter pilot, and frank admissions that "advanced pilot status" played a key role in deciding what kind of plane one flew—are eerily reminiscent of assertions, arguments, and accusations made throughout the four decades immediately following World War II. Now, as then, most of the pilots involved in these discussions would vehemently disavow any sexist intent in their celebration of power, skill, or the practicality of wearing a military flight suit to ward off grease. But that's exactly the point. The tradition of masculinity is so embedded in aviation that it is almost invisible to most individuals who have successfully run the gauntlet and been acculturated into the community of pilots. And men are not the only ones who take this culture for granted. Some women involved in aviation either do not see or choose not to acknowledge the gendered culture of flying. For instance, one recent study on professional pilots reported: "While there is a strong anti female pilot bias, i.e. hostile sexism, within the occupation, surprisingly, several female pilots reported that they had not experienced or were unaware of any prejudice directed towards them in their flying career. This may well be that these female pilots have embraced the masculinity of the occupation through self-socialisation [*sic*] and that they ae a product of the company they keep."[16] And if longtime pilot, aviation writer, and editor emeritus of *Flying* magazine Richard Collins is indeed correct regarding the relationship between masculinity and aviation, it may just be that the predominantly male "company" that this small number of women fliers keeps has indeed created and continues to perpetuate "a fraternal bond in flying that largely excludes females."[17]

Ultimately, the story of masculinity in postwar private flying transcends this particular pastime and instead reflects the continued (and often effectively invisible) gendering of many technologies and activities within our society. Only by stepping back from the here and now can we begin to fully comprehend how seemingly innocuous everyday practices establish and reinforce power relationships according to who is, and who is not, expected to pursue specific careers, purchase particular items, or engage in certain activities. Thus, the story of "Janie" Nelson getting tossed into a hotel swimming pool because "she learned to fly without letting her husband know" is not merely some quaint anecdote from a gathering of white, middle-class, male pilots in the early 1960s.[18] Instead, it reflects the nature of gender, technology, and power relationships in post–World War II America and challenges us to consider the persistence of these relationships into the early twenty-first century.

Introduction

1. Charles P. Miller, "Biggest Plantation Party," *AOPA Pilot* 6, no. 12 (Dec. 1963): 85.

2. Ibid., 85; Don Downie, "A New Flying Concept: The 'Pinch-Hitter,'" *AOPA Pilot* 6, no. 12 (Dec. 1963): 23–28; Jane Nelson, "Why Be an Airport Widow?" *AOPA Pilot* 6, no. 8 (Aug. 1963): 81–82, 87.

3. Miller, "Biggest Plantation Party," 85. During World War II, military aviation cadets often celebrated a classmate's first solo by throwing him into a pool or other body of water. However, after the war, civilian pilots created a new tradition based on cutting off a new pilot's shirttails (for more on these two traditions, see chapter 2). This 1963 account of dunking Jane Nelson contains no reference to the older military tradition (which never made the transition to civilian flying), and the author's references to "punishment for a wife" and "the dunking chair for shrewish wives" indicate good-natured revenge rather than celebration of a milestone.

4. Lois C. Philmus, "Women Banish Fear of Flying," *AOPA Pilot* 6, no. 5 (May 1963): 29.

5. The proportion of women military pilots is even lower than it is in civilian aviation. According to a 2013 *Washington Times* article, women account for 2 percent of air force pilots, 4 percent of the navy's carrier-based jet fighter and radar jamming aircraft pilots, and less than 1 percent of Marine Corps fixed-wing (airplane) pilots. Rowan Scarborough, "She Has the Right Stuff: Female Combat Pilots Have Been Flashing Their Skills for 20 Years," *Washington Times*, Mar. 25, 2013, http://www.washingtontimes.com/news/2013/mar/25/more-than-a-sound-barrier-female-pilots-have-displ/#ixzz2hqtTq7EG.

6. The percentage of women pilots reached a new high of 6.73 percent in 2010, but then leveled off again over the next several years. Unless otherwise noted, all statistics on pilots in this book include student pilots. Data on total pilots and women pilots from tables found in U.S. Department of Commerce, CAA, *Statistical Handbook of Civil Aviation* (Washington, DC: U.S. GPO, published annually 1944–58); U.S. Department of Transportation, FAA, *Statistical Handbook of Aviation* (Washington, DC: U.S. GPO, published annually starting in 1959); U.S. Department of Transportation, FAA, *U.S. Civil Airmen Statistics* (annual tables for 1999–2013), http://www.faa.gov/data_research/aviation_data_statistics/civil_airmen_statistics/.

7. Margaret A. Weitekamp, *Right Stuff, Wrong Sex: America's First Women in Space Program* (Baltimore: Johns Hopkins University Press, 2004).

8. This statement is supported by the statistical fact that only 1 in 20 pilots were women. A 1986 study on young women who chose careers in aviation found that a significant proportion had personally known, and been encouraged by, real pilots while they were growing up. These young women also reported that most of the pilots who encouraged them to pursue this "nontraditional" career path were men. Charles L. Rodriguez, "A Descriptive Study of Females Preparing for a Nontraditional Career in Aviation" (MS thesis, Southern Illinois University at Carbondale, Dec., 1986), 23–24, 60–62, 73.

9. Helen Richey was hired in 1934 as a copilot for Central Airlines. Although she was an accomplished pilot, she was hired as a ploy to generate publicity for the airline. Blackballed by male pilots and frustrated by new regulations that limited women (but not men) to flying only in good weather, she resigned in protest a year later. Many aviation historians consider Emily Warner, hired by Frontier Airlines in January 1973, to be the first female airline pilot in the United States, although American Airlines claims that it was "the first *major* airline to hire a female pilot" (emphasis added) when it admitted Bonnie Tiburzi to its ranks a few weeks later. Deborah G. Douglas, *American Women and Flight since 1940* (Lexington: University Press of Kentucky, 2004), 9, 176; Joseph J. Corn, *The Winged Gospel: America's Romance with Aviation, 1900–1950* (New York: Oxford University Press, 1983), 80; American Airlines, "Female Pilots Make History," revised Jan. 2011, http://www.aa.com/i18n/amrcorp/corporateInformation/facts/femalepilots.jsp.

10. Douglas, *American Women and Flight*, 186–87, 189; Amy E. Foster, *Integrating Women into the Astronaut Corps: Politics and Logistics at NASA, 1972–2004* (Baltimore: Johns Hopkins University Press, 2011), 2, 106.

11. Richard [L.] Collins, "The Great Debate: Boy v. Girl Pilots," Debate, *Air Facts Journal*, Jan. 9, 2012, http://www.airfactsjournal.com/2012/01/the-great-debate-boy-v-girl-pilots/.

12. Richard L. Collins, "A Long and Satisfying Journey," On Top, *Flying*, Oct. 2008, 33.

13. Electronics Inc., "Hot to Go in 70 Seconds!" [advertisement], *AOPA Pilot* 4, no. 9 (Sept. 1961): 42; "There Wasn't Any Emergency . . . " [cartoon], *AOPA Pilot* 12, no. 1 (Jan. 1969): 74.

14. Margaret J. Ringenberg, with Jane L. Roth, *Girls Can't Be Pilots* (Fort Wayne, IN: Daedalus Press, 1998), 147–48.

15. Robert Peterson, "Not Her Cup of Tea: How to Handle a Woman When She Won't Fly," *Flying*, June 1969, 57; Mrs. Stephen S. De Clue and Katy Dance [two letters], "Feminine Funk," Flying Mail, *Flying*, Aug. 1969, 93; Linda Rickles, Edgar R. Mycroft, and Mrs. R. K. Fleming [three letters], Flying Mail, "Drink Tea and See," *Flying*, Sept. 1969, 90.

16. Joseph R. Novello and Zakhour I. Youssef, "Psycho-social Studies in General Aviation I: Personality Profile of Male Pilots," *Aerospace Medicine* 45, no. 2 (Feb. 1974): 187–88.

17. Joseph R. Novello and Zakhour I. Youssef, "Psycho-social Studies in General Aviation II: Personality Profile of Female Pilots," *Aerospace Medicine* 45, no. 6 (June 1974): 631, 632–33.

18. See, for instance, Jim Mitchell et al., "How Pink Is the Sky? A Cross-national Study of the Gendered Occupation of Pilot," *Employment Relations Record* 5, no. 2 (July

2005): 43–60; John A. Wise, V. David Hopkin, and Daniel J. Garland, eds., *Handbook of Aviation Human Factors*, 2nd ed. (Boca Raton, FL: CRC Press, 2010), 5-24, 10-8.

19. *Top Gun*, film, directed by Tony Scott (Hollywood, CA: Paramount Pictures, 1986); *The Right Stuff*, film, directed by Philip Kaufman (Burbank, CA: Warner Bros., 1983); Tom Wolfe, *The Right Stuff* (New York: Farrar, Straus and Giroux, 1979).

20. Claude Grahame-White, "No Place in the Air for Women," *Mobile (AL) Register*, Sept. 3, 1911.

21. Roger Bilstein, "The Airplane and the American Experience," and Dominick A. Pisano, "The Greatest Show Not on Earth: The Confrontation between Utility and Entertainment in Aviation," both in *The Airplane in American Culture*, ed. Dominick A. Pisano (Ann Arbor: University of Michigan Press, 2003), 16–35, 39–74.

22. Robert Wohl, *The Spectacle of Flight: Aviation and the Western Imagination, 1920–1950* (New Haven, CT: Yale University Press, 2005), 37–38.

23. David T. Courtwright, *Sky as Frontier: Adventure, Aviation, and Empire* (College Station: Texas A&M University Press, 2005), 46–50, 58–59.

24. Dean C. Smith, *By the Seat of My Pants: A Pilot's Progress, from 1917 to 1930* (Boston: Little, Brown, 1961), 139.

25. Walt Ballard, quoted in PBS, *The American Experience: Lindbergh* (Boston: WGBH and Insignia Films, 1990), documentary film transcript, http://www.pbs.org/wgbh/amex/lindbergh/filmmore/transcript/transcript1.html.

26. Charles A. Lindbergh, with William Jovanovich and Judith A. Schiff, eds., *Autobiography of Values* (New York: Harcourt Brace Jovanovich, 1978), 121.

27. *Oxford English Dictionary Online*, "aviator," http://www.oed.com.

28. Corn, *Winged Gospel*, 71–90; Courtwright, *Sky as Frontier*, 5–88.

29. A. Scott Berg, *Lindberg* (New York: J. P. Putnam's Sons, 1998), 278.

30. Corn, *Winged Gospel*, 91–111; William F. Trimble, "Collapse of a Dream," in *From Airships to Airbus: The History of Civil and Commercial Aviation*, vol. 2, *Pioneers and Operations*, ed. William F. Trimble (Washington, DC: Smithsonian Institution Press, 1995), 128–45.

31. Richard Thruelsen, "Fly It Yourself," *Saturday Evening Post*, Dec. 21, 1940, quotes from 20, 52.

32. *Tomorrow's Customers for Aviation* (New York: Crowell-Collier, Aug. 1944), 22–23.

33. *Survey on Aviation: February 1945* (Princeton, NJ: Benson & Benson, 1945), 8.

34. "John H. Geisse Sees Roadable Plane as Big Boon to Private Aviation," *Civil Aeronautics Journal*, Nov. 15, 1945, 122. A 1946 Gallup Poll survey asked the same question to a broader audience. The results suggested some decline in enthusiasm but still reflected widespread interest in learning to fly: 27 percent of all respondents said yes, with men outnumbering women by a small margin. Again, age was the most important factor, with 50 percent of those aged 21 to 29 years answering in the affirmative. George H. Gallup, ed., *The Gallup Poll: Public Opinion, 1935–1971*, vol. 1, *1935–1948* (New York: Random House, 1972), 568.

35. Cessna Aircraft Co., "To the Family That Can't Take a Vacation Trip This Year" [advertisement], *Flying*, July 1943, cover.

36. Cessna Aircraft Co., "To the Golfer Who Used to Drive over 8800 Yards" [advertisement], *Time*, July 5, 1943, 67. The same advertisement appeared one month later in at least one other magazine. See *Flying*, Aug. 1943, inside front cover.

37. Cessna Aircraft Co., "To the Family That Can't Take a Vacation."

38. Based on the author's survey of all Cessna advertisements published in *Flying* magazine between January 1942 and December 1945. This also follows a prewar pattern of industry attempts to make flying "thinkable" to a skeptical public by employing female pilots to prove that "almost anyone" could fly. Corn, *Winged Gospel*, 71–90, quote from 75.

39. *Air Travel Survey, 1998* (Washington, DC: Air Transport Association of America, 1998), III-2, III-8; "Table 7: Resident Population by Sex and Age: 1980 to 2010," in U.S. Department of Commerce, U.S. Census Bureau, *Statistical Abstract of the United States: 2012*, 11, http://www.census.gov/prod/2011pubs/12statab/pop.pdf; "Table 7.1, Estimated Active Pilot Certificates Held: December 31, 1976–1985," in FAA, *Statistical Handbook of Aviation*, 1985 ed., 134, table; "Table 1, Estimated Active Airmen Certificates Held: December 31, 1990–1999," FAA, *U.S. Civil Airmen Statistics* (1999).

40. U.S. adult population (aged 18 and over) from U.S. Census Bureau, "Resident Population—Estimates by Age, Sex, and Race: July 1, 1939," and "Resident Population Plus Armed Forces Overseas—Estimates by Age, Sex, and Race: July 1, 1951," both at http://www.census.gov/popest/data/national/asrh/pre-1980/PE-11.html; data on pilots from "Certificated Civil Aircraft and Airmen, 1927–54," in CAA, *Statistical Handbook of Civil Aviation*, 1955 ed., 27, table.

41. Total active civilian pilots for various years from CAA, *Statistical Handbook of Civil Aviation* (1950); FAA, *Statistical Handbook of Aviation* (1990); U.S. Department of Transportation, FAA, *U.S. Civil Airmen Statistics* (2000). Number of pilots who flew for a living by decade, 1950–2000, from "Airplane Pilots and Navigators" [listed under "Professional, Technical, and Kindred Workers"], "Table 3: Race of the Experienced Civilian Labor Force and of Employed Persons, by Detailed Occupation and Sex, for the United States: 1950," in U.S. Department of Commerce, Bureau of the Census, *Occupational Characteristics*, Special Reports: Report P-E No. 1B, 1950 Census of Population (Washington, DC: U.S. GPO, 1956), 1B-29; "Airplane Pilots and Navigators" [listed under "Professional, Technical, and Kindred Wkrs"], "Table 3: Race of the Experienced Civilian Labor Force and of the Employed, by Detailed Occupation and Sex, for the United States: 1960," in U.S. Department of Commerce, Bureau of the Census, *Occupational Characteristics*, Subject Reports: Final Report PC(2)-7A, 1960 Census of Population (Washington, DC: U.S. GPO, 1963), 21; "Airplane Pilots" [listed under "Technicians, Except Health, and Engineering and Science"], "Table 223: Detailed Occupation of Employed Persons by Race and Sex: 1970," in U.S. Department of Commerce, Bureau of the Census, *Detailed Occupation of Employed Persons by Race and Sex for the United States: 1970*, Supplemental Report PC(S1)-32, 1970 Census of Population (Washington, DC: U.S. GPO, Mar. 1973): 1–740; "Airplane Pilots and Navigators" [listed under "Technicians; Exc(ept) Health, Engineering, & Science"], "Table 1: Detailed Occupation of the Civilian Labor Force by Sex, Race, and Spanish Origin: 1980," U.S. Department of Commerce, Bureau of the Census, *Detailed Occupation and Years of School Completed by Age, for the Civilian Labor Force by Sex, Race, and Spanish Origin: 1980*, Supplementary Report PC80-S1-8, 1980 Census of Population (Washington, DC: U.S. GPO, Mar. 1983), 9; U.S. Department of Commerce, U.S. Census Bureau, "Census '90 Detailed Occupation by Race, Hispanic Origin and Sex," http://censtats.census.gov/eeo/eeo.shtml; U.S. Department of

Commerce, U.S. Census Bureau, "Census 2000 EEO Data Tool," http://www.census.gov/eeo2000/.

42. General aviation fatal accidents per 100,000 hours flown dropped more or less steadily from 3.5 per 100,000 hours flown in 1959 to 1.38 in 1989. However, this statistic is somewhat misleading, since the number of hours flown by professional pilots flying corporate-owned executive transport aircraft increased considerably during this time period, and this type of general aviation has a much better safety record than private flying. According to a National Transportation Safety Board investigator who spoke at a 2012 forum, private aviation flights were 12 times more likely to crash than other kinds of general aviation flights. General aviation accident statistics for 1959–89 from FAA, *Statistical Handbook of Aviation*, 1968, 1972, 1980, and 1990 eds.; comparison of safety record for private aviation to other forms of general aviation from Alan Levin, "Deadly Private-Plane Crashes Prompt U.S. Call for Basics," Bloomberg.com, June 19, 2012, http://www.bloomberg.com/news/2012-06-19/pilots-deadly-private-plane-crashes-prompt-u-s-call-for-basics.html.

43. J. Jefferson Miller, "Diminishing Returns: A Complex Set of Issues with No Easy Answers and Many Adverse Consequences," *AOPA Pilot* 28, no. 4 (Apr. 1985): 38–39, 41–44, 46, 48; Dick McKenzie, "Motion Sickness," *Plane & Pilot*, Feb. 1985, 24, 70.

44. News in the Air, *Aviation Consumer* 10, no. 8 (Apr. 15, 1980): 1.

45. "Cessna Single Engine Production Totals by Model and Year," in John Frank, *Cessna 172 Skyhawk Buyers Guide* (Santa Maria, CA: Cessna Pilots Association, 2009): 26; Cessna Aircraft Co., "Skyhawk: The Best-Selling and Most-Flown Aircraft Ever Built," http://cessna.txtav.com/en/single-engine/skyhawk.

46. Patricia Cayo Sexton, *The Feminized Male: Classrooms, White Collars, and the Decline of Manliness* (New York: Random House, 1969), 15.

47. Janet T. Spence and Camille Buckner, "Masculinity and Femininity: Defining the Undefinable," in *Gender, Power, and Communication in Human Relationships*, ed. Pamela J. Kalbfleisch and Michael J. Cody (Hillsdale, NJ: Lawrence Erlbaum, 1995), 105–38.

48. R. W. Connell, *Masculinities*, 2nd ed. (Berkeley: University of California Press, 2005); R. W. Connell and James W. Messerschmidt, "Hegemonic Masculinity: Rethinking the Concept," *Gender & Society* 19, no. 6 (Dec. 2005): 829–59, quote from 832.

49. Novello and Youssef, "Psycho-social Studies in General Aviation I," 187.

50. Gail Bederman, "Why Study 'Masculinity,' Anyway? Perspectives from the Old Days," *Culture, Society & Masculinities* 3, no. 1 (Spring 2011): 14–16.

51. Ibid., 18.

52. Novello and Youssef, "Psycho-social Studies in General Aviation II," 632–33.

53. Connell and Messerschmidt, "Hegemonic Masculinity," 829–59, quote from 837.

54. Examples include Janet R. Daly Bednarek, with Michael H. Bednarek, *Dreams of Flight: General Aviation in the United States* (College Station: Texas A&M University Press, 2003); Donald M. Pattillo, *A History in the Making: 80 Turbulent Years in the American General Aviation Industry* (New York: McGraw-Hill, 1998); Roger E. Bilstein, *Flight in America: From the Wrights to the Astronauts*, 3rd ed. (Baltimore: Johns Hopkins University Press, 2001), 191–92, 195–203, 239–44, 257, 292–98, 356–60.

55. David N. Lucsko, *The Business of Speed: The Hot Rod Industry in America, 1915–1990* (Baltimore: Johns Hopkins University Press, 2008), 8; Robert C. Post, *High Performance: The Culture and Technology of Drag Racing, 1950–2000*, rev. ed. (Baltimore: Johns Hopkins University Press, 2001).

56. Like private pilots, most Americans who built and operated amateur two-way "ham radio" stations in their homes were white middle-class men. Indeed, male "hams," as they called themselves, outnumbered women in the hobby by 19 to 1 in the postwar era, a ratio remarkably similar to that found in aviation. Despite these demographic similarities, however, the "technical culture" of ham radio was somewhat different from that of private aviation. Although both ham operators and pilots were licensed by the federal government, most hams taught themselves the skills needed to pass the exam, whereas all fliers were required to complete standardized training with a certified flight instructor before taking their private pilot check ride with a government-designated examiner. As a matter of pride, most hams built their own equipment from kits rather than purchase ready-to-use radios, while most postwar pilots flew factory-built, government-certificated aircraft (there are exceptions in aviation; chapter 6 addresses the nascent homebuilt or experimental aircraft movement of the early postwar years, and the conclusion briefly describes how it gained momentum after 1985). Finally, aside from falling from a roof when working on an antenna or the occasional (and seldom fatal) electric shock, ham radio operators rarely risked physical injury in pursuit of their pastime but instead endured ridicule or disdain from family, friends, neighbors, and co-workers who could not understand the appeal of the activity. Pilots, on the other hand, made decisions with very real life-or-death consequences every time they flew, and once safely on the ground, they generally enjoyed a certain degree of the prestige and mystique that the American public had long accorded to fliers. Kristen Haring, *Ham Radio's Technical Culture* (Cambridge, MA: MIT Press, 2007), xii, xvii, 8, 36, 49–73.

57. Technological enthusiasm as an "ends" to justify an expensive and time-consuming pastime carries more explanatory power than previous attempts to understand why people engage in hobbies. For instance, historian Stephen Gelber's argument that hobbies reinforce the capitalistic work ethic holds up well with the examples of "productive leisure" he marshals in its defense, such as collecting objects (stamps, coins, etc.), woodworking, and "household do-it-yourself" projects. However, his thesis is less persuasive when applied to other activities deliberately excluded from his study but that many participants insist are hobbies, including mountaineering (which he specifically mentions in his introduction), as well as other leisure activities such as hunting, skiing, and private flying. Stephen M. Gelber, *Hobbies: Leisure and the Culture of Work in America* (New York: Columbia University Press, 1999), x, 6, 8–9, references to mountaineering on 8–9, 17.

58. Joan Wallach Scott, *Gender and the Politics of History*, rev. ed. (New York: Columbia University Press, 1999), 41–50.

59. Nina E. Lerman, Arwen Palmer Mohun, and Ruth Oldenziel, "Versatile Tools: Gender Analysis and the History of Technology" and "The Shoulders We Stand On and the View from Here: Historiography and Directions for Research," *Technology and Culture* 38, no. 1 (Jan. 1997): 1–8, 9–30, quote from 30.

Chapter 1 · Who Is "Mr. General Aviation"?

1. Lew Townsend, "AOPA Members Cast Large Shadow," *AOPA Pilot* 12, no. 5 (May 1969): 94–97, quotes and graph from 94. Townsend admitted in the article that only about one-fifth of active U.S. pilots belonged to the AOPA and that roughly 47,000 individuals—slightly less than one-third of the organization's roughly 150,000 members—actually completed the survey. Thus, the AOPA's 1969 Pilot Poll represents about 7 percent of all active U.S. civilian pilots at the time. This was not the first time that the AOPA had surveyed its members, but past polls concentrated on what type of flying they did (business vs. pleasure) and revealed little about members' socioeconomic status, educational background, and professional lives outside of aviation. See, for instance, "AOPA and General Aviation," editorial, *AOPA Pilot* 3, no. 9 (Sept. 1960): 28.

2. Townsend, "AOPA Members Cast Large Shadow," 94–95, 97. Information about U.S. household income and assets for 1968 from "Table H-1: Income Limits for Each Fifth and Top 5 Percent of All Households: 1967 to 2010," in U.S. Department of Commerce, U.S. Census Bureau, Historical Income Tables: Income Inequalities, http://www.census.gov/hhes/www/income/data/historical/inequality/index.html; "[Table] No. 648, Family Units, by Liquid Asset Holdings: 1968 to 1971," in U.S. Department of Commerce, Bureau of the Census, *Statistical Abstract of the United States: 1974*, 95th ed. (Washington, DC: U.S. GPO, 1974), 397, table.

3. Townsend, "AOPA Members Cast Large Shadow," 95. According to a 1969 survey by the U.S. Department of Transportation, 48 percent of U.S. households owned one car, 26 percent owned two, 5 percent had three or more cars, and the remaining 21 percent did not own a car. At 53 percent, AOPA members enjoyed twice the national rate of two-car-ownership. U.S. Department of Transportation, Office of Highway Policy Information, Federal Highway Administration, "Introduction to the 1969 Nationwide Personal Transportation Survey," Feb. 4, 1999, http://www.fhwa.dot.gov/ohim/1969/back.htm.

4. "Table 7.5—Women Actively Engaged in Aviation: December 31, 1963–71," and "Table 7.11—Active Pilots by Type of Certificate [as of January 1, 1970]," in U.S. Department of Transportation, FAA, *Statistical Handbook of Aviation*, 1972 ed. (Washington, DC: U.S. GPO, 1972), 168, 177, tables.

5. For a concise, firsthand retrospective on prewar private flying, see Alice S. Fuchs, "U.S. Civil Aviation 1939," *AOPA Pilot* 7, no. 5 (May 1964): 60–62. For descriptions of how pilot-hero worship affected an entire generation of young Americans, especially boys, during the interwar years, see "Those Missing Kids," editorial, *AOPA Pilot* (July 1953): 50a; Donald S. Lopez, *Into the Teeth of the Tiger* (New York: Bantam Books, 1986), 9–18. For a broader scholarly analysis of how the American public viewed famous fliers of the interwar era, see Joseph J. Corn, *The Winged Gospel: America's Romance with Aviation, 1900–1950* (New York: Oxford University Press, 1983); A. Bowdoin Van Riper, *Imagining Flight: Aviation and Popular Culture* (College Station: Texas A&M University Press, 2004).

6. Janet R. Daly Bednarek, "The Flying Machine in the Garden: Parks and Airports, 1918–1938," *Technology and Culture* 46, no. 2 (Apr. 2005): 354–55, 367–68, quote from 372.

7. U.S. adult population (aged 18 and older) from U.S. Census Bureau, "Resident Population—Estimates by Age, Sex, and Race: July 1, 1939," and "Resident Population Plus Armed Forces Overseas—Estimates by Age, Sex, and Race: July 1, 1951," both at http://www.census.gov/popest/data/national/asrh/pre-1980/PE-11.html; data on pilots from "Certificated Civil Aircraft and Airmen, 1927–54," in U.S. Department of Commerce, CAA, *Statistical Handbook of Civil Aviation*, 1955 ed. (Washington, DC: U.S. GPO, 1955), 27, table.

8. For contemporary descriptions of the dream of universal postwar private flying, see Historical Service Board, American Historical Association, *Will There Be a Plane in Every Garage?*, GI Roundtable Series EM 37 (Washington, DC: U.S. GPO, Aug. 1945), Internet Archive, http://www.archive.org/details/WillThereBeAplaneInEveryGarage; Daniel R. Zuck, *An Airplane in Every Garage* (New York: Vintage Press, 1958). For historical analysis of this phenomenon, see Dominick A. Pisano, *To Fill the Skies with Pilots: The Civilian Pilot Training Program, 1939–1946* (Washington, DC: Smithsonian Institution Press, 2001), 131–37; William F. Trimble, "Collapse of a Dream," in *From Airships to Airbus: The History of Civil and Commercial Aviation*, vol. 2, *Pioneers and Operations*, ed. William F. Trimble (Washington, DC: Smithsonian Institution Press, 1995), 128–45.

9. Combined, women and black men made up just over 1 percent of the program's graduates. Pisano, *To Fill the Skies with Pilots*, 76–77, 131.

10. SMSgt. Timothy Tellgren, "A Brief History of Air Education and Training Command," U.S. Air Force, Air Education and Training Command, Office of History and Research, Randolph Air Force Base, TX, ca. 2002, http://www.aetc.af.mil/library/history/historyofaetc.asp; "Appendix 1: The History of Naval Aviator and Naval Aviation Pilot Designations and Numbers, the Training of Naval Aviators and the Number Trained (Designated)," in *United States Naval Aviation, 1910–1995*, Department of the Navy, Naval History and Heritage Command, Naval Aviation History Office, ca. 1995, 414, http://www.history.navy.mil/avh-1910/APP01.PDF.

11. Leaders of the U.S. Army Air Forces resisted any wartime programs that threatened to undermine their long-term efforts to establish an independent air force after the war. This may account for their steadfast insistence that the CPTP failed to adequately prepare pilot candidates for military flight school. The navy apparently harbored no such reservations and accepted many CPTP graduates into its own flight training programs, but it trained far fewer total pilots than the army. Pisano, *To Fill the Skies with Pilots*, 77–82, 86–93, 130–31.

12. "According to FAA, over 787,000 pilots were trained by the armed forces during World War II." Jack Wilson, "'Old Rip' Flies Again," *AOPA Pilot* 10, no. 5 (May 1967): 35. Unfortunately, the author did not cite the specific source for this number. It is likely that this figure, if correct, included all aviators trained by the military services plus the CPTP, which, as noted earlier, was essentially militarized after the United States entered the war, even though many CPTP graduates did not go on to actually fly for the U.S. military.

13. "Your License," *AOPA Pilot* (June 1944): 66d; "For Military Pilots," *AOPA Pilot* (Feb. 1945): 62b–62c; "Servicemen Require Civilian Medical," *AOPA Pilot* (Aug. 1945): 64b; U.S. Department of Commerce, Program Planning Staff, Office of the Administrator of Civil Aeronautics, *Staff Study: The Outlook for Private Flying, 1950–1955* (Washington, DC: U.S. GPO, Mar. 1951), 19.

14. Douglas D. Bond, M.D., *The Love and Fear of Flying* (New York: International Universities Press, 1952), 9–13; Frank Coviello, "Sold on Flying—AGAIN," *AOPA Pilot* 9, no. 4 (Apr. 1966): both quotes from 58; "Insurance Company Pamphlet 'Briefs' Ex-Military Pilots," *AOPA Pilot* (Apr. 1946): 54b; Janet R. Daly Bednarek with Michael H. Bednarek, *Dreams of Flight: General Aviation in the United States* (College Station: Texas A&M University Press, 2003), 89.

15. Donn Munson, "Report from a Retread Tiger," *AOPA Pilot* 5, no. 11 (Nov. 1962): 34; Wilson, " 'Old Rip' Flies Again," 35, 37, quotes from 35. For contemporary sources that indicate that former military pilots were a significant presence in civil aviation during the immediate postwar years, see Howard Mingos, ed., *The Aircraft Yearbook for 1947* (New York: Lanciar Publishers, 1947), 113; *Outlook for Private Flying*, 19.

16. U.S. Department of Veterans Affairs, "Born of Controversy: The GI Bill of Rights," http://www.gibill.va.gov/gi_bill_info/history.htm; Servicemen's Readjustment Act of 1944, Public Law 345, Chapter 268, 78th Cong., 2nd sess. Note that many recent secondary sources, including those published by the federal government, anachronistically employ the gender-neutral term *servicemember* (in place of *servicemen*) when citing the title of the original bill.

17. Estimated expenditure from "Survival in the Air Age: A Report by the President's Air Policy Commission, 1947" (Washington, DC: U.S. GPO, Jan. 1, 1948), 125; quote from U.S. Department of Commerce, CAA Research Division and Office of Aviation Information, *Aircraft Use in 1947* (Washington, DC: U.S. GPO, Feb. 1949), 11. According to figures published one year later by the CAA, a total of 345,149 veterans had applied their GI Bill benefits toward flight training as of November 1, 1947 (339,300 at for-profit and 5,849 at nonprofit schools). Of those who had signed up for lessons at for-profit schools, 107,200 had completed their training, 118,400 were still enrolled, and 113,700 had dropped out. "Veterans Enrolled in Veteran's Administration Flight Training Program under Public Law 346," in CAA, *Statistical Handbook of Civil Aviation*, 1948 ed. (Washington, DC: U.S. GPO, 1948), 33, table.

18. The number of government-approved civilian flight schools rose and fell in the immediate postwar era in direct relationship to the status of GI Bill flight training. From 403 such schools at the start of 1946, this increased steadily to a high point of 3,287 in July 1948. That summer, Congress passed a bill that significantly curtailed GI Bill flight training, and by June 1950 there were only 2,237 flight schools still in operation. *Outlook for Private Flying*, 24.

19. Harry Burnett, "Spins," VOX Pilot, *AOPA Pilot* (Dec. 1948): 58h; "Gosh, It's Grandma," *AOPA Pilot* (Sept. 1946): 50g; "Hot Pilot," *AOPA Pilot* (Jan. 1949): 50d. In addition to sources cited in previous notes, a 1950 government report further reinforces arguments that most new student pilots in the mid- to late 1940s funded their lessons with the GI Bill. Specifically, the report explains: "Indications are that flight training activities . . . declined drastically after the Veterans' [*sic*] Administration tightened up the eligibility requirements for G.I. training in July 1948." U.S. Department of Commerce, CAA Aviation Statistics Division, *Aircraft Use in 1948* (Washington, DC: U.S. GPO, Feb. 1950), 4. Pilots in the field reported the same thing in anecdotal observations, often in the form of complaints by non–GI Bill recipients who claimed that they had "learned to fly the hard way" and resented "the free flying lessons handed out to

many who are not particularly interested in flying." Harold S. Cover, "Something for Nothing," VOX Pilot, *AOPA Pilot* (Oct. 1948): 50h.

20. Pilot ages from Zuck, *An Airplane in Every Garage*, 10, table I-A. Census Bureau data indicate that 70 percent of male World War II veterans were younger than 29 when the war ended in 1945, and that 94 percent were under the age of 39. "Table No. 251, Male Veterans of World War II by Age, Marital Status, and Number of Own Children under 18 Years Old: April 1951," in U.S. Department of Commerce, Bureau of the Census, "Chapter 8: Labor Force, Employment, and Earnings," *Statistical Abstract of the United States: 1952*, 73rd ed. (Washington, DC: U.S. GPO, 1952), 205.

21. The CAA first published data on pilots' ages beginning with calendar year 1954. Data on pilot ages for 1954 through 1958 from tables titled "Airmen by Categories in Each Age Group," in CAA, *Statistical Handbook of Civil Aviation* (Washington, DC: U.S. GPO); data for 1959 through 1996 from similar tables in FAA, *Statistical Handbook of Aviation* (Washington, DC: U.S. GPO).

22. Coviello, "Sold on Flying—AGAIN," 58.

23. Munson, "Report from a Retread Tiger," 34–35, 63–65; Wilson, "'Old Rip' Flies Again," 35–37.

24. Pisano, *To Fill the Skies with Pilots*, 76–77.

25. Deborah G. Douglas, *American Women and Flight since 1940* (Lexington: University Press of Kentucky, 2004), 97–103, 192–94; General H. H. Arnold, "To Each Member of the WASP," official letter from Headquarters of the Army Air Forces, Washington, DC, Oct. 1, 1944, reproduced in Gail M Gutierrez, *Forgotten Wings: Oral History of Women Air Service Pilots, the WASPs* (Fullerton: California State University Oral History Program, 1992), 333.

26. [WISPS], Women in the News, *Skylady*, Sept.–Oct. 1946, 9. Note that WASP is commonly misspelled "WASPS" or "WASPs" in popular literature. Because "Women Airforce Service Pilots" is already plural, adding an "s" to the acronym WASP is redundant.

27. For instance, a young flight instructor described in 2005 how the autobiography of former WASP Margaret J. Ringenberg, *Girls Can't Be Pilots*, had inspired and sustained her through her own sometimes frustrating efforts to break into the male-dominated field of aviation. From Keri T. Wiznerowicz, interviewed by the author at the EAA Headquarters, Oshkosh, WI, July 31, 2005, audio recording in author's collection; Margaret J. Ringenberg with Jane L. Roth, *Girls Can't Be Pilots* (Fort Wayne, IN: Daedalus Press, 1998).

28. Douglas, *American Women and Flight*, 91–92. At least one young woman with no prior flight experience was inspired to learn how to fly during the war in order to qualify for the WASP program (she succeeded), but she appears to be a rare exception to the rule that most applicants held at least a private pilot's license before the United States entered World War II. Melinda Pradarelli, "Suzanne Parish: In the Pink," *Aviation for Women*, Sept.–Oct. 1999, 30–33, obtained via WASP on the Web, http://www.wingsacrossamerica.us/wasp/resources/parish.pdf.

29. Evelyn Monahan and Rosemary Neidel-Greenlee, *A Few Good Women: America's Military Women from World War I to the Wars in Iraq and Afghanistan* (New York: Anchor Books, 2011), 64; L.S. Willis, "Flight Training and the G.I. Bill," *Flying*, Aug. 1946, 17–18; "Born of Controversy: The GI Bill of Rights."

30. Data for women pilots from Howard Mingos, *Aircraft Yearbook* (New York: Aeronautical Chamber of Commerce of America), 460 (1940 ed.), 660 (1942 ed.); Mingos, *Aircraft Yearbook* (New York: Lanciar Publishers), 232 (1946 ed.); "Women in Aviation," in CAA, *Statistical Handbook of Civil Aviation*, 1944 ed., 80, table. Data on all pilots (men and women) for this period from "Certificated Civil Aircraft and Airmen, 1927–54," in CAA, *Statistical Handbook of Civil Aviation*, 1955 ed., 27, table.

31. J. Todd Moye, *Freedom Flyers: The Tuskegee Airmen of World War II* (New York: Oxford University Press, 2010), 99–104.

32. Many secondary sources quote different figures for the total number of black pilots in the United States in 1940. Some apparently rely on a document published by the U.S. Bureau of the Census in September 1940, which lists 231 total black pilots, without acknowledging that the Bureau amended this number to 269 in an update released on September 20, 1940. Regardless of which figure one uses, the total number was minuscule in comparison to the overall pilot population. U.S. Department of Commerce, U.S. Bureau of the Census, "Negro Aviators (Supplemental List)," appended to "Negro Aviators," *Negro Statistical Bulletin* No. 3 (Washington, DC: U.S. Government, Sept. 1940), Connecticut State Library Digital Collections (for the amended number of black pilots, see the first page of the "Supplemental List," which is found on page 23 of the complete downloaded document), http://cslib.cdmhost.com/cdm/singleitem/collection/p4005coll11/id/511/rec/6.

33. U.S. Department of Commerce, Bureau of the Census, "Table II—Race and Nativity, by Sex, for the United States: 1940," in *Sixteenth Census of the United States—1940—Population, Volume II: Characteristics of the Population-Part 1* (Washington, DC: U.S. GPO, 1943), 9.

34. Pisano, *To Fill the Skies with Pilots*, 76.

35. About half of the roughly 1,000 black pilots who earned their wings in the U.S. Army Air Forces during World War II served in combat. Moye, *Freedom Flyers*, 12, 39.

36. Aside from the 53,000-plus Puerto Ricans who served between 1940 and 1946, the U.S. government did not track the number of Hispanic Americans in the military during World War II. Office of the Deputy Assistant Secretary of Defense for Military Manpower and Personnel Policy, *Hispanics in America's Defense* (Washington, DC: U.S. GPO, 1990), 27.

37. Lopez retired from the U.S. Air Force as a lieutenant colonel, then went on to serve as an engineer for the U.S. space program and later as director of the Smithsonian's National Air and Space Museum. Lopez, *Into the Teeth of the Tiger*, especially 15–18; U.S. Department of Transportation, FAA, William J. Hughes Technical Center, "Nuestra Herencia," *Intercom* (Sept. 1999): 3, http://www.tc.faa.gov/intercom/sep99a.PDF; Adam Bernstein, "Donald Lopez; Fighter Ace, Museum Official," obituaries, *Washington Post*, Mar. 5, 2008, http://www.washingtonpost.com/wp-dyn/content/article/2008/03/04/AR2008030402792.html.

38. *Hispanics in America's Defense*, 45.

39. Lawrence Gonzales, *Deep Survival: Who Lives, Who Dies, and Why* (New York: W. W. Norton, 2003), 9–12; Henry "Hank" Cervantes, oral history interview with Maggie Rivas-Rodríguez and Bruce Ashcroft, Washington, DC, May 30, 2004, as quoted in Carry-Anne Olsen, "Henry 'Hank' Cervantes," written summary of

interview, n.d., University of Texas Libraries VOCES Oral History Project, http://www.lib.utexas.edu/voces/template-stories-indiv.html?work_urn=urn%3Autlol%3Awwlatin.456&work_title=Cervantes%2C+Enrique%22Hank%22.

40. Bill Robinet, *By the Skin of My Teeth: A Cropduster's Story* (Veneta, OR: Billville Press, 1997), 59 (Vercellino is misspelled "Vercillino" in this source); Gary K. Nelson, Arizona State Attorney General, Opinion Letter 68-024-L to James Vercellino, Director, Arizona Department of Aeronautics, Sept. 24, 1968, Arizona State Library, Archives and Public Records, http://azmemory.azlibrary.gov/cdm/singleitem/collection/agopinions/id/10970/rec/1.

41. Keith Connes, "Can a Black Man Fly?" *Flying*, July 1969, 54–55.

42. "Roy Comeaux: Summary of a 1989 Oral History by Debbi Jones; 1996," In the Steps of Esteban: Tucson's African American Heritage, http://parentseyes.arizona.edu/esteban/bios_otheroccupations_comeaux.html.

43. U.S. Department of Commerce, Bureau of the Census, "Airplane Pilots and Navigators" (listed under "Professional, Technical, and Kindred Workers"), "Table 3. Race of the Experienced Civilian Labor Force and of Employed Persons, by Detailed Occupation and Sex, for the United States: 1950," in *Occupational Characteristics*, Special Reports: Report P-E No. 1B, 1950 Census of Population (Washington, DC: U.S. GPO, 1956), 1B-29.

44. Cessna Aircraft Co., "Cessna's New Program of Flight Training Starts with a New Kind of Instructor" [advertisement], *Flying*, Jan. 1971, inside front cover. The same advertisement appeared elsewhere, including *AOPA Pilot* 13, no. 12 (Dec. 1970): 79.

45. Connes, "Can a Black Man Fly?" 53–57; "Black Pilots," Flying Mail, *Flying*, Sept. 1969, 7, 86; "Black Pilots," Flying Mail, *Flying*, Oct. 1969, 6–7. In the September issue of *Flying*, the editor published one explicitly hostile letter (from "signature illegible") and four positive reader responses. The next month, the editor published seven more, and noted "With the exception of the two letters from the gentleman in Tallulah, Louisiana [submitted by Tommy E. Wixson and Merle B. Gustafson], reader response to 'Can a Black Man Fly?' was heavy and favorable." This assessment chose to ignore a third letter to the editor that month (from Thomas E. Baur) that can be read as either neutral or negative.

46. U.S. Department of Commerce, U.S. Census Bureau, "Census '90 Detailed Occupation by Race, Hispanic Origin and Sex," http://censtats.census.gov/eeo/eeo.shtml; U.S. Department of Commerce, U.S. Census Bureau, "Census 2000 EEO Data Tool," http://www.census.gov/eeo2000/.

47. Derald Wing Sue et al., "Racial Microaggressions in Everyday Life: Implications for Clinical Practice," *American Psychologist* 62, no. 4 (May–June 2007): 271–86, quote from 271.

48. Connes, "Can a Black Man Fly?" 56–57.

49. Interview with Roosevelt Lewis by the author at the Tuskegee Memorial Day Fly-in, Moton Field, Tuskegee, AL, May 28, 2011, notes in author's files.

50. Jesse Lee Brown and Rodney Brown, interviewed by the author at the EAA Headquarters, Oshkosh, WI, July 29, 2005, audio recording in author's collection.

51. Ibid.

52. Chuck Yeager and Leo Janos, *Yeager: An Autobiography* (New York: Bantam Books, 1985), 3–10.

53. Before World War II, a civilian pilot certificate (better known as a pilot's license) expired in the same manner as a modern-day driver's license. Individuals who neglected to renew their certificate or failed to meet medical or flight proficiency standards were stricken from the rolls of active fliers. During World War II, the Civil Air Regulations (CAR) were changed so that pilot certificates (with the exception of those issued to student pilots) no longer expired. Instead, the CAA now relied on a separate medical certificate, which had to be renewed periodically via a physical exam, to determine whether a licensed pilot could fly. However, instead of estimating the number of active pilots by counting current (unexpired) medical certificates, the CAA continued its old practice of counting unexpired pilot certificates. Since pilot certificates no longer expired, this led to increasingly inflated annual figures that failed to account for individuals who had given up flying since the rule change. By the mid-1940s, the CAA acknowledged this problem even as it failed to take corrective action. For instance, a 1945 table titled "Certification Summary, 1927–45" included a footnote that cautioned readers: "The count of certificated pilots after 1941 is not directly comparable with the previous years as the CAR were amended to permit pilot certificates currently effective on April 1, 1942, to continue in effect indefinitely." It took the federal government until 1952—an entire decade after the rule change—to start using current medical certificates (which *did* expire periodically) to track the population of active pilots. On the one hand, this issue makes the boom-and-bust period of the immediate postwar era seems even more dramatic than it truly was, since the CAA's estimate of the total pilot population became increasingly inflated as time went on, and part of the sudden drop in pilot population in the early 1950s reflects the switch to new—and more accurate—accounting practices. However, this is not to say that private aviation did not experience a very real downturn starting in the late 1940s. For instance, the CAA issued 193,000 new student certificates in 1947 at the height of postwar GI Bill flight training. By 1951, the number of new student certificates issued in the past year had plummeted to 45,000, indicating that the number of people heading to the airport to sign up for flying lessons had dropped by more than 75 percent. Civil Aeronautics Authority, *Civil Air Regulations Part 20: Pilot Certificates*, effective May 1, 1940 (Washington, DC: U.S. GPO, 1940), 10–11; Civil Aeronautics Board, *Civil Air Regulations Amendment 20-37* (elimination of periodic expiration of pilot certificates), effective Apr. 1, 1942 (Washington, DC: CAB, 1942), 1; quote describing effects of this rule change from "Certification Summary, 1927–45," in CAA, *Statistical Handbook of Civil Aviation*, 1945 ed., 107, table; statistics for 1945–51 from "Certificated Civil Aircraft and Airmen, 1927–54," in CAA, *Statistical Handbook of Civil Aviation*, 1955 ed., 27, table; statistics for 1956 from "Active Airmen Certificates Held—1953–61, as of Jan 1 each year," in FAA, *Statistical Handbook of Aviation*, 1961 ed., 43, table.

54. *Outlook for Private Flying*, 28–29.

55. Several historians and present-day aviation industry observers agree that little had changed regarding the practical and economic shortfalls of private flying in the half century since World War II ended. Trimble, "Collapse of a Dream," 141–42; Tom D. Crouch, "An Airplane for Everyman: The Department of Commerce and the Light Airplane Industry, 1933–37," in *Innovation and the Development of Flight*, ed. Roger D. Launius (College Station: Texas A&M University Press, 1999), 167; Bill Sweetman, "I Fly," *Popular Science*, June 2000, 54.

56. "Table No. 251, Male Veterans of World War II by Age, Marital Status, and Number of Own Children under 18 Years Old," in *Statistical Abstract of the United States: 1952*, 205; Zuck, *An Airplane in Every Garage*, 32.

57. "Things *Aren't* Tough All Over," editorial, *AOPA Pilot* (Oct. 1952): 50a.

58. AOPA, "Wanted: Do You Recognize This Man?" [advertisement], *AOPA Pilot* 7, no. 8 (Aug. 1964): 46.

59. Jim Chanault, "Student Pilots on the GI Bill," transcript of oral history interview, n.d., Wessels Living History Farm, York, NE, "Farming in the 1940s," http://www.livinghistoryfarm.org/farminginthe40s/life_20.html.

60. "GI Flight Training," *AOPA Pilot* (Aug. 1948): 50d; *Outlook for Private Flying*, 19–22, quotes from 21.

61. Pisano, *To Fill the Skies with Pilots*, 131–32.

62. This figure does not include FIP graduates. Unlike the CPTP during World War II, many new private pilots produced by the FIP went on to complete military flight training, so adding them to the total number of U.S. Air Force flight school graduates would amount to counting them twice. "Appendix B: Training Production," in Thomas A. Manning et al., *History of Air Education and Air Training Command: 1942–2002* (Randolph Air Force Base, TX: Office of History and Research Headquarters, Air Education and Training Command. 2005), 335–38, table.

63. Based on total number of naval aviators trained between 1946 and 1985. *United States Naval Aviation, 1910–1995*, 414.

64. Ann Krueger Hussey, "Air Force Flight Screening: Evolutionary Changes, 1917–2003," U.S. Air Force, Air Education and Training Command, Office of History and Research, Randolph Air Force Base, TX, Dec. 2004, 25–27, http://www.aetc.af.mil/shared/media/document/AFD-061109-020.pdf.

65. After the U.S. Air Force split from the U.S. Army in 1947 to become its own separate service, the army shifted its focus to helicopters and trained just a handful of fixed-wing (airplane) pilots for liaison, battlefield reconnaissance, and executive transport duties over the next several decades. Although the army would train thousands of helicopter pilots throughout the postwar era, there is little indication that many of these chopper pilots bothered to complete the additional training—at their own expense—needed to obtain their civilian airplane rating. Student pilot figures from U.S. Department of Commerce, CAA, *Statistical Handbook of Civil Aviation*, 1944–58 eds., FAA, *Statistical Handbook of Aviation*, 1959–85 eds.

66. Frank A. Burnham, *Hero Next Door* (Fallbrook, CA: Aero Publishers, 1974).

67. "CAP Goes Aloft," *Flying*, Aug. 1965, 86; Civil Air Patrol, *Civil Air Patrol Annual Report to Congress: 1965* (Ellington Air Force Base, Houston: Civil Air Patrol, 1966), 5, 13; Norm Abell, "The Old/Young CAP," *Plane & Pilot*, Oct. 1971, 45–46; Civil Air Patrol, *Civil Air Patrol Annual Report to Congress: 1972* (Maxwell Air Force Base, Montgomery, AL: Civil Air Patrol, 1972) 8, 29; Civil Air Patrol, *Civil Air Patrol Annual Report to Congress: 1976* (Maxwell Air Force Base, Montgomery, AL: Civil Air Patrol, 1976), 2, 13, 24; Civil Air Patrol, *Civil Air Patrol Annual Report to Congress: 1986* (Maxwell Air Force Base, AL: Civil Air Patrol, 1986), 7, 8, 18, 23, 40.

68. Per capita participation in private flying reached an all-time high in 1951 when the CAA counted 1 out of 183 adult Americans as pilots. In 1960, with civil aviation on the mend after bottoming out in the mid-1950s, the ratio was only 1 in 334. By 1970 the

industry had recovered, and the ratio of American adults who were actively flying had rebounded to 1 in 185. But then private aviation entered a long period of decline: 1 in 197 Americans were pilots in 1980; dropping to 1 in 263 by 1990; 1 in 334 in 2000; and only 1 in 374 by 2010. Information about the effects of economic trends on flying from "General Aviation in 1974: No Year of Tears," *AOPA Pilot* 17, no. 3 (Mar. 1975): 28. Ratios derived from statistics in CAA, *Statistical Handbook of Civil Aviation*, 1944–58 eds.; FAA, *Statistical Handbook of Aviation*, 1959–96 eds.; U.S. Department of Transportation, FAA, *U.S. Civil Airmen Statistics* (annual tables for 1999–2013), http://www.faa.gov/data_research/aviation_data_statistics/civil_airmen_statistics/; U.S. Census Bureau, "Resident Population Plus Armed Forces Overseas—Estimates by Age, Sex, and Race: July 1, 1951"; Resident Population plus Armed Forces Overseas—Estimates by Age, Sex, and Race: July 1, 1960 and July 1, 1970, both at http://www.census.gov/popest/data/national/asrh/pre-1980/PE-11.html; "Table 7—Resident Population by Sex and Age: 1980 to 2010," in U.S. Department of Commerce, U.S. Census Bureau, "Section 1: Population," *Statistical Abstract of the United States: 2012*, 11, http://www.census.gov/compendia/statab/2012/tables/12s0007.pdf.

69. "Who Pays? You!" editorial, *AOPA Pilot* 19, no. 10 (Oct. 1976): 6; Patrick Bradley, "Crashes and the Court: Product Liability's Wide-Ranging Effects on Airplanes, the Aviation Industry and the Cost of Flight," *Flying*, Nov. 1984, 51–52, 54; Steven Thomson, "The Great American Pilot Shortage," *Air & Space Smithsonian*, Oct.–Nov. 1986, 62–63.

70. M. J. Walker, "The Bumpy Road of a Free-Lance Instructor: Or How You Can Teach Flying and Still Exist," *AOPA Pilot* 9, no. 5 (May 1966): 62.

71. Arnold L. Winter, "Good Old Days," VOX Pilot, *AOPA Pilot* 1, no. 9 (Nov. 1958): 10.

72. Thomas Caldwell, "Where Are the Kids?" VOX Pilot, *AOPA Pilot* 8, no. 4 (Apr. 1965): 20–21; Barry J. Sullivan, "Us Kids," VOX Pilot, *AOPA Pilot* 8, no. 9 (Sept. 1965): 20, 22; Len Morgan, "Small Fry," Vectors, *Flying*, Oct. 1984, 10; Paul Jennings, "Dire Straits . . . ," Flying Mail, *Flying*, Dec. 1984, 2.

73. Richard [L.] Collins, "Better Side to the 'Bottom End,'" editorial, *Flying*, Mar. 1981, 34; Danny S. Wright, "Ready to Eat, Sleep, and Drink Aviation . . . ," Flying Mail, *Flying*, June 1981, 4.

74. Estimated cost of learning to fly between 1950 and 1985 assumes 25 hours of dual instruction (aircraft rental combined with instructor's fee), plus 25 additional hours of solo flight (aircraft rental only), but does not include ground school, books, or examination fees. In cases where more than one source is available for a particular year, figures for that year are averaged. Cost information for various years from *Outlook for Private Flying*, 47; Leslie [Les] and Helen Greene, interviewed by the author at the Greene residence, Three Rivers, MI, Aug. 3, 2005, audio recording in author's collection; Harland Wilson, "Answers for Armchair Pilots," *Flying*, July 1951, 54; Frank Kingston Smith, *Week-End Pilot* (New York: Random House, 1957), 182; U.S. Dept. of Commerce, CAA Office of Planning, Research and Development, *The Airplane at Work for Business and Industry: A 1955 Survey, Covering 1954 Activity* (Washington, DC: U.S. GPO, Dec. 1955), 23; Donald Chase, "For Sale—Pilot Training Only," *AOPA Pilot* 5, no. 1 (Jan. 1962): 38; Wayne Thoms, *Flying for Fun or Business* (New York: Arco, 1967), 9–10; William D. Price, "G.I. Bill Recommendation," VOX Pilot, *AOPA Pilot* 10, no. 4

(Apr. 1967): 18; Chester Peterson Jr., "Financial Advice for Fledgling Pilots," *AOPA Pilot* 10, no. 6 (June 1967): 76; George E. Burns, "Put Yourself in the Pilot's Seat," Travel, *New York Times*, Apr. 10, 1977, sec. 10, pp. 1, 18; "Aircraft Rental Rates (per hour) Available to Members of the Stanford Flying Club," in "Learn to Fly with the Stanford Aviation Association," advertising mailer/newspaper insert, "Stanford Flying Club," Jan. 1978, quote from 1, table from 2 (author's collection); Walter J. Boyne, *Flying: An Introduction to Flight, Airplanes, and Aviation Careers* (Englewood Cliffs, NJ: Prentice-Hall, 1980), 92; Gordon Baxter, "First Encounter with Flight," *Flying*, July 1980, 67; B. R. Cullumber, "Part 141 or Part 61: Which Flight School?" *Plane & Pilot*, Nov. 1980, 34; "Getting the Most from Your Flight Instruction Dollar," *Aviation Consumer*, Mar. 1, 1980, 11–12; author's personal experience learning to fly with Kalamazoo High Flyers in Kalamazoo, Michigan, during July–Aug. 1984; Jennings, "Dire Straits . . . ," 2; Cessna Aircraft Co. [ad], "It Took You a Year to Learn to Walk. In Six Months, You Can Learn to Fly," *Flying*, July 1984, 10–11; Ted Spitzmiller, "Who Needs an IFR Rating?" Jobs & Schools, *Plane & Pilot*, Sept. 1984, 56. Costs adjusted to 1950 dollars using U.S. Department of Labor, Bureau of Labor Statistics, "CPI [Consumer Price Index] Inflation Calculator," http://www.bls.gov/data/inflation_calculator.htm.

75. Average pretax household income for 1950 through 1985 from U.S. Department of Labor, U.S. Bureau of Labor Statistics, *100 Years of U.S. Consumer Spending: Data for the Nation, New York City, and Boston*, U.S. Bureau of Labor Statistics Report 991, May 2006, 22–42, http://www.bls.gov/opub/uscs/report991.pdf. Prices of Cadillac four-door sedans model years 1950–90 from National Automobile Dealers Association, NADA Guides, "Classic Car Values and Specs," http://www.nadaguides.com/Classic-Cars. Models used are 62 four-door sedan for 1950–59; Deville four-door sedan for 1960 through 1990. All figures are the manufacturer's suggested retail price (MSRP) for base model with standard equipment.

76. For examples of the cost of earning an instrument rating in 1951 and 1984, respectively, see *Outlook for Private Flying*, 47; Spitzmiller, "Who Needs an IFR Rating?" 56.

77. Model 120 and 140 production figures from Mitch Mayborn and Bob Pickett, *Cessna Guidebook*, vol. 1, 2nd ed. (Dallas: Flying Enterprise Publications, 1976), 38–40; model 170 production figures from Cleo M. Bickford and Buddy McGown, eds., *The 170 Book*, 4th ed. (Lebanon, MO: International Cessna 170 Association, 1999), 3–4, 10; quote from Cessna Aircraft Co. "Skyhawk: The Best-Selling and Most-Flown Aircraft Ever Built," http://cessna.txtav.com/en/single-engine/skyhawk (accessed on Jan. 26, 2015); model 150, 152, and 172 production figures from "Cessna Single Engine Production Totals by Model and Year," in John Frank, *Cessna 172 Skyhawk Buyers Guide* (Santa Maria, CA: Cessna Pilots Association, 2009): 26 (note: figure for model 172 includes model 172 RG [produced 1980–85] but does not include roughly 5,000 new planes built after Cessna resumed production of an updated version of the 172 in 1997).

78. Smith, *Week-End Pilot*, 211.

79. Edward H. Phillips, *Beechcraft: Pursuit of Perfection* (Eagan, MN: Flying Books, 1992), 40, 78. See also my more detailed description of the Beechcraft Bonanza in chapter 4.

80. New plane prices from Bickford and McGown, *The 170 Book*, 11; Bill Clarke, *The Cessna 172*, 2nd ed. (Blue Ridge Summit, PA: TAB Books/McGraw-Hill, 1993), 18;

Mayborn and Pickett, *Cessna Guidebook*, 38–40; Phillips, *Beechcraft: Pursuit of Perfection*, 34, 36–37, 40, 78; "The *AOPA Pilot* Bonus Supplement: General Aviation Aircraft and Engines 1965," *AOPA Pilot* 8, no. 3 (Mar. 1965): 56–59; "General Aviation Aircraft—1970," *AOPA Pilot* 13 no. 3 (Mar. 1970): 118–21; "*AOPA Pilot*'s March Bonus Supplement: General Aviation Aircraft 1975," *AOPA Pilot* 18, no. 3 (Mar. 1975): 84, 86; "1980 General Aviation Aircraft Directory," *AOPA Pilot* 23, no. 3 (Mar. 1980): 68–69, 73; Bernice D. Gambell, Judy B. Haines, and Dawn E. Tamkin, "Shifting Sands: 1985 General Aviation Aircraft Directory," *AOPA Pilot* 28, no. 3 (Mar. 1985): 53–54, 57. Information from contemporary aircraft reviews suggests that prices for personal planes in 1985 were even higher than the figures used here; thus the dramatic price increases between 1980 and 1985 may have in fact been greater than described in this chapter. Keith Connes, "Plane Vanilla," *Plane & Pilot*, Mar. 1985, 33; Bill Cox, "Eclipsing the Skyhawk," *Plane & Pilot*, Apr. 1985, 34.

81. Bradley, "Crashes and the Court," 52, 54; J. Jefferson Miller, "Diminishing Returns: A Complex Set of Issues with No Easy Answers and Many Adverse Consequences," *AOPA Pilot* 28, no. 4 (Apr. 1985): 38–39, 41–44, 46, 48.

82. Cadillac prices from National Automobile Dealers Association (NADA) Guides, "Classic Car Values and Specs." Home values from U.S. Department of Commerce, U.S. Census Bureau, "Historical Census of Housing Tables: Home Values," http://www.census.gov/hhes/www/housing/census/historic/values.html.

83. Matt Thurber, "Affordable Classics: The Recreational Airplane as an Investment," *Flying*, Sept. 1985, 58–64.

84. New plane prices from John M. Frank, *150/152 Model History* (Santa Maria, CA: Cessna Pilots Association, July 17, 2006), 6, 8, http://www.cessna.org/documents/public_info/samplemodel.pdf; "*AOPA Pilot* Bonus Supplement [1965]," 56–59; used plane prices from *Trade-a-Plane*, first April issue, 1965, 13–15.

85. For representative examples from the 1940s through 1980s with break-even figures expressed in flight hours per year, see Neil E. Berboth, "Should You Own or Rent a Plane?" *Flying*, Aug. 1946, 31 (125–200 hours); *Outlook for Private Flying*, 32 (200 hours); "Figure 8: Airplane A, Graphical Cost Comparison," in Stacy Weislogel, "Rent or Buy?" *AOPA Pilot* 8, no. 9 (Nov. 1965): 84, graph (230 hours); "Figure 3: Graph of Equal Cost, Rent versus Buy," in Gene Holman, "The Rent-or-Buy Decision . . . How to Graph the Economics," *AOPA Pilot* 16, no. 4 (Apr. 1973): 40, graph (108 hours); Boyne, *Flying*, 171 (200 hours).

86. J. L. Howard, "A Mechanic Bites Back," *AOPA Pilot* 4, no. 4 (Apr. 1961): 59.

87. "Median Family Income by race/ethnicity of head of household: 1950 to 1993," in Thomas Snyder and Linda Shafer, U.S. Department of Education, National Center for Education Statistics, *Youth Indicators, 1996*, NCES 96-027 (Washington, DC: 1996), 44, table, http://nces.ed.gov/pubs98/yi/; "Table F-1—Income Limits for Each Fifth and Top 5 Percent of Families (All Races): 1947 to 2010," in U.S. Department of Commerce, U.S. Census Bureau, Historical Income Tables: Income Inequalities, http://www.census.gov/hhes/www/income/data/historical/inequality/index.html.

88. Quotes from Archie Trammell, "Fun-Plane," *Flying*, Nov. 1970, 41; typical rental rate based on data from Jerrold R. Williams, "A Flying Vacation in Hawaii," *AOPA Pilot* 10, no. 6 (June 1967): 127; aircraft rental rates adjusted from 1967 to 1970 prices using "CPI Inflation Calculator."

89. U.S. Department of Labor, Wage and Hour Division, "History of Federal Minimum Wage Rates under the Fair Labor Standards Act, 1938–2009," http://www.dol.gov/whd/minwage/chart.htm#.UOpbJax36Hi; Trammell, "Fun-Plane," 41.

90. Robert Blodget, "How to Run a Flying Club," *Flying*, Sept. 1971, 55.

91. Paul Poberezny, The Homebuilder's Corner, *Sport Aviation* 33, no. 2 (Feb. 1984): 2.

92. Boyne, *Flying*, 29.

93. Berl Brechner, "The Price Is Flight," From the Tower, *Flying*, May 1981, 9.

94. Howard F. Haines, "Back to Sailing," VOX Pilot, *AOPA Pilot* (Aug. 1950): 50h.

95. Leonard J. Redding, "Mechanic Gripe," VOX Pilot, *AOPA Pilot* (July 1951): 50h.

96. J.M. [Jane Mahon?], "Pinch-Hitter Comes Home to AOPA," *AOPA Pilot* 8, no. 8 (Aug. 1965): 101.

97. George H. Moloney Jr., "The Middle-Aged Student vs. 'The System,'" *AOPA Pilot* 15, no. 9 (Sept. 1972): 78.

98. Burns, "Put Yourself in the Pilot's Seat," sec. 10, p. 1.

99. Vincent Gallagher, "Little Guys Speak Up," VOX Pilot, *AOPA Pilot* 7, no. 12 (Dec. 1964): 16–17.

100. Donald Chase, "Fun Is a Thing with Wings," *AOPA Pilot* 11, no. 9 (Sept. 1968): 64.

101. Nosewheel Nick [pseud.], "No Nylons for Mama," *AOPA Pilot* 3, no. 2 (Feb. 1960): 23, 48.

102. Lois C. Philmus, "Women Banish Fear of Flying," *AOPA Pilot* 6, no. 5 (May 1963): 29.

103. Lois C. Philmus, "Beware the Skirt Barrier," *AOPA Pilot* 3, no. 6 (June 1960): 39, 61–63; Robert Peterson, "Not Her Cup of Tea: How to Handle a Woman When She Won't Fly," *Flying*, June 1969, 56–59.

104. William B. Murphy, "They Build What They Fly," *AOPA Pilot* 8, no. 7 (July 1965): 49.

105. Mrs. R. K. Fleming, "Drink Tea and See," Flying Mail, *Flying*, Sept. 1969, 90.

106. "Only for the Rich?" editorial, *AOPA Pilot* 18, no. 12 (Dec. 1975): 6; letters from John W. Keski and L. C. Blauvelt, "Only for the Rich?" VOX Pilot, *AOPA Pilot* 19, no. 2 (Feb. 1976): 13–14.

107. Pilot ages from CAA, *Statistical Handbook of Civil Aviation*; FAA, *Statistical Handbook of Aviation*.

108. Paul H. Poberezny, interviewed by the author at the EAA Headquarters, Oshkosh, WI, Aug. 3, 2004, audio recording in author's collection.

109. T. P. Wright, [Chief] Administrator of the CAA, "Personal Aircraft: An American Appraisal" (paper presented at the Anglo-American Conference under the joint auspices of the Royal Aeronautical Society and the Institute of Aeronautical Sciences in London, Sept. 6, 1947), 7.

110. Lee L. Schiek, "Disillusioned," VOX Pilot, *AOPA Pilot* 11, no. 2 (Feb. 1968): 21.

111. W. R. Caswell, "Disillusioned," VOX Pilot, *AOPA Pilot* 11, no. 2 (Feb. 1968): 21–22.

112. AOPA member 235,167, "Pilot Shortage Myth," VOX Pilot, *AOPA Pilot* 11, no. 7 (July 1968): 18.

113. AOPA member 295,947, "Pilot Shortage," VOX Pilot, *AOPA Pilot* 11, no. 12 (Dec. 1968): 25.

Chapter 2 · Shouting, Shirttails, and Spins

1. Biographical information about Frank Kingston Smith from, "Longtime Aviation Author Frank Kingston Smith Dies," What's New, Sept. 3, 2003, http://www.aopa.org/whatsnew/newsitems/2003/03-3-126x.html; EAA, "'Weekend Pilot' Frank Kingston Smith Dies," *EAA News*, Sept. 4, 2003, http://www.eaa.org/communications/eaanews/030904_kingston.html; Jim Campbell, "R.I.P. Frank Kingston Smith," *Aero-News Network*, Sept. 5, 2003, http://www.aero-news.net/index.cfm?do=main.textpost&id=21cffd85-f2e3-4811-8926-c3b613aadd01. Information about Diane Ackerman from *International Who's Who in Poetry, 2005* (London: Europa Publishers, 2004), 6; Jeremiah Tax, "Booktalk," *Sports Illustrated*, Oct. 28, 1985, 10.

2. AOPA, "Longtime Aviation Author Frank Kingston Smith Dies"; Frank Kingston Smith, *Week-End Pilot* (New York: Random House, 1957), 4–5; Diane Ackerman, *On Extended Wings* (New York: Atheneum, 1985), 3.

3. Comparing flight instruction syllabi from 1939, 1956, and 1964 reveals that except for the addition of two-way radio communication, a few hours on the fundaments of instrument flight, and an introduction to flying at night, very little about flight instruction changed over this 25-year period. Descriptions of flight training in the 1970s and 1980s indicate few additional changes. Throughout the postwar era, the private pilot curriculum called for a minimum of 40 hours of total flight time (35 hours for government approved schools), although most individuals took longer than this to complete their training. "Controlled Private Flying Course: CPTP Demonstration Program," *Air Commerce Bulletin*, Mar. 15, 1939, 232–33, reproduced in Dominick A. Pisano, *To Fill the Skies with Pilots: The Civilian Pilot Training Program, 1939–1946* (Washington, DC: Smithsonian Institution Press, 2001), app. B; U.S. Department of Commerce, CAA, *Flight Instructors' Handbook: CAA Technical Manual No. 105* (Washington, DC: U.S. GPO, Jan. 1956); U.S. Department of Transportation, FAA, Flight Standards Service (FSS), *Flight Instructor's Handbook* (Washington, DC: U.S. GPO, 1964); U.S. Department of Transportation, FAA, *A Flying Start* (Washington, DC: U.S. GPO, ca. 1977); Walter J. Boyne, *Flying: An Introduction to Flight, Airplanes, and Aviation Careers* (Englewood Cliffs, NJ: Prentice-Hall, 1980), 83–85.

4. In a 1971 book for potential pilots, aviation writer Barry Schiff included *Stick and Rudder* in his list of "Books to Read" and noted that it was "a classic." *Stick and Rudder* was still in print as of 2014, and the dust jacket of a recent copy claims "Over 250,000 copies in print." This is not to suggest that Langewiesche's book is currently used (or has been lately) as the primary text for flight instruction; rather, its continued popularity indicates that the fundaments of flying a small airplane have changed little since World War II. Barry Schiff, *Flying* (New York: Golden Press, 1971), 158; Wolfgang Langewiesche, *Stick and Rudder: An Explanation of the Art of Flying* (New York: McGraw Hill, 1944, 1972, 1990), dust jacket.

5. Patricia Cayo Sexton, *The Feminized Male: Classrooms, White Collars, and the Decline of Manliness* (New York: Random House, 1969), 15.

6. Stanley R. Mohler, M.D., U.S. Department of Transportation, FAA Office of Aviation Medicine, *Recent Findings on the Impairment of Airmanship by Alcohol* (Springfield, VA: Clearinghouse for Federal Scientific and Technical Information, Sept. 1966), 3.

7. For a succinct description of the mental "demands of flight" above and beyond those required for safe driving, see George I. LeBaron Jr., M.D., "Pilot Error: A Discussion of Causes and Some Cures," *AOPA Pilot* 17, no. 9 (Sept. 1974): 39–42, quote from 39.

8. Smith, *Week-End Pilot*, 29–31.

9. Ibid., 30, 79–81.

10. Ackerman, *On Extended Wings*, 66.

11. Ibid., 58–59.

12. Ibid., 66, 69–71.

13. George I. LeBaron Jr., M.D., "Pussy Cat in the Cockpit: Part I," *AOPA Pilot* 15, no. 12 (Dec. 1972): 70; LeBaron, "Pussy Cat in the Cockpit: Part II," *AOPA Pilot* 16, no. 1 (Jan. 1973): 33.

14. Pisano, *To Fill the Skies with Pilots*, 63–65, including quote on 65 from Morris S. Viteles et al., *A Course in Training Methods for Pilot Instructors*, Report no. 20 (Washington, DC: Civil Aeronautics Administration, Division of Research, Sept. 1943), viii, emphasis in original.

15. Army Air Forces Training Command, *Instructors' Manual: Basic Flying* (U.S. Army, ca. 1944), 10.

16. David T. Courtwright, *Sky as Frontier: Adventure, Aviation, and Empire* (College Station: Texas A&M University Press, 2005), 115, see also n. 10 on 243.

17. Lieut. John R. Hoyt, *Manual for Aviation Cadets* (New York: McGraw Hill, 1943); John R. Hoyt, "When You Learn to Fly—Get a Tough Instructor," *Flying*, June 1945, quotes from 56, 138.

18. Hoyt, "When You Learn to Fly," 138.

19. Don Downie, "The Flight Instructor: Aviation's Weakest Link?" *AOPA Pilot* 7, no. 11 (Nov. 1964): 30.

20. Ibid.

21. George W. Glacius, "More on Flight Instructors," VOX Pilot, *AOPA Pilot* 8, no. 2 (Feb. 1965): 15.

22. William D. Strohmeier, "When You Teach Flying, Understand the Student," *Flying*, June 1945, 57, 144.

23. Duane Cole, "Do Unto Others . . . ," Pro's Nest, *Flying*, Jan. 1983. 21.

24. Len Morgan, "A Better Breed of Instructor," Vectors, *Flying*, May 1981, 36.

25. Boyne, *Flying*, 81–82.

26. CAA, *Flight Instructors' Handbook*, 1–11, quote from 7.

27. FAA, *Flight Instructor's Handbook*, 1–17, 21, 101, 102. For other examples from the same manual that stressed the centrality of the student-instructor relationship, see "Professionalism in Flight Instruction," 21; "The Instructor-Student Relationship," 21–23; "Watch Your Language," 33; "Emphasize the Positive," 33–35; and "The Flight Instructor Image," 100–102.

28. John Pennington, "A Rookie Talks Instructing," *AOPA Pilot* 1, no. 9 (Nov. 1958): 26, 56–57; George H. Moloney Jr., "The Middle-Aged Student vs. 'The System,' " *AOPA Pilot* 15, no. 9 (Sept. 1972): 77–79.

29. Ackerman, *On Extended Wings*, 66.

30. Claude Grahame-White, "No Place in the Air for Women," *Mobile (AL) Register*, Sept. 3, 1911.

31. Marion Hart, "If I Can Do It, You Can," *Air Facts* 9, no. 12 (Dec. 1, 1946), 57–60; "Know How," *Air Facts* 10, no. 1 (Jan. 1, 1947), 72.

32. Ruth Downie and Don Downie, "She Wouldn't Be 'Washed Out,'" *AOPA Pilot* 2, no. 3 (Mar. 1959): 33.

33. Frank A. Tinker, "Where Are All the Gals?" *AOPA Pilot* 6, no. 7 (July 1963): 42; Lois C. Philmus, "Women Banish Fear of Flying," *AOPA Pilot* 6, no. 5 (May 1963): 29.

34. Boyne, *Flying*, 15.

35. Moloney, "The Middle-Aged Student," 77.

36. Boyne, *Flying*, 8.

37. Gordon Baxter, "First Encounter with Flight," *Flying*, July 1980, 68–69.

38. Boyne, *Flying*, 9.

39. G. S. Frissel, "Pride Goeth before a Fall," Never Again! *AOPA Pilot* 21, no. 9 (Sept. 1978): 88; Smith, *Week-End Pilot*, 80.

40. Robert T. Shaw, "Requiem for Four Zero Kilo," *AOPA Pilot* 3, no. 8 (Aug. 1960): 32.

41. Boyne, *Flying*, 10.

42. Smith, *Week-End Pilot*, 72–73.

43. *Duckhook* is a golfing term for a flubbed drive that hooks to the left (for a right-handed golfer). By "duckhook the takeoff," Smith meant that the pilot had neglected to apply the required additional pressure to the right rudder pedal to counter increased propeller torque when adding power for takeoff, thus allowing the plane to swerve to the left. Frank Kingston Smith, "The Spirit Moved Me," Off and Winging, *AOPA Pilot* 22, no. 5 (May 1979): 142.

44. Boyne, *Flying*, 8–10.

45. Tinker, "Where Are All the Gals?" quotes from 42.

46. Virginia M. Schmidt, "A Funny Thing Happened on the Way from the Airport," *AOPA Pilot* 12, no. 4 (Apr. 1969): 55.

47. Jamie Becket, "Mind Your Manners," *General Aviation News*, June 27, 2012, http://www.generalaviationnews.com/2012/06/mind-your-manners/.

48. "Madam, You're Landing on the Sky" [cartoon], *AOPA Pilot* 12, no. 5 (May 1969): 107; Demchuck, "'His' and 'Hers'" [cartoon], *AOPA Pilot* 10, no. 5 (May 1967): 96; Janice Crisp, "A Woman Objects," VOX Pilot, *AOPA Pilot* 10, no. 8 (Aug. 1967): 24.

49. For instance, in a 1975 cartoon, a man talking on a telephone inside a building marked "Flying School" says, "Yes, Dr. Morley, your wife is coming in from her first solo now" as a woman parachutes onto the lawn. "Yes, Dr. Morley . . . " [cartoon], *AOPA Pilot* 18, no. 12 (Dec. 1975): 15.

50. George Christian, "Out of the Pattern without a Chart," Never Again! *AOPA Pilot* 19, no. 4 (Apr. 1976): 76–78, quotes from 78.

51. Patricia Houtz, "Female Logic," Never Again! *AOPA Pilot* 8, no. 10 (Oct. 1965): 78.

52. Jennifer Ortiz, "I Can't Be Lost!" Never Again! *AOPA Pilot* 19, no. 11 (Nov. 1976): 78–80, quote from 79.

53. Philmus, "Women Banish Fear of Flying," 28–30, quote from 29.

54. Tinker, "Where Are All the Gals?" 40–44.

55. For details about how various women coped, from acting like "one of the boys" at one extreme to playing the role of "girly girl" at the other, see chapter 6.

56. Barry Schiff, "Clipping Shirttails," Proficient Pilot, *AOPA Pilot* 40, no. 12 (Dec. 1997): 113.

57. Ibid.

58. Daryl Murphy, "How Did the Post-Solo Ritual of Cutting Our Shirttails Begin?" *General Aviation News* 53, no. 18 (Aug. 31, 2001): 33.

59. Schiff, "Clipping Shirttails," 113.

60. Doyle Earl Bradford, "War Time Memories of a Man-Child" (copyright 1994 by the Bradford family), 15, www.91stbombgroup.com/91st_tales/78_doyle_bradford.pdf. The earliest reference I've found that ties shirttails to soloing was written by a naval aviator who learned to fly in 1946. In addition to getting thrown into a "slimy worm-infested mudhole" at Naval Air Station Hensley Field near Dallas, Texas, he reported having his tie clipped ("a common tradition") and his uniform shirttail torn, not clipped, off his back as part of the post-solo celebrations. Fred "Crash" Blechman, "Earning the Mudhole," *Airport Journals*, Apr. 2007, http://www.airportjournals.com/Display.cfm?varID=0704013.

61. This is not entirely true: during the interwar period fledging naval aviators tossed the last member of their class to solo a seaplane into Pensacola Bay, fully clothed; however, this ritual was meant to single out the most ham-fisted student, not as a universal rite of passage into the world of fliers. Richard K. Smith and Cargill R. Hall, *Five Down, No Glory: Frank G. Tinker, Mercenary Ace in the Spanish Civil War* (Annapolis: Naval Institute Press, 2011), 43–44.

62. Amy Harr, "Oh Solo Mio," Training, *Flying*, Dec. 1989, 88.

63. Charles A. Lindbergh, *"We"* (Guilford, CT: Lyons Press, 2002; originally published New York: G. P. Putnam's Sons, 1927), 43 (page references are to the 2002 edition).

64. Boyne, *Flying*, 84.

65. Gordon Baxter, *How to Fly, for People Who Are Not Sure They Want To* (New York: Summit Books, 1981), 18.

66. Courtwright, *Sky as Frontier*, 50, 115.

67. Janet D. Davis, "That's Mama Up There," *AOPA Pilot* 6, no. 11 (Nov. 1963): 69.

68. Sally Buegeleisen, "Abbott's Rib," *Flying*, Aug. 1965, 108.

69. In 1938 candidates for a private pilot certificate had to demonstrate to a flight instructor their ability to recover from spins. By the end of World War II, the Civil Air Regulations required student pilots to demonstrate this skill to a designated federal flight examiner. This rule remained in effect until 1949. U.S. Department of Commerce, Bureau of Air Commerce, *Civil Air Regulations Part 20: Pilot Rating*, as amended to May 31, 1938 (Washington, DC: U.S. GPO, 1938), 4; Civil Aeronautics Authority, *Civil Air Regulations Part 20: Pilot Certificates*, effective May 1, 1940 (Washington, DC: U.S. GPO, 1940), 4; Civil Aeronautics Board, *Civil Air Regulations Amendment 20-0: Pilot Certificates*, adopted Apr. 26, 1945, effective July 1, 1945 (Washington, DC: CAB, 1945), 3; Civil Aeronautics Board, *Civil Air Regulations Amendment 20-3: Elimination of Spin Test Requirements*, effective Aug. 15, 1949 (Washington, DC: CAB, 1949), 1–3.

70. Schiff, *Flying*, 22.

71. Neil D. Van Sickle, ed., *Modern Airmanship*, 3rd ed. (Princeton, NJ: D. Van Nostrand, 1966), quotes from 318, 319. For similar reminders in other manuals, see Langewiesche, *Stick and Rudder*, 18–24; FAA, *Flight Instructor's Handbook*, 46–48.

Warnings like this were published to counteract the apparently widespread belief that low airspeed was the sole reason for stalls and the corresponding myth that simply staying above an airplane's published stalling speed would thus guarantee safety. These and other manuals argued that pilots needed to understand that low airspeed was just one of many flight conditions that could lead to a stall.

72. Langewiesche, *Stick and Rudder*, 3.

73. According to a 1985 article on spin training, "Early airworthiness requirements provided that all airplanes of less than 4,000 pounds recover from a six-turn spin within 1½ additional turns, at all certificated CG [center of gravity] positions, simply by releasing the flight controls. Current airworthiness standards on spins (FAR 23.221) require a single-engine normal-category airplane to be able to recover from a one-turn spin or a three-second spin, whichever takes longer, in not more than one additional turn, with the controls used in a normal manner for recovery." In lay terms, this means that by law, early postwar aircraft designs had to self-recover from a spin without any input from the pilot (beyond simply letting go of the controls); whereas more recent designs, produced under amended federal regulations, require quick application of proper spin recovery techniques by the pilot to ensure success. Don Dwiggens, "Aeronautic Breakdancing," *Plane & Pilot*, Apr. 1985, 69.

74. CAA, *Flight Instructors' Handbook*, 41.

75. Peter Garrison, "What's Wrong with Stall Practice?" *Flying*, May 1991, 59.

76. Langewiesche, *Stick and Rudder*, 195.

77. Smith, *Week-End Pilot*, 34–35.

78. Joseph Bourque, "The Spin Debate," *Air & Space Smithsonian*, Oct.–Nov. 2003, 38.

79. This universally accepted spin recovery procedure appears in almost every instruction manual for private pilot candidates published throughout the postwar era. See, for example, sec. 26, "Normal Spins," and sec. 27, "Accidental Spins," in *Civil Air Regulations and Flight Standards for Pilots*, 24th ed. (Los Angeles: Aero Publishers, Mar. 1962), n.p.

80. "I'll Have to Flunk You on That Spin" [cartoon], *AOPA Pilot* 4, no. 9 (Sept. 1961): 43.

81. Peter Derfner, "Get the Flaps Off!" Never Again! *AOPA Pilot* 21, no. 12 (Dec. 1978): 68–69.

82. Dwiggens, "Aeronautic Breakdancing," 73.

83. Safety Bureau, Civil Aeronautics Board, *Proposed Parts 20, 43 and 60 of the Civil Air Regulations* (Washington, DC: CAB, Nov. 10, 1944), 2; Civil Aeronautics Board, *Civil Air Regulations Amendment 20-0: Pilot Certificates* (1945), 3.

84. Accident statistics from 1948–49 from "CAB-CAA Expected to Boost Plane Safety Minimums," *AOPA Pilot* (July 1950): 50a–50b; "CAB Gives AOPA Summary of 1949 Civil Accidents," *AOPA Pilot* (Sept. 1950): 50e; Frederick Graham, "Safety Engineer Warns Non-commercial Fliers to Reduce Their Accident Rate," Aviation: Private Pilots, *New York Times*, Feb. 1, 1948, sec. 2, p. 19.

85. "Spins Unpopular," and "General Comments," both in *AOPA Pilot* (Oct. 1944): 70c–70d, quotes from 70d.

86. Harry Burnett, "Spins," VOX Pilot, *AOPA Pilot* (Dec. 1948): 58h.

87. U.S. Department of Commerce, Program Planning Staff, Office of the Administrator of Civil Aeronautics, *The Outlook for Private Flying, 1950–1955* (Washington, DC: U.S. GPO, Mar. 1951), 54.

88. For an overview of the AOPA's official stance during the mid-1940s through mid-1950s regarding mechanical stall-warning devices, spin-proof airplanes, and other technological solutions to safety problems, see "Safety in Plane Design Lacking," *AOPA Pilot* (Dec. 1945): 74b; "Safety and 1948," *AOPA Pilot* (Feb. 1948): 50c; "Do You *Want* a Safer Airplane?" editorial, *AOPA Pilot* (Aug. 1949): 50a; "Okay Boys, What Now?" editorial, *AOPA Pilot* (Sept. 1949): 50a; "We *Will* Get More Safety," editorial, *AOPA Pilot* (Aug. 1950): 50a; "Do You *Really* Want Safety?" editorial, *AOPA Pilot* (Jan. 1952): 50a; "You Can't Sell Safety," editorial, *AOPA Pilot* (Aug. 1956): 50b.

89. T. P. Wright, Administrator, CAA, excerpt of letter to the AOPA, reprinted in "Pilot Opinion Solicited: Are Low Altitude Maneuvers Needed?" *AOPA Pilot* (Dec. 1947): 58b. Just three months before it published Wright's letter, the *AOPA Pilot* carried a thumbnail reproduction of the CAA's latest antibuzzing poster: a caricature of a low-flying airplane about to collide with power lines because the distracted pilot is busy waving to a beautiful woman on the ground. CAA, "Hot Rock Pilots" [image], *AOPA Pilot* (Sept. 1947): 50a. For examples of the AOPA's opinion on the matter, see "Low Flying Must Stop!" *AOPA Pilot* (May 1947): 50b; "An Editorial," *AOPA Pilot* (Sept. 1947): 50a. Although Wright and the AOPA leadership disagreed on many matters—including spin training and the advisability of low flying—historian John R. M. Wilson credits Wright with simplifying regulations regarding private flying during his tenure as CAA administrator, a philosophy that met with AOPA approval. John R. M. Wilson, *Turbulence Aloft: The Civil Aeronautics Administration amid Wars and Rumors of Wars, 1938–1953* (Washington, DC: U.S. GPO, 1979), 169.

90. Al Knouff, "Hi-Low?" [and editor's comments], VOX Pilot, *AOPA Pilot* (Feb. 1948): 50h.

91. Civil Aeronautics Board, *Civil Air Regulations Amendment 20-3: Elimination of Spin Test* (1949).

92. "New No-Spin Law Eases Requirements for Pilot License," *AOPA Pilot* (Oct. 1949): 50a–50b.

93. "Okay Boys, What Now?" 50a.

94. "Do You *Want* a Safer Airplane?" 50a.

95. Nadine G. Tatum, "Spin Requirement," VOX Pilot, *AOPA Pilot* (Apr. 1950): 50h.

96. Truitt Vinson, "Student Days," VOX Pilot, *AOPA Pilot* 1, no. 7 (Sept. 1958): 6.

97. Frank A. Tinker, "Your FAA Check Ride: Why?" *AOPA Pilot* 5, no. 3 (Mar. 1962): 28–29, 59–60 (including postscript), quote from 28.

98. W. C. Heller, "Look at the Record," VOX Pilot, *AOPA Pilot* 5, no. 5 (May 1962): 10.

99. Downie, "The Flight Instructor," 30–32.

100. Robert J. Traister, *The Joy of Flying* (Blue Ridge Summit, PA: Tab Books, 1981), 34–35.

101. Gene [no last name], "'Spins' in a Cardinal RG," Pilots of America Message Board, Apr. 7, 2009, http://www.pilotsofamerica.com/forum/showthread.php?t=28329.

102. Van Sickle, *Modern Airmanship*, 3rd ed., 321. In later editions of the same book, this qualifier was changed to sound less derogatory toward the average "Sunday pilot," instead simply suggesting that "the teaching of spin recovery should remain a firm part of pilot training." John F. Welch, ed., *Van Sickle's Modern Airmanship*, 5th ed. (New York: Van Nostrand Reinhold, 1981), 407.

103. Smith, *Week-End Pilot*, 31–32.

104. Spencer F. Houghton, "Spin Recovery Instruction," VOX Pilot, *AOPA Pilot* 12, no. 5 (May 1969): 25.

105. Few question the statistic that stalls and spins were a leading cause of fatal accidents involving small planes in the 1940s, but evidence that practice spins "gone wrong" killed more pilots than accidental spins is elusive. However, the flying community has repeated this second claim so often that it is now widely accepted as fact. For examples of this, see Russel F. Holdren, "Build Better Planes," *AOPA Pilot* (June 1948): 58d; Bruce Landsberg, "A Little Too Real," Safety Pilot, *AOPA Pilot* 46, no. 8 (Aug. 2003): 56; Bourque, "The Spin Debate," 38. In 2003, Rich Stowell, a nationally known flight instructor and outspoken supporter of spin training, challenged the AOPA, which had just published a report on spin training, "to provide credible and properly cited references (or better yet, copies of the text) dated ca. 1949" to support the "myth" that the federal government had dropped mandatory spin training because more pilots were dying as a result of intentional practice spins than from unintentional spins. Rich Stowell, "Review of New AOPA Stall/Spin Study," *Fighter Combat International (FCI): Flight Training and Safety Newsletter,* no. 21 (Aug. 2003), Aviation Performance Solutions (APS), LLC, http://apstraining.com/Docs/FCI_Newsletter_Sep03_AOPA_Spin_Study.pdf.

106. Sec. 27, "Accidental Spins," in *Civil Air Regulations and Flight Standards for Pilots,* n.p.

107. CAA, *Flight Instructors' Handbook,* 44; FAA, *Flight Instructor's Handbook.*

108. David D. Pierce, "A Different Kind of Flight School," *AOPA Pilot* 11, no. 1 (Jan. 1968): 53–56. Interestingly, this description of the psychological dangers of spins, used in this case to advocate spin training, is almost identical to one made years before by opponents who argued for designing spin-proof airplanes in lieu of training pilots in spin recovery. See Leighton Collins, "The Dangers of the Air," in Langewiesche, *Stick and Rudder,* 343.

109. Don Downie, "Pilot Flight Check: The Bellanca Citabria," *AOPA Pilot* 19, no. 4 (Apr. 1976): 38.

110. "Stall/Spin: Reducing the Sting," Safety Corner, *AOPA Pilot* 21, no. 10 (Oct. 1978): 106.

111. Garrison, "What's Wrong with Stall Practice?" 56.

112. Bourque, "The Spin Debate," 42–43, quote from magazine cover.

113. Philmus, "Women Banish Fear of Flying," 29. Later that same year, *AOPA Pilot* editor Max Karant described "the fearful woman" as "one of general aviation's greatest enemies." Max Karant, "Welcome Aboard, Honey!" editorial, *AOPA Pilot* 6, no. 12 (Dec. 1963): 20.

114. Max Karant, "Two Thousand Attend AOPA Event," *AOPA Pilot* 5, no. 12 (Dec. 1962): 33.

115. Karant, "Welcome Aboard, Honey!" 20; Don Downie, "A New Flying Concept: The 'Pinch-Hitter,'" *AOPA Pilot* 6, no. 12 (Dec. 1963): 28.

116. "Table 4.9—Active Airman Certificates Held: Jan. 1, 1955–63," and "Table 4.11—Women Actively Engaged in Aviation: Jan. 1, 1961–63," in U.S. Department of Transportation, FAA, *Statistical Handbook of Aviation,* 1963 ed. (Washington, DC: U.S. GPO, 1964), 51–52, tables.

117. Downie, "A New Flying Concept," 26–27.

118. Ibid., 27.

119. Thomas A. Horne, "ASF at 50: The AOPA Air Safety Foundation's Legacy of Safety, and a View to the Future," *AOPA Pilot* 43, no. 11 (Nov. 2000): 63.

120. Douglas D. Bond, M.D., *The Love and Fear of Flying* (New York: International Universities Press, 1952), 11.

121. Downie, "A New Flying Concept," 27.

122. Robert L. Parrish, "Flight Training at a Busy Airport," *AOPA Pilot* 9, no. 6 (June 1966): 111.

123. Ibid., 110; Mrs. Bernie Shapiro, "Group Therapy," VOX Pilot, *AOPA Pilot* 7, no. 6 (June 1964): 16.

124. *The AOPA Pinch-Hitter Course Pilot Training Manual*, 1st ed., special insert in *AOPA Pilot* 7, no. 4 (Apr. 1964): 93.

125. Ibid., 94.

126. A common challenge for student pilots is to learn how the airplane's control wheel is *different* from the steering wheel on a car. Unless the airplane is equipped with interconnected controls (e.g., the ERCO Ercoupe and Piper Tri-Pacer, both described in chapter 3), the control wheel will not turn an airplane on the ground. Instead, rudder pedals and differential braking accomplish this, and on windy days using the control wheel like a steering wheel is not only ineffective but can actually cause a small plane to flip over during taxiing. In the air, turns are properly accomplished by using the wheel or stick and the rudder pedals in unison. Using the control wheel alone results in slow, skidding ("uncoordinated") turns, sloppy and inefficient but workable in an emergency. Welch, *Van Sickle's Modern Airmanship*, 5th ed., 377–79, 391–93.

127. Downie, "A New Flying Concept," 27.

128. *Pinch-Hitter Course Pilot Training Manual*, 1st ed., 100. By 1973, the revised Pinch-Hitter Course manual continued to teach a nontechnical approach to radio communication but had dropped the condescending comparison between talking to an air traffic controller and "telling your problems on the phone to your closest friend." *The AOPA Pinch-Hitter Course Pilot Training Manual*, rev. ed., special insert in *AOPA Pilot* 16, no. 5 (May 1973): 65–67.

129. *Pinch-Hitter Course Pilot Training Manual*, 1st ed., 96–101.

130. Ibid., 96.

131. Ibid., 97–98. As with the original comparison between using the radio and talking on the telephone, by 1973 the revised Pinch-Hitter manual had dropped the housewife-specific exercise that used the kitchen sink to help teach aerial navigation. In its place, the manual instructed students to "Practice other examples of this kind until you become as familiar with the heading numbers and directions as with the streets in your hometown." *The AOPA Pinch-Hitter Course Pilot Training Manual*, rev. ed., 63.

132. Quote from AOPA Foundation, Inc., "Pinch-Hitter Course Sets Stage for a New Air Age . . . of Women!" [advertisement], *AOPA Pilot* 6, no. 12 (Dec. 1963): 44.

Chapter 3 · The Family Car of the Air versus the Pilot's Airplane

1. Fred Weick and James R. Hansen, *From the Ground Up: The Autobiography of an Aeronautical Engineer* (Washington, DC: Smithsonian Institution Press, 1988), 188–91.

2. Not only did the Ercoupe perform as advertised (stall proof, spin proof, and easier to fly and land than other airplanes of the era), but as of 2013 a respectable one-sixth of

all Ercoupe variants ever produced were still flying, most of them more than 50 years old. This is the hallmark of a technical success that failed in the marketplace, much like the gas-powered home refrigerator. Ruth Schwartz Cowan, "How the Refrigerator Got Its Hum," in *The Social Shaping of Technology: How the Refrigerator Got Its Hum*, ed. Donald MacKenzie and Judy Wajcman (Philadelphia: Open University Press, 1985), 202–18; U.S. Department of Transportation, FAA, Make/Model Inquiry of the FAA Registry database for "Ercoupe," "Forney," "Alon," and "Mooney," http://registry.faa.gov/aircraftinquiry/AcftRef_Inquiry.aspx.

3. *Official Gazette of the United States Patent Office*, vol. 259, no. 3. (Washington, DC: U.S. GPO, Tuesday, Feb. 18, 1919), 402; Peter M. Bowers, *Curtiss Aircraft, 1907–1947* (London: Putnam, 1979), 75–76.

4. Ford Richardson Bryan, *Beyond the Model T: The Other Ventures of Henry Ford*, rev. ed. (Detroit: Wayne State University Press, 1997), 99, 168–70.

5. Tom D. Crouch, "An Airplane for Everyman: The Department of Commerce and the Light Airplane Industry, 1933–37," in *Innovation and the Development of Flight*, ed. Roger D. Launius (College Station: Texas A&M University Press, 1999), 168, 172, 178–80.

6. Contrary to a persistent myth within the postwar aviation community, the W-1 did not take first prize in the federal government's $700 Airplane for Everyman competition. In fact, it was never even a contestant. This misconception likely stems from the fact that the government used wind tunnel test results from the W-1 to help set the standards for the contest, then later purchased Weick's experimental plane for further testing. Weick and Hansen, *From the Ground Up*, 137–40, 145–48; Crouch, "An Airplane for Everyman," 178.

7. Richard Thruelsen, "Fly It Yourself," *Saturday Evening Post*, Dec. 21, 1940, quotes from 20, 52.

8. Weick and Hansen, *From the Ground Up*, 148–49.

9. Wolfgang Langewiesche, *Stick and Rudder: An Explanation of the Art of Flying* (New York: McGraw Hill, 1944, 1972, 1990), 164–65, 180–85, 336–45.

10. Weick also utilized differential ailerons, which decreased the drag induced by the downward deflected aileron, to help eliminate adverse yaw and mounted the Ercoupe's engine several degrees downward and to the right of centerline to do away with the need to use the right rudder pedal to overcome the effect of propeller torque during takeoff and climb at full power. Weick and Hansen, *From the Ground Up*, 174–76; Thomas A. Horne, "The Ercoupe," *AOPA Pilot* 23, no. 3 (Mar. 1980): 124.

11. Prewar Ercoupes used a single brake handle; postwar models used a car-style pedal instead. Thruelsen, "Fly It Yourself," 20; Max Karant, "The Ercoupe," *Flying*, Aug. 1946, 34, 78, 80.

12. Engineering and Research Corp., "When You Drive a Car, You Have to Coordinate Footwork and Handwork" [advertisement], *Aviation*, Dec. 1945, 301.

13. Maxwell Hamilton, "I Flew My First Time Up!" *Mechanix Illustrated*, Sept. 1941, 35.

14. Karant, "The Ercoupe," 78.

15. See chapter 2 for a more detailed discussion of the causes and outcomes of stalls and spins.

16. Karant, "The Ercoupe," 78, 80. Later articles describe how determined pilots can force the Ercoupe into a "whip stall" by purposely diving to pick up speed, then pulling

back abruptly on the controls, but even then only part of the wing stalls, and the pilot remains in full control of the airplane. Thomas A. Horne, "Friendly Flier: Fred Weick's Classic Straddles the Generations," Budget Buy, *AOPA Pilot* 49, no. 3 (Mar. 2006): 58.

17. As early as 1940, the federal government made special allowances for individuals who learned to fly "in aircraft fully determined to be characteristically incapable of spinning," on the condition that these pilots were restricted to flying only this kind of aircraft until they received additional instruction in a conventional airplane capable of stalls and spins. Civil Aeronautics Authority, *Civil Air Regulations Part 20: Pilot Certificates*, effective May 1, 1940 (Washington, DC: U.S. GPO, 1940), 3–4.

18. "Arrested Development," News Sidelights, *Aviation Week*, July 26, 1948, 7; Weick and Hansen, *From the Ground Up*, 227–29.

19. Aeronca Aircraft Corp. built two prototypes of the two-control Chum under license from ERCO, but this model never entered production. "Have You Seen?" *Flying*, Nov. 1946, 53.

20. "Sell Airplanes over the Counter," Marketing, *Aviation*, Nov. 1945, 111; Karant, "The Ercoupe," 80; Horne, "The Ercoupe," 132; MSRP for a new 1946 Cadillac model 62 from National Automobile Dealers Association, NADA Guides, "Classic Car Values and Specs," http://www.nadaguides.com/Classic-Cars.

21. William F. Trimble, "Collapse of a Dream," in *From Airships to Airbus: The History of Civil and Commercial Aviation*, vol. 2, *Pioneers and Operations*, ed. William F. Trimble (Washington, DC: Smithsonian Institution Press, 1995), 135–36, 139, 141–42.

22. Weick and Hansen, *From the Ground Up*, 234.

23. At least two other companies have owned the Ercoupe design rights, Universal Aircraft Industries during the early 1950s and UNIVAIR after 1974, but they manufactured only spare parts (not complete airplanes) to support the existing fleet. Horne, "Friendly Flier," 61–62.

24. "New Airplane from Alon," Traffic Pattern, *Flying*, Dec. 1965, 113. Although the historical record is sparse on this subject, presumably this airplane was based on Weick's earlier "Ercoupe Four" design that he worked on in the 1940s while employed by ERCO.

25. Total production numbers for Ercoupes and later variants depend on source. For instance, see Ed Burkhead, "Ercoupers FAQ," Ercoupe Owners Club, http://www.ercoupe.org/ercoupers_faq.htm (5,591 total aircraft); Matt McDaniel, "My Other Plane Is an Ercoupe," *CIRRUS Pilot*, Jan.–Feb. 2010, 22, http://www.progaviation.com/articles/MoP_ercoupe.pdf (5,685 total aircraft).

26. Cessna Aircraft Co., "Bride's Eye View of Tomorrow" [advertisement], *Flying*, July 1942, inside front cover.

27. Cessna Aircraft Co., "To the Family That Can't Take a Vacation Trip This Year" [advertisement], *Flying*, July 1943, inside front cover.

28. Cessna Aircraft Co., "One Letter . . . To Answer Thousands from You" [advertisement], *Flying*, Oct. 1945, 15.

29. Cessna Aircraft Co., "We Did It Before and We're Planning to Do It Again" [advertisement], *Flying*, Nov. 1945, 19.

30. Cessna Aircraft Co., "One Letter . . . ," 15.

31. Weick and Hansen, *From the Ground Up*, 211–12.

32. Fred Weick held patents on tricycle landing gear, simplified aircraft controls, and several design features that made airplanes unlikely to stall or spin. Before World War II he offered to license other manufacturers who wished to incorporate any of these innovations. Whether the prospect of paying license fees had any bearing in Cessna's decision to abandon its plans to build an easy-to-fly Family Car of the Air is unclear. Fred E. Weick, Airplane, U.S. Patent 2,110,516, filed Jan. 18, 1934, and issued Mar. 8, 1938; correspondence between Fred Weick & Associates and various aircraft manufacturers, Apr. 1938 to Mar. 1939; reproduced as documents 3-16a through 3-16e in James R. Hansen et al., eds., *Reinventing the Airplane*, vol. 2, *The Wind and Beyond: A Documentary Journey into the History of Aerodynamics in America* (Washington, DC: NASA, 2007), 321–39, 340–43.

33. Mitch Mayborn and Bob Pickett, *Cessna Guidebook*, vol. 1, 2nd ed. (Dallas: Flying Enterprise Publications, 1976), 38–40; Cleo M. Bickford and Buddy McGown, eds., *The 170 Book*, 4th ed. (Lebanon, MO: International Cessna 170 Association, 1999) 3, 10; Bill Cox, "The Littlest Cessna," Featured Aircraft, Cessna Owner Association, Mar. 31, 2009, http://www.cessnaowner.org/features/5-cessna-120-140-littlest-cessna.html.

34. These numbers do not include the nearly 65,000 tricycle gear Cessna 150/152 and 172 aircraft manufactured between 1956 and 1986, which if combined with the company's first generation of postwar entry-level planes would make the ratio nearly 14:1. "Cessna Single Engine Production Totals by Model and Year," in John Frank, *Cessna 172 Skyhawk Buyers Guide* (Santa Maria, CA: Cessna Pilots Association, 2009): 26, table.

35. Gordon Baxter, *How to Fly, for People Who Are Not Sure They Want To* (New York: Summit Books, 1981), 264–65.

36. Thruelsen, "Fly It Yourself," 20, 51, 52.

37. Karant, "The Ercoupe," 34.

38. John F. Welch, ed., *Van Sickle's Modern Airmanship*, 5th ed. (New York: Van Nostrand Reinhold, 1981), 412–13, 425–26.

39. Karant, "The Ercoupe," 78, 80; J. Mac McClellan, "Stalling: The Good, Bad and Ugly," Left Seat, *Flying*, May 1995, 11; Horne, "The Ercoupe," 124; Horne, "Friendly Flier," 60.

40. Thruelsen, "Fly It Yourself," 20.

41. Karant, "The Ercoupe," 78.

42. Horne, "The Ercoupe," 127.

43. Max Karant, "The Ryan Navion," Flying's Check Pilot, *Flying*, Mar. 1948, 73.

44. W. L. Jones, The Forum, *Air Pilot and Technician*, Sept. 1943, 30.

45. Swanee Taylor, "Simplified Control: Eliminating the Rudder Is a Major Issue in Light Plane Design," *National Aeronautics* 23, no 10 (Oct. 1945): 16.

46. A. Naughton Lane, "Simplified Control," Contact, *National Aeronautics* 23, no 12 (Dec. 1945): 4.

47. Frank Kingston Smith, *Week-End Pilot* (New York: Random House, 1957), 206.

48. Ray Withman, "Pilots Have Argued for Years—Now Decide for Yourself: How Good Is the Aircoupe?" *Science and Mechanics*, Apr. 1958, 91, 94.

49. Alice S. Fuchs, "*Flying's* Check Pilot Report on the Forney Aircoupe," *Flying*, Apr. 1957, 43.

50. Patricia Cayo Sexton, *The Feminized Male: Classrooms, White Collars, and the Decline of Manliness* (New York: Random House, 1969), 15.

51. Lloyd Mangrum, as told to Don Downie, "From Fairways to Airways," *AOPA Pilot* 1, no. 10 (Dec. 1958): 26.

52. Weick and Hansen, *From the Ground Up*, 300.

53. "Boys, 11 and 12, Fly Stolen Plane 120 Miles on Knowledge Obtained from Comic Books," *New York Times*, May 22, 1948, 17.

54. "Boy of 13 Who Watched a Flier Fly Does It Himself," *New York Times*, Sept. 15, 1956, 1, 15.

55. Fred Weick, "Development of the ERCOUPE . . . An Airplane for Simplified Private Flying," reprint of paper presented at the Society of Automotive Engineers meeting in Washington, DC, Mar. 13, 1941, *SAE Transactions* 49, no. 6 (Dec. 1941): 529.

56. Weick and Hansen, *From the Ground Up*, 183–84.

57. "More Fog," The Accident Corner, *AOPA Pilot* (Apr. 1953): 50f.

58. "Meet the Missus," Safety Corner, *AOPA Pilot* 1, no. 1 (Mar. 1958): 47.

59. "Big Night," The Accident Corner, *AOPA Pilot* (May 1955): 50g.

60. "Bottle Confidence," The Accident Corner, *AOPA Pilot* (Aug. 1955): 50g.

61. "'Damn-fool' proof" quote from Alfred L. Wolf, "What Private Flying Needs from You," preprint of paper presented to the Philadelphia Section of the Institute of the Aeronautical Sciences, Sept. 24, 1946, 16, Smithsonian National Air and Space Museum, NASM Tech. Files P0000100, GA General no. 1, folder no. 1.

62. For examples of manufacturer's claim for "world's safest airplane," see Engineering and Research Corp., "When You Drive a Car," 301; Sanders Aviation, Inc., "Big News for '48! The Improved ERCOUPE has More Power . . . and Is Still the World's Safest Airplane" [advertisement], *Flying*, Mar. 1948, 75.

63. I. W. Cole, "Flying Professors," *Flying*, Jan. 1953, 34, 46.

64. "Michigan Ercoupers," *Flying*, Sept. 1954, 32–33.

65. Gordon Baxter, "The $5,000 Airplane," *Flying*, Apr. 1979, 66.

66. Horne, "The Ercoupe," 126–27; Marc E. Cook, "Fifty Years of Ercoupe," *AOPA Pilot* 33, no. 10 (Oct. 1990): 95; Bill Werner, "So WHOSE [*sic*] the sissy????" discussion thread posted on Ercoupers.com forum, Feb. 12, 2001, http://www.mail-archive.com/coupers@topica.com/msg00691.html.

67. Physically disabled pilots, particularly paraplegics and amputees who relied on the Ercoupe's lack of rudder pedals to allow them to fly without legs, were exempt from negative labels in articles and written comments about the Ercoupe. Rus Walton, "Wheel-Chair Aces," *Flying*, Mar. 1947, 46–47, 99.

68. Arthur L. Klein, "The Future of Aircraft Engineering," speech delivered to the Society of Automotive Engineers in Detroit, Jan. 10, 1955, reprinted in *Engineering and Science* 18, no.4 (Jan. 1955): 22. Klein also revealed his dismissive attitude toward private flying during an oral history interview years later when he declared: "Aircraft are capital goods—except the small ones, the little private airplanes; I was never interested in that class of airplane." Arthur L. Klein, transcript of oral history interviews with Harriett Lyle and John Greenberg, Pasadena, CA, Feb. 1979, and Palos Verdes Estates, CA, Apr. 1982, Oral History Project, California Institute of Technology Archives, 21, http://oralhistories.library.caltech.edu/102/01/OH_Klein_A.pdf.

69. "You Can't Sell Safety," editorial, *AOPA Pilot* (Aug. 1956): 50b. For an overview of the AOPA's earlier efforts to promote safety through better aircraft design, see "Safety in Plane Design Lacking," *AOPA Pilot* (Dec. 1945): 74b; "Safety and 1948," *AOPA Pilot* (Feb. 1948): 50c; "Do You *Want* a Safer Airplane?" editorial, *AOPA Pilot* (Aug. 1949): 50a; "Okay Boys, What Now?" editorial, *AOPA Pilot* (Sept. 1949): 50a; "We *Will* Get More Safety," editorial, *AOPA Pilot* (Aug. 1950): 50a; "Do You *Really* Want Safety?" editorial, *AOPA Pilot* (Jan. 1952): 50a.

70. National Highway Traffic Safety Administration, Presidential Initiative for Increasing Seat Belt Use Nationwide, "America's Experience with Seat Belt and Child Seat Use," http://www.nhtsa.gov/people/injury/airbags/Archive-04/PresBelt/america_seatbelt.html; U.S. Department of Transportation, National Highway Traffic Safety Administration, *Quick Reference Guide (2010 Version) to Federal Motor Vehicle Safety Standards and Regulations* (Washington, DC: National Highway Traffic Safety Administration, Feb. 2011), ii; Ralph Nader, *Unsafe at Any Speed: The Designed-In Dangers of the American Automobile* (New York: Grossman Publishers, 1965), 295–346.

71. James T. Bruce III and John B. Draper, *Crash Safety in General Aviation Aircraft* (copyright James Bruce and John Draper, Jan. 1970), quotes from 87, 48. Excerpts (including these two quotes) also appeared in "The Nader Report," *AOPA Pilot* 13, no. 3 (Mar. 1970): 58–59.

72. Quote from Bruce and Draper, *Crash Safety in General Aviation Aircraft*, 4–5. The Air Commerce Act of 1926 called for the federal government to certify new aircraft in a process that included federal inspection and approval of aircraft designs as well as flight testing a sample aircraft prior to final certification. Minimum engineering standards for aircraft were first described in detail in a handbook issued by the Department of Commerce in October 1927. Nick A. Komons, *Bonfires to Beacons: Federal Civil Aviation Policy under the Air Commerce Act, 1926–1938* (Washington, DC: Smithsonian Institution Press, 1989), 98–99.

73. Bruce and Draper, *Crash Safety in General Aviation Aircraft*, 12.

74. Hamilton, "I Flew My First Time Up!" 35.

75. Wesley Price, "Look Out! She's Stalling!" *Saturday Evening Post*, Oct. 25, 1947, 12; Safe Flight Instrument Corp., "Safe Flight Celebrates 60th," News and Press Releases, ca. 2006, http://www.safeflight.com/mmain.php?px=1&cm=3&cs=109&css=223.

76. Clarence H. Oskey, "Quit Stalling," The Mail Box, *Flying*, July 1948, 83.

77. William A. Scott, "Quit Stalling," The Mail Box, *Flying*, July 1948, 84.

78. Ed Hoadley, "Back-Seat Pilot," *Flying*, Jan. 1952, 31.

79. Edwin D. Merry, "Float Plane Trolling," *Flying*, Sept. 1958, 77.

80. Norman Jacobshagen, "Beech Debonair," Check Pilot Report, *Flying*, June 1960, 90.

81. Richard L. Collins, "Standards," *Flying*, May 1969, 97.

82. Charles L. Black, "AOPA's First 10 Years," *Flying*, Aug. 1949, 74.

83. "AOPA's 1949 Annual Report," *AOPA Pilot* (July 1950): 50c.

84. General aviation aircraft estimated by subtracting the number of airline aircraft in service in 1949 (913) from the total number of civilian aircraft in service (92,622). "Aircraft Certification and Airmen Approvals, 1927–50," and "Annual Data—Aircraft in Service, by Make and Model, 1943–1950 [scheduled air carriers]," in U.S. Department

of Commerce, CAA, *Statistical Handbook of Civil Aviation, 1950* (Washington, DC: U.S. GPO, 1950), 25, 52, tables.

85. Not all general aviation aircraft produced in the United States between 1949 and 1974 were equipped with stall warning devices, and some of Safe Flight's 200,000 stall warning devices were almost certainly installed on aircraft produced by non-U.S. companies for overseas use. Still, these figures provide an idea of how widespread stall warning devices were by 1974. Safe Flight Instrument Corp., "Don't Fly without Our Kind of Information" [advertisement], *Flying*, June 1974, 97; "Shipments of Airplanes Manufactured in U.S.," General Aviation Manufacturers Association, *General Aviation Statistical Databook, 2004* (Washington, DC: GAMA, 2005), 10, table.

86. On its sixtieth anniversary in 2006, the company boasted that "Safe Flight products are installed on over two-thirds of the world's aircraft in the general aviation, commercial and military sectors." Safe Flight Instrument Corp., "Safe Flight Celebrates 60th."

87. Weick and Hansen, *From the Ground Up*, 267; Piper Aircraft Co. Service Memo No. 21, "Model PA-22 Rigging Procedure (Supplement to PA-20 Rigging Procedure)," UNIVAIR Aircraft Corp., Aircraft Bulletins, PA-20 & PA-22, http://univairparts.com/pa20_22.php; *Comanche Flight Training Manual* (Vero Beach, FL: Piper Training Center, ca. 1989), 3.2, 3.6; Piper Aircraft Corp., "Anywhere" [advertisement], *Flying*, Mar. 1965, back cover.

88. H. F. King and John W. R. Taylor, *Milestones of the Air: Jane's 100 Significant Aircraft* (New York: McGraw-Hill, 1969), 134; David Thurston, *Design for Flying* (New York: McGraw Hill, 1995), 67.

89. U.S. Department of Transportation, FAA, Type Certification Sheet A18SO, Revision 4, Piper PA-38-112, Jan. 16, 1996.

90. Gordon Baxter, "First Encounter with Flight," *Flying*, July 1980, 68.

91. Max Karant, "'Three-Control' Modification for Ercoupes Permits Only Shallow Slips," AOPA Report, *AOPA Pilot* (Aug. 1950): 50d; Horne, "Friendly Flier," 58; McDaniel, "My Other Plane is an Ercoupe," 24.

92. Edward B. Oliver, "Disservice," VOX Pilot, *AOPA Pilot* (Sept. 1957): 66h.

93. Archie Trammell, "First Flight: The Cadet," *Flying*, Dec. 1968, 34.

94. Weick and Hansen, *From the Ground Up*, 307, 328–52.

95. Thomas A. Horne, "Training Wings," New Pilot, *AOPA Pilot* 37, no. 11 (Nov. 1994): 90; Bruce Landsberg, "Tomahawk Safety Review," Safety Pilot, *AOPA Pilot* 40, no. 2 (Feb. 1997): 115.

96. The interconnected wing tanks in the Cessna 150 (and its replacement, the model 152) simultaneously feed the carburetor using gravity alone, so there is no need (nor is there the option) to switch from one tank to the other, and also no need for an auxiliary fuel pump. Landsberg, "Tomahawk Safety Review," 116.

97. J. Mac McClellan, "Why People Become Pilots," editorial, *Flying*, June 1991, 50.

98. Smith, *Week-End Pilot*, 185–86.

99. Diane Ackerman, *On Extended Wings* (New York: Atheneum, 1985), 6, 164–65.

100. Harty [pseud.], comment posted Mar. 21, 2002, re: discussion thread "Tomahawk or TRAUMAHAWK?" Rising Up Aviation, http://www.risingup.com/forums/airplane-ownership/611-tomahawk-traumahawk.html.

101. Virginia Scharff, *Taking the Wheel: Women and the Coming of the Motor Age* (Albuquerque: University of New Mexico Press, 1991), 64–66, 119–21, 146–64.

102. Thruelsen, "Fly It Yourself," 52.

Chapter 4 · The "Right Stuff" Syndrome

1. Richard B. Weeghman, "The Wrong Stuff," Editor's Memo, *Aviation Consumer*, Jan. 1, 1981, 2.

2. David Thurston, *Design for Flying* (New York: McGraw Hill, 1995), 41.

3. Richard L. Collins, "Airplane Evolution: Cosmetic Touches or Real Improvements?" *Flying*, Aug. 1984, 61.

4. Although the first model 35 prototype flew in late 1945, the aircraft did not actually go on sale until 1947. The last new V-tail Bonanza was delivered to a customer in 1984. Edward H. Phillips, *Beechcraft: Pursuit of Perfection* (Eagan, MN: Flying Books, 1992), 36–40, 78.

5. The V-tail did reduce the plane's overall weight and therefore increased its potential payload. However, published performance figures indicate that an otherwise similar model 33 Debonair straight-tail Bonanza was just as fast as the V-tail model 35. For instance, the 285 hp model 33 V-tail Bonanza enjoyed a 3 mph advantage over an otherwise identical straight-tail model 33. With a published maximum cruising speed of 203 mph (compared to 200 mph), this gave the V-tail a statistically insignificant 1.5 percent advantage. All other things being equal, for a 500-mile flight, a V-tail Bonanza would arrive 2 minutes and 13 seconds sooner than its straight-tail counterpart. By 1980, Beechcraft was advertising that except for a 15-pound difference in useful load, "the performance numbers are identical" for the V-tail and straight-tail versions, effectively renouncing earlier claims that the V-tail made the plane faster by reducing aerodynamic drag. "V-Tail Has Several Advantages, NACA Report Discloses," *AOPA Pilot* (May 1949): 50f; performance figures from Bill Clarke, *The Illustrated Buyer's Guide to Used Airplanes*, 5th ed. (New York: McGraw-Hill, 2000), 266, 270, 276–77; quote from Beechcraft Corp., "Anything Else is Something Less" [advertisement], *AOPA Pilot* 23, no. 3 (Mar. 1980): 28–29.

6. Speeds from Clarke, *Illustrated Buyer's Guide to Used Airplanes*, 270–77; quote from Nigel Moll, "Prized Possessions," *Flying*, July 1985, 58.

7. Published figures indicate that a 1956 Ercoupe 415-D cruises at 112 mph at 75 percent power and 2,300 rpm; this drops to an "economy cruise speed" of 80 mph at 55 percent power and 2,000 rpm. Thomas A. Horne, "The Ercoupe," *AOPA Pilot* 23, no. 3 (Mar. 1980): 132. Cessna 150, Piper J-3 Cub, and Piper PA-38 Tomahawk cruise speeds from Clarke, *Illustrated Buyer's Guide to Used Airplanes*, 167–70, 188, 198–99.

8. "Beech Bonanza," in *The Aviation Consumer Used Aircraft Guide*, ed. Richard B. Weeghman (New York: Aviation Consumer, 1981), 98–99; Thurston, *Design for Flying*, 112; Peter Garrison, "Telling Tails," *Flying*, July 1985, 76; quote from Moll, "Prized Possessions," 55.

9. Clarke, *Illustrated Buyer's Guide to Used Airplanes*, 266.

10. For a more detailed comparison of new plane prices, see chapter 1. Regarding used planes: for example, at $13,125, a brand-new Cessna 172 cost less in 1970 than an 11-year-old 1959 model K35 Bonanza with an estimated price of $15,000. "General

Aviation Aircraft—1970," *AOPA Pilot* 13 no. 3 (Mar. 1970): 118–21; "The Bonanza," in "Summer Sale: Used-Plane Pilot Reports," *Flying*, July 1970, 55.

11. Moll, "Prized Possessions," 55–56.

12. For one insider's views regarding why "professional men, as a group, may be more accident prone in the air than other flyers," see Robert L. Wick Jr., M.D., "A Professional Confession," *AOPA Pilot* 6, no. 12 (Dec. 1963): 28–30. Opposing viewpoints tended to come from the very type of person whom Wick implicated. See letters from Robert W. Carver, D.O., and Mark S. Kochevar, M.D., "Professional Confession," VOX Pilot, *AOPA Pilot* 7, no. 3 (Mar. 1964): 19–20.

13. Captain "X" [pseud.] and Reynolds Dodson, *Unfriendly Skies: Revelations of a Deregulated Airline Pilot* (New York: Doubleday, 1989), 71–72.

14. Richard L. Collins, "V-Tail," *Flying*, Mar. 1982, 63.

15. For an early description of the general Bonanza in-flight breakup scenario written by someone who was decidedly biased in favor of the airplane, see Frank Kingston Smith, *Week-End Pilot* (New York: Random House, 1957), 211–14. Three decades and numerous fatal crashes later, Bonanza supporters continued to describe the problem in essentially the same terms. See J. Mac McClellan, "V-Tail Bonanza: A Tail of Controversy," *Flying*, Feb. 1985, 37–43.

16. Kurt Hoover, Wallace T. Fowler, and Ronald O. Stearman, "Studies in Ethics, Safety, and Liability for Engineers: The V-tail Bonanza," University of Texas at Austin, n.d., 2, www.engr.sjsu.edu/nikos/courses/ae171/V-Tail%20Bonanza.doc; Brent Silver, "The V-Tail Bonanza—Breaking of a Legend," *Aviation Consumer*, Feb. 1980, 4–15. In 1968, Beechcraft began officially marketing the model 33 Debonair under the Bonanza name, but purists insisted for decades that only the V-tail model 35 was a "real" Bonanza. A longtime pilot and aviation commentator captured this attitude when he wrote, "The Bonanza Society is so exclusive that the V-tail people slight the straight-tail people with the whispered curse 'Debonair,' a designation Beechcraft dropped long ago. At the factory today they are all Bonanzas, but within the cult, a true Bonanza must have a rudder that shares the same bed as your elevators, toe to toe in an elegant vee." Gordon Baxter, "Who Flies Bonanzas? The Inside Scoop on This Flying Cuisinart," Bax Seat, *Flying*, Feb. 1984, 16. Information about the origins and evolution of the model 33 Debonair from R. L. C. [Richard L. Collins], "Beechcraft Debonair," Used Airplane Report, *Flying*, Dec. 1994, 86.

17. Russell Munson, "The Longest Run," *Flying*, May 1980, 126–27; McClellan, "V-Tail Bonanza," 38.

18. Hoover, Fowler, and Stearman, "Studies in Ethics," 2.

19. Bert Shivers, "More Bonanza Comments," Notes & Letters, *Aviation Consumer*, June 1, 1980, 3.

20. J. Jefferson Miller, "Tri-Tail Bonanza: Playing It Straight with a Nagging Question of Structural Integrity," *AOPA Pilot* 28, no. 11 (Nov. 1985): 70–74, quote from 71.

21. Bill Prymak, "Butterfly Bonanza," Notes & Letters, *Aviation Consumer*, May 1, 1980, 3.

22. Edward W. Constant II, "A Tale of Two Bonanzas: How Knowledgeable Communities Think about Technology," *Technology and Culture* 47, no. 2 (Apr. 2006): 271, 283. On November 28, 1994, FAA Airworthiness Directive AD 94-20-04 became

effective, mandating the installation of new tail reinforcements and other measures to prevent potential in-flight problems related to the V-tail. For details, see "V-Tail Fact Sheet—C35 through V35B," American Bonanza Society, ca. 1994, http://www.bonanza.org/downloads/VTailFactShtC35-V35B.pdf; Department of Transportation, FAA, [Docket No. 93-CE-37-AD; Amendment 39-13147; AD 94-20-04 R2] "Airworthiness Directives; Raytheon Aircraft Company Beech Models C35, D35, E35, F35, G35, H35, J35, K35, M35, N35, P35, S35, V35, V35A, and V35B Airplanes" (Washington, DC: U.S. GPO, 2003), http://rgl.faa.gov/Regulatory_and_Guidance_Library/rgAD.nsf/0/21b1b224dff2889286256d2600546895/$FILE/942004R2.pdf.

23. Bruce Landsberg, "Bonanza Safety Review," *AOPA Pilot* 37, no. 2 (Feb. 1994): quote from 106; Constant, "A Tale of Two Bonanzas," 271.

24. Constant's belief that statistical analysis alone can explain not only human activity, but also deeply held cultural attitudes, proves unsatisfactory to some who are familiar with the aviation community. For a thoughtful challenge to Constant's article, see Stephen L. Thompson, "Bonanza Belief-O-Meter," Communications, *Technology and Culture* 48, no. 3 (July 2007): 680–83.

25. In this case the left wing, not the tail, reportedly broke off as the pilot apparently attempted to pull out of a dive at low altitude after losing control in the clouds. Since later accidents revealed that a wing (or both wings) often separated from the plane after the tail had failed, it is impossible, with the scant details provided, to determine whether the tail was a factor in this case. Regardless of the tail's role, it is clear that the AOPA blamed the pilot, not the plane. "Case Histories of Crashes Contain Lessons for All Pilots," The Accident Record . . . , *AOPA Pilot* (Nov. 1948): 50d.

26. "Thunderstorm" and "More Weather," both in The Accident Corner, *AOPA Pilot* (Sept. 1954): 50f–50g.

27. Smith, *Week-End Pilot*, 211–13.

28. "Circuit Court Upholds Beech in Bonanza Weather Accident Case," Legally Speaking, *AOPA Pilot* 1, no. 9 (Nov. 1958): 12–13, 43–44, quote from 13.

29. Bruce Landsberg, "Landmark Accidents: The Day the Music Died," Safety Pilot, *AOPA Pilot* 52, no. 2 (Feb. 2009): 64–69. For chilling accounts of narrow escapes from this sort of accident spanning two decades, see George E. Hickey, "Terror in the Soup," Never Again! *AOPA Pilot* 5, no. 10 (Oct. 1962): 58–59; W. G. Bradley, M.D., "To Thine Own Self Be True," Never Again! *AOPA Pilot* 21, no. 11 (Nov. 1978): 94–95; Gary Ramelli, "Just What the Doctor Didn't Order," I Learned about Flying from That! *Flying*, Jan. 1981, 92–93. Given that these near-accidents involved the "doctor killer" Bonanza, it is noteworthy that two of these three pilots identified themselves as doctors (the third neglected to mention his occupation).

30. Vern Ballenger, "V-Tail Views," Flying Mail, *Flying*, May 1985, 10, 12.

31. "Speed Control in Dives," On the Airways, *AOPA Pilot* 8, no. 4 (Apr. 1965): 91; H. A. Slingsby, *Technical Report ADS-12: Limiting of Aircraft Speed by Automatic Engine Thrust Control* (Washington, DC: FAA, 1963).

32. Moll, "Prized Possessions," 55–56.

33. Edward G. Tripp, "The Bonanza Mystique," *AOPA Pilot* 25, no. 2 (Feb. 1981): 35, 37–38.

34. Ibid., 37–38.

35. Ibid., 38.

36. McClellan, "V-Tail Bonanza," 37–38.

37. Ibid., 43.

38. Percentage of all pilots who held an instrument rating does not include student pilots, who by law cannot be instrument rated. Richard L. Collins, On Top, *Flying*, Nov. 1970, 20; "Table 31—U.S. Civil Aircraft by Manufacturer and Model, Number of Seats, and Powerplant Data: Piston Aircraft (Fixed Wing)," in U.S. Department of Transportation, FAA, Office of Management Systems, *Census of U.S. Civil Aircraft*, Calendar Years 1970 and 1971 (Washington, DC: U.S. GPO, 1971), 86, table; "Table 7.11—Active Pilots by Type of Certificate, as of January 1, 1971," U.S. Department of Transportation, FAA, *Statistical Handbook of Aviation* (Washington, DC: U.S. GPO, 1972), 177, table.

39. Norbert Aubuchon, "Wisdom of the Solomons," *Flying*, Mar. 1974, 97; "Table 24—U.S. Civil Aircraft by Manufacturer and Model, Number of Seats and Powerplant Data: Piston Aircraft (Fixed Wing)," in U.S. Department of Transportation, FAA, Office of Management Systems, *Census of U.S. Civil Aircraft*, Calendar Year 1974 (Washington, DC: U.S. GPO, 1974), 146, table; Harry G. Hadler, "Letter from the President," *American Bonanza Society Newsletter*, May 1985, 1600; "Young Eagles Win Bonanza," Reporting Points, *Flying*, Apr. 1994, 28; Carl Chance, "About the American Bonanza Society History," Wings Over Kansas, Dec. 7, 2009, http://www.wingsoverkansas.com/legacy/a298/.

40. Aubuchon, "Wisdom of the Solomons," 96, 97.

41. Ira B. Hartzog, President, Hartzog Aviation, Inc., "A Response to the *Aviation Consumer* Article," *American Bonanza Society Newsletter*, Dec. 1978, 848; Virgil Fisher, "To Speak Up *for* the V-Tail," *American Bonanza Society Newsletter*, May 1986, 1748.

42. J. Norman Colvin, "News and Views: FAA Approves V-Tail Beef-Up C35 through V35B," Colvin's Corner, *American Bonanza Society Newsletter*, Aug. 1981, 1097.

43. Description from "ABS/ASF Bonanza Pilot Proficiency Program," *American Bonanza Society Newsletter*, Feb. 1985, 1573. The ABS introduced this course under the name Bonanza Pilot Checkout Program in 1983. By early 1985 it had changed the name to Bonanza Pilot Proficiency Program, perhaps to remove the potential stigma attached to the term *checkout* (within the aviation community, pilots are "checked out" in a plane before flying it for the first time; by this definition, signing up for a "checkout" would not seem appropriate to experienced Bonanza pilots). "ABS Bonanza Pilot Checkout Program, Rockford, Illinois, October 7th thru 9th," *American Bonanza Society Newsletter*, Aug. 1983, 1360.

44. Scott Mathews, "BPPP Gets Better and Better," Forum, *American Bonanza Society Newsletter*, Dec. 1997, 5028; Ray L. Leadabrand, "BPPP Testimonial," *American Bonanza Society Newsletter*, Jan. 1993, 3135; Jim Slavik, "More Praise for the BPPP!" For the Record, *American Bonanza Society Newsletter*, July, 1990, 2549.

45. Just as "standard" (manual) transmissions are no longer "standard equipment" in automobiles sold in the U.S., having been replaced by automatic transmissions, "conventional landing gear" ceased to be the industry standard for personal planes by the mid-1950s. Even so, the label remains in use to this day to officially differentiate "taildraggers" from airplanes equipped with tricycle landing gear.

46. B. Holland, "No, a Jury of Your Peers . . . " [cartoon], *AOPA Pilot* 19, no. 10 (Oct. 1976): 109.

47. Smithsonian National Air and Space Museum, "Langley Aerodrome A," http://www.nasm.si.edu/collections/artifact.cfm?id=A19180001000.

48. Glenn Research Center, NASA, "Center of Gravity—CG," http://www.grc.nasa.gov/WWW/K-12/airplane/cg.html.

49. H. S. Plourde, *The Compleat [sic] Taildragger Pilot* (Goffstown, NH: Mugguette B. Plourde, 2004), 18–19.

50. Walter G. Vincenti, "The Retractable Airplane Landing Gear and the Northrop 'Anomaly': Variation-Selection and the Shaping of Technology," *Technology and Culture* 35, no. 1 (Jan. 1994): 1–33.

51. Most early pilots learned to fly without benefit of manuals. W. David Lewis, "Historical Introduction," in *Fighting the Flying Circus*, by Captain Edward V. Rickenbacker (originally published in 1919; edited edition with a historical introduction by W. David Lewis published in Chicago: R. R. Donnelley & Sons, 1997), xliv–xlv (references are to the 1997 edition); Fred Weick and James R. Hansen, *From the Ground Up: The Autobiography of an Aeronautical Engineer* (Washington, DC: Smithsonian Institution Press, 1988), 21–31; Roscoe Turner and Jean H. Dubuque, *Win Your Wings: Book One* (Chicago: Frederick J. Drake, 1941), 233–62, groundloops described on 242.

52. Although a few earlier aircraft like the Curtiss D-3 "Headless Pusher" (ca. 1909–12) were equipped with fixed (nonsteerable) nosewheels, Weick insisted that the designers of these planes had not sought to solve the groundloop problem but were instead simply trying to make their aircraft planes less likely to flip over during landing. In his autobiography, Weick also related how Orville Wright took him aside during the 1936 NACA conference to tell him that the tricycle gear innovation was (in Weick's words) "a new line of thought and a worthwhile improvement." Weick and Hansen, *From the Ground Up*, 149; Smithsonian National Air and Space Museum, "Curtiss D-III 'Headless Pusher,'" http://www.nasm.si.edu/collections/artifact.cfm?id=A19280009000.

53. Weick and Hansen, *From the Ground Up*, 148–49.

54. Joseph J. Corn, *The Winged Gospel: America's Romance with Aviation, 1900–1950* (New York: Oxford University Press, 1983), 74–75.

55. Dominick A. Pisano, *To Fill the Skies with Pilots: The Civilian Pilot Training Program, 1939–1946* (Washington, DC: Smithsonian Institution Press, 2001), 66–67, 126.

56. Max Karant, "Piper Tri-Pacer Has Rugged Gear, Good Performance," *AOPA Pilot* (June 1951): 50c–50d.

57. "New Membership Contest Launched; Prizes Galore," *AOPA Pilot* (Feb. 1956): 50a. The Tri-Pacer represents a transitional phase in aircraft design, and by the early 1960s Max Karant had reconsidered his assessment of the model and noted that it could be difficult to handle on the ground because the former taildragger's ad hoc factory conversion to tricycle landing gear resulted in an airplane that sat high above the runway on a relatively narrow wheelbase. In an article about the Tri-Pacer's successor, the Piper Cherokee, Karant stated: "It's no secret that the Tri-Pacer is a 'hot' airplane, has a high ground accident rate, and is as likely as not to blow over while standing still if you're not careful where and how you park." Max Karant, "Cherokee: The Plane with a Future," *AOPA Pilot* 4, no. 8 (Aug. 1961): 36.

58. Max Karant, "Piper's 1953 Pacers Have Bigger Cabins, Excellent Sound Proofing," *AOPA Pilot* (Jan. 1953): 50d–50e.

59. "Pacer Discontinued, More Powerful New Tri-Pacer Unveiled," *AOPA Pilot* (Jan. 1955): 50c; "This Cessna Features a 'Land-o-Matic' Gear," *AOPA Pilot* (Jan. 1956): 50e; Smith, *Week-End Pilot*, 217.

60. Production of the two-seat Cessna 140 taildragger ended in 1950, and it was not replaced until Cessna introduced the tricycle model 150 in 1959. There was one exception to the company's otherwise universal switch from conventional to tricycle gear: Cessna continued to produce the taildragger model 180 Skywagon alongside its tricycle gear 182 Skylane up until 1981. However, the Cessna 180 catered mainly to a niche market of backcountry bush pilots who preferred tailwheel-equipped airplanes for landing on rough terrain. Clarke, *Illustrated Buyer's Guide to Used Airplanes*, 166, 289, 347.

61. An earlier government study of all general aviation accidents that occurred in 1964 reported similar findings for the same aircraft models, which, the report noted, had been " 'transitioned' from tail wheel configurations to tricycle gear configurations" by the manufacturer (e.g., Cessna's models 170 and 172). This suggests that the high incidence of taildraggers involved in groundloop accidents in the 1970s was not simply a result of fewer pilots having originally learned to fly with conventional landing gear. NTSB figures for the 1970s from Clarke, *Illustrated Buyer's Guide to Used Airplanes*, 473, 477–78; 1960s NTSB findings from National Transportation Safety Board, Bureau of Safety, Civil Aeronautics Board, *Aircraft Design-Induced Pilot Error* (Washington, DC: U.S. Department of Transportation, July 1967), xvii, 56.

62. Dennis W. Newton, "Some Can, Some Can't: Tame a Taildragger," *Plane & Pilot*, Aug. 1970, 44–49.

63. Max Karant, "Tricycle-Gear Does Wonders for Cub, Already Has CAA Approval," *AOPA Pilot* (Sept. 1949): 50f; "Piper Devises Tricycle-Gear for Pacer, Calls New Model 'Tri-Pacer,' " *AOPA Pilot* (Mar. 1951): 50b.

64. "Tri Cessna," *AOPA Pilot* (July 1955): 50e; "Cessna 180 Gets Tricycle Gear, Performs Well," *AOPA Pilot* (Aug. 1956): 50c; Met-Co-Aire, "Tri-Gear Conversion for the Cessna 120, 140, 140A" [advertisement], *AOPA Pilot* 3, no. 2 (Feb. 1960): 50.

65. George Stone Cary, "The Value of Speed," *AOPA Pilot* 4, no. 4 (Apr. 1961): 33.

66. Jim Weisen, "What Do You Mean . . . ?" [cartoon], *AOPA Pilot* 7, no. 7 (July 1964): 75.

67. "Don't Park it There . . . " [cartoon], *AOPA Pilot* 15, no. 7 (July 1972): 14.

68. Newton, "Some Can, Some Can't," 44–49.

69. Don Downie, "Taming the Taildragger: Toil, Tears, and Sweat—But Worth the Trouble," *AOPA Pilot* 17, no. 9 (Sept. 1975): 41–44, quotes from 41–42.

70. See letters from John G. Tietz and Joseph Mancusi in "Taildragger Techniques," VOX Pilot, *AOPA Pilot* 18, no. 12 (Dec. 1975): 12–13; Glen Kezer, Alan Bramson, John G. Tietz, and Ross A. McLean, plus comments by aerodynamicist Irve Culver, in "Taildragger Techniques—Continued," VOX Pilot, *AOPA Pilot* 19, no. 2 (Feb. 1976): 12–13.

71. Richard Bach, "What Is an Antiquer?" *Flying*, Jan. 1964, 56.

72. "Nosedragger" nickname from Downie, "Taming the Taildragger," 41; "training wheels" from Weeghman, "The Wrong Stuff," 2.

73. K. O. Eckland, "The Old Aerodrome," *AOPA Pilot* 19, no. 8 (Aug. 1976): 60.

74. Roger Rozelle, "Taylorcraft: The Basic Taildragger Lives!" *AOPA Pilot* 22, no. 8 (Aug. 1979): 36–42; Don Downie, "Pilot Flight Check: The Great Great Lakes," *AOPA Pilot* 19, no. 7 (July 1976): 32–36.

75. American Champion resumed production after acquiring the rights in 1990. Clarke, *Illustrated Buyer's Guide to Used Airplanes*, 145–46.

76. Production of the PA-18 Super Cub (including the militarized L-21) began in 1949 and ended in 1982, then resumed in limited numbers between 1988 and 1994. Piper Cub Forum, "Cub Production, 1931–2008," http://www.pipercubforum.com/cubprod.htm; Clarke, *Illustrated Buyer's Guide to Used Airplanes*, 187.

77. Budd Davidson, "Biplanes You Can Build: Affordable Designs for Those Who Want a Pair of Wings," *Sport Aviation* 49, no. 1 (Jan. 2000): 24–25, quote from 24.

78. Lazair, "Kitfox Aircraft History," http://kitfox.lazair.com/skystar/Aircraft_History.htm; Glasair Aviation, "A Bit of History," http://www.glasairaviation.com/history.html; Van's Aircraft, "RV-6/6A," http://www.vansaircraft.com/public/rv-6int.htm.

79. Light Plane Components, "PA22-20 Conversions: Convert Your Tri-Pacer to a Tail Dragger" [advertisement], *29th Annual International EAA Convention & Sport Aviation Exhibition* [convention program] (Oshkosh, WI: EAA, 1981), 113, Smithsonian National Air and Space Museum, NASM Tech. Files J1-1981-600-1; U.S. Department of Transportation, FAA, "Supplemental Type Certificate No. SA45RM," issued to Univair Aircraft Corp. on Feb. 20, 1975, reissued July 11, 1981 (photocopy of s/n 22-4829 in author's files).

80. Donald Chase, "The Pendulum Swings . . . and an Ohio Firm Thrives on Converting Tri-Gear Lightplanes Back into Taildraggers," *AOPA Pilot* 21, no. 11 (Nov. 1978): 90–93, quote from 93.

81. Chase, "The Pendulum Swings," quotes from 91, 93.

82. Mark Twombly, "Cessna 140: Flying This Classic Airplane Is Only Half the Fun," *AOPA Pilot* 28, no. 7 (July 1985): 56.

83. This began to change in the 1990s (after the period covered by this book) when Cirrus Designs became the first light-plane manufacturer to receive FAA approval to install airframe parachutes in personal planes. Activated by the pilot in an emergency, the system uses a rocket to rapidly deploy the parachute, which then lowers the entire aircraft safely to the ground while its occupants remain strapped in their seats. Since that time, the FAA has issued Supplemental Type Certificates so owners can retrofit certain used aircraft with airframe parachutes, but adoption remains far from universal. Cirrus Aircraft, "Cirrus Backgrounder," ca. 2011, http://cirrusaircraft.com/static/pricesheets/press/Press%20backgrounder.pdf; BRS Aviation, "Manufacturers of Whole Aircraft Parachute Systems," http://brsparachutes.com/brs_aviation_home.aspx.

84. Richard L. Collins, "Is Twin-Engine Safety a Myth?" *Flying*, Aug. 1971, 71 [S11].

85. Col. Charles A. Lindbergh, "Lindbergh on Flying: Aircraft for Private Owners and How to Choose One," *New York Times*, Nov. 25, 1928, sec. 11, p. 10.

86. Dr. Paul Nichols [Ph.D.], "Twins vs. Singles?" Pilot Opinion, *Plane & Pilot*, Nov. 1985, 14, 67, quote from 14.

87. Fatality rate based on accidents per 100,000 hours flown in light twins vs. single-engine planes. Richard L. Collins, "Trouble with Twins: Two Engines Don't Guarantee Safety," editorial, *Flying*, Feb. 1985, 20. This disparity in fatality rates likely reflects the fact that when a single-engine plane suffers an engine failure, the pilot has few options but to attempt an off-airport emergency landing. In such cases, a landing in a farmer's field that damages the plane counts as a crash, even though the occupants might walk away uninjured. On the other hand, when an engine fails on a light

twin, generally the pilot either makes a successful emergency landing on one engine at an airport or loses control of the plane at some point during the flight and suffers a catastrophic accident. Berl Brechner, "When the Engine Goes . . . ," *AOPA Pilot* 16, no. 5 (May 1973): 73.

88. Actual figures vary depending on year and source. For instance, two contemporary new-plane reviews indicate that a 1978 Seminole has a single-engine ceiling of either 3,200 feet or 4,100 feet, while a sales brochure for the 2011 Seminole (which has the same maximum weight and horsepower as the 1978 model) lists a single-engine ceiling of 3,800 feet. Roger Rozelle, "Pilot Flight Check: Piper Seminole," *AOPA Pilot* 21, no. 9 (Sept. 1978): 34; Richard L. Collins, "Seminole: Light Twin from Piper," *Flying*, Oct. 1978, 40; "2011 Seminole Pricing and Performance Specifications" (Vero Beach, FL: Piper Aircraft, Inc., 2011).

89. Performance data for Cessna 310R from Richard B. Weeghman, "310 Romeo," *Flying*, Jan. 1975, 53; quote from Richard L. Collins, "Understanding a Twin's Bad Manners," On Top, *Flying*, Apr. 1982, 114.

90. Neil D. Van Sickle, ed., *Modern Airmanship*, 3rd ed. (Princeton, NJ: D. Van Nostrand, 1966), 480. For discussions of the myriad challenges facing light twin pilots in the event of an engine failure, see Collins, "Understanding a Twin's Bad Manners," 114; J. R. Williams, "VMC Myths," *Plane & Pilot*, Feb. 1985, 20–23.

91. Archie Trammell, "Cessna Super Skymaster," Pilot Report, *Flying*, Jan. 1971, 84.

92. Collins, "Trouble with Twins," 20.

93. Richard Wyeroski, "Twin Training," Flying Mail, *Flying*, Apr. 1985, 6.

94. Henry Appleby, "Twins Again," Flying Mail, *Flying*, May 1985, 10.

95. Matt Thurber, J. Mac MacClellan, and Tom Hammitt. "The Odd Squad," *Flying*, Apr. 1985, 38.

96. Cessna Aircraft Co., "The Radically New Cessna Skymaster . . . with CLT" [advertisement], *Flying*, July 1963, 2–3.

97. "New 'Push-Pull' Twin, the Skymaster, Scores Hit at Cessna Sales Meet," *AOPA Pilot* 5, no. 11 (Nov. 1962): 86–87; Max Karant, "CLT—Cessna Breakthrough," *AOPA Pilot* 6, no. 3 (Mar. 1963): 38.

98. Trammell, "Cessna Super Skymaster," 52–57, 84; Marc E. Cook, "Cessna 337 Skymaster: Just Say Three-Three-Seven," *AOPA Pilot* 36, no. 7 (July 1993): 82–91; "Cessna 337 Skymaster: One of the Most Affordable Twins, Skymasters Are Easy to Fly and Offer Good Payload/Range," *Aviation Consumer*, July 2009, 24–31.

99. Richard L. Collins, "The Virtuous Twin," *Flying*, Oct. 1975, 28.

100. Cook, "Cessna 337 Skymaster," 83.

101. R. Kirk Mauldin, "The Role of Humor in the Social Construction of Gendered and Ethnic Stereotypes," *Race, Gender and Class* 9, no. 3 (Jan. 2002): 76–95.

102. Trammell, "Cessna Super Skymaster," 52.

103. Cook, "Cessna 337 Skymaster," 83.

104. Collins, "Virtuous Twin," 28.

105. Thurber et al., "The Odd Squad," 40.

106. Robert I. Stanfield, "Cessna Skymaster," Pilot Report, *Flying*, May 1963, 49.

107. Cessna Aircraft Co., "What's So Funny about 18,000 Feet on One Engine?" [advertisement], *AOPA Pilot* 11, no. 3 (Mar. 1968): inside front cover; Cessna Aircraft Co., "Some People Like to Kid a Winner" [advertisement], *AOPA Pilot* 12, no. 2 (Feb. 1969):

6; Cessna Aircraft Co., "'Flying Tiger' or 'Pussy Cat'?" [advertisement], *AOPA Pilot* 12, no. 8 (Aug. 1969): inside front cover.

108. Cessna Aircraft Co., "Stay Twin Proficient and Still Have Time to Swing" [advertisement], *AOPA Pilot* 13, no. 1 (Jan. 1970): inside front cover; Cessna Aircraft Co., "If You Want to Buy a Skymaster . . . " [advertisement], *Flying*, Sept. 1974, inside front cover.

109. "How Dangerous Is the Skymaster?" *Aviation Consumer*, Nov. 1, 1979, 18–19; George Spacek, "Where There's Smoke, There Are Skymasters?" Queries, *Aviation Consumer*, Sept. 1, 1980, 23; Cook, "Cessna 337 Skymaster," 90.

110. Trammell, "Cessna Super Skymaster," 54–55.

111. Details about rear-engine-out procedures from Collins, "The Virtuous Twin," 98.

112. Richard L. Collins, "Rejected Takeoff," Safety Check, *Flying*, Feb. 1972, 80.

113. "Engine-Out Sensor for Skymasters," What's New? *AOPA Pilot* 16, no. 1 (Jan. 1973): 18.

114. "How Dangerous Is the Skymaster?" 18–19; U.S. Department of Transportation, FAA, "Airworthiness Directive 77-08-05, CESSNA Models 337 Series, T337 Series and M337 Series Airplanes," effective date April 28, 1977, http://www.airweb.faa.gov /Regulatory_and_Guidance_Library/rgAD.nsf/0/40B73D072FD60CCE86256A34006 8370C?OpenDocument&Highlight=77-08-05.

115. Cook's "yaw-spin-boom" cleverly condenses the typical sequence of an asymmetric-thrust light twin accident into just three words. Cook, "Cessna 337 Skymaster," 84.

116. "Cessna 337 Skymaster: One of the Most Affordable Twins," 26; Thurber et al., "The Odd Squad," 40.

117. Collins, "Trouble with Twins," 20.

118. Newton, "Some Can, Some Can't," 49.

119. Robert J. Hamp III, "Cream of the Crop," Flying Mail, *Flying*, July 1979, 6.

Chapter 5 · Hog Wallow Airports, Hangar Flying, and Hundred-Dollar Hamburgers

1. Quotes from Walter J. Boyne, *Flying: An Introduction to Flight, Airplanes, and Aviation Careers* (Englewood Cliffs, NJ: Prentice-Hall, 1980), 15, 8.

2. Gordon Baxter, "First Encounter with Flight," *Flying*, July 1980, 67–69.

3. Ibid., 69.

4. Ibid.

5. Definition of fixed-base operator from Ira Harkey, "Some FBOs I Have Known," *AOPA Pilot* 9, no. 11 (Nov. 1966): 62.

6. Lynn L. Bollinger and Arthur Houghton Tully Jr., *Personal Aircraft Business at Airports* (Boston: Division of Research, Graduate School of Business Administration, Harvard University, 1948), 69.

7. Ibid., 69–70.

8. Ibid., 74.

9. "How Not to Run an Airport," *Flying*, June 1945, 35–42, 107; "How Not to Run an Airport—One Year Later," *Flying*, July 1946, 45–52.

10. "Here's How to Run an Airport," *Flying*, July 1945, 150.

11. "How Not to Run an Airport," 35.

12. "Fewer than Eight Per Cent of Fields Rate Excellent," *AOPA Pilot* (Apr. 1946): 54c.

13. "Help! Help!" editorial, *Skylady*, Sept.–Oct. 1946, 2.

14. Grace Jones, "Keeping UP with the Joneses," *Skylady*, Mar.–Apr. 1946, 5–6.

15. The report cover is marked "RESTRICTED!" and a statement on the first page indicates that the company considered the material to be "confidential." However, there is no indication that the U.S. government assigned any formal level of classification to this document. Henry J. Kaiser, Co., "Kaiser Terminals: 5,000 Airports for Private Flying," Apr. 1944, quotes from 11, 20, 23, 27, 31, Smithsonian Libraries, NASM Library / ALM Special Collections.

16. Ibid., 5, 9, 34.

17. Ibid., 38; entry for Nov. 28, 1944, in U.S. Department of Transportation, FAA, "FAA Historical Chronology, 1926–1996," 36, http://www.faa.gov/about/history/chronolog_history/.

18. U.S. Department of Commerce, CAA, *Statistical Handbook of Civil Aviation*, 1944 ed. (Washington, DC: U.S. GPO, 1944), 11, 12, 16.

19. U.S. Department of Commerce, CAA, *Small Airports* (Washington, DC: U.S. GPO, September 1, 1945), 3. This booklet follows Kaiser's lead by providing standardized templates for airports of various sizes, but it emphasizes local planning, execution, and management with no mention of a Kaiser-style nationwide company to franchise and supervise these facilities.

20. CAA, *Statistical Handbook of Civil Aviation*, 1955 ed. (Washington, DC: U.S. GPO, 1955), 3, 4.

21. J. C. D. Blaine, "The Development of a National Airport Plan," Reports and Comments, *Land Economics* 30, no. 3 (Aug. 1954): 273–74.

22. Henry J. Kaiser, Co., "Kaiser Terminals," 5.

23. Dudley M. Steele, [no title], VOX Pilot, *AOPA Pilot* (Aug. 1946): 50h.

24. "Two-Gal Field," *Flying*, July 1947, 40–41.

25. Jack Rushing, M.D., Address, in *Proceedings of the 1950 Conference [on] Airport Management and Operations Held at the Extension Study Center, University of Oklahoma, October 12–13, 1950* (published ca. 1950 by *Flight* magazine and the University of Oklahoma Extension Division, in cooperation with the CAA), 52, Smithsonian Libraries, NASM Library / ALM Special Collections.

26. "The Men from the Boys," editorial, *AOPA Pilot* (May 1952): 50a.

27. Lois C. Philmus, "A Red Carpet for Flyers," *AOPA Pilot* 1, no. 8 (Oct. 1958): 27.

28. Martin Caidin, "Dream Flight," *AOPA Pilot* 4, no. 2 (Feb. 1961): 61.

29. Joseph Whitehill, "As Rare as Oz," VOX Pilot, *AOPA Pilot* 7, no. 7 (July 1964): 16.

30. C. Pfeiffer Trowbridge, "Flyer's Dream Trip," *AOPA Pilot* 9, no. 10 (Oct. 1966): 22.

31. "Between Adolescence and Adulthood?" editorial, *AOPA Pilot* 4, no. 10 (Oct. 1961): 28.

32. Nola Mae McFillen, "DANGER! Female Airport Operators at Work," *AOPA Pilot* 4, no. 9 (Sept. 1961): 30.

33. Lois C. Philmus, "A Revolution in California," *AOPA Pilot* 4, no 10 (Oct. 1961): 96, 98.

34. Ibid., 98.

35. American Mercury Insurance Co., "AMERICO Salutes Bill Whitesell and His Flying W Ranch" [advertisement], *AOPA Pilot* 5, no. 7 (July 1962): 11.

36. Matthew Dolan, "A Flight of Fancy for Airport Dude Ranch: The New Owners of the Flying W Ranch Airport Hope to Recapture Its '60s Glory Days," *Philadelphia Inquirer*, May 6, 1996, http://articles.philly.com/1996-05-06/news/25624850_1_caves-new-owners-small-corporate-jet.

37. Lois C. Philmus, "Erasing the Old to Make Way for the New," *AOPA Pilot* 8, no. 11 (Nov. 1965): 43–44; "Kaye Group Buys Airport Base Concern," *Los Angeles Times*, Dec. 14, 1966, SF8. The company currently operating under the name Van Nuys Skyways was established in 1985. Dun & Bradstreet Credibility Corp., "Credibility Review: Van Nuys Skyways, Inc.," http://www.dandb.com/businessdirectory/vannuysskywaysinc-burbank-ca-19870789.html.

38. Sally Buegeleisen, "Skirts Flying," *Flying*, May 1967, 94.

39. "Willful Women," Traffic Pattern / News and Views, *Flying*, Aug. 1966, 20, 120; entry for June 6, 1967, in "FAA Historical Chronology, 1926–1996," 123. Sources disagree about who received the inaugural FAA Airport Beautification Award, with several airports claiming this distinction. For instance, a 1971 article claimed that Skaneateles, New York, was "a field so fine that last year, it won the first FAA Airport Beautification Award." Stephan Wilkinson, "Tour New York by Air," *Flying*, Aug. 1971, S16.

40. "Junk Airplanes New FAA Concern," On the Airways, *AOPA Pilot* 13, no. 6 (June 1970): 124–125; "Beautification Award to South Texas Airport," On the Airways, *AOPA Pilot* 13, no. 6 (June 1970): 125–26. For background about the Highway Beautification Act of 1965, see U.S. Department of Transportation, Federal Highway Administration, "How the Highway Beautification Act Became a Law," Highway History, Apr. 7, 2011, http://www.fhwa.dot.gov/infrastructure/beauty.cfm.

41. "Demolition Derby," Reporting Points, *Flying*, Dec. 1972, 21.

42. "The New American Domesticated Male: A Boon to the Household and a Boom to Industry," *LIFE*, Jan. 4, 1954, 42–45.

43. Kristen Haring, *Ham Radio's Technical Culture* (Cambridge, MA: MIT Press, 2007), 138; "A Room of His Own: Havens Reflecting Men's Special Interests," *McCall's*, Aug. 1967, 82–87.

44. A growing literature regarding the "politics of dirt" analyzes the socially constructed relationship between cleanliness, gender, class, and race. See, for instance, Ruth Schwartz Cowan, *More Work for Mother: The Ironies of Household Technology from the Open Hearth to the Microwave* (New York: Basic Books, 1983); Phyllis Palmer, *Domesticity and Dirt: Housewives and Domestic Servants in the United States, 1920–1945* (Philadelphia: Temple University Press, 1990); Judith A. McGaw, "Reconceiving Technologies: Why Feminine Technologies Matter," in *Gender and Archeology*, ed. Rita P. Wright (Philadelphia: University of Pennsylvania Press, 1996), 52–75; Roger Horowitz, "'Where Men Will Not Work': Gender, Power, Space, and the Sexual Division of Labor in America's Meatpacking Industry," *Technology and Culture* 38, no. 1 (Jan. 1997): 187–213.

45. Quote from McFillen, "DANGER! Female Airport Operators at Work," 30.

46. Quote from Boyne, *Flying*, 15. For extensive descriptions and analysis of the so-called masculinity crisis in postwar America, see Michael S. Kimmel, *Manhood in*

America: A Cultural History, 2nd ed. (New York: Oxford University Press, 2006); Miriam G. Reumann, *American Sexual Character: Sex, Gender, and National Identity in the Kinsey Reports* (Berkeley: University of California Press, 2005), 54–85.

47. Jones, "Keeping UP with the Joneses," 6.

48. From 1947 to 1960, the federal government spent $66.7 million on general aviation airports and $506.1 million on airports served by airlines; between July 1, 1970, and December 31, 1971, the ratio was $18.6 million vs. $264.7 million. "1947–61 Federal Aid Program—Number of Airports and Federal Funds Allocated—as of January 1, 1961," in U.S. Department of Transportation, FAA, *Statistical Handbook of Aviation*, 1961 ed. (Washington, DC: U.S. GPO, 1961), 11, table; FAA, *Statistical Handbook of Aviation*, 1972 ed. (Washington, DC: U.S. GPO, 1972), 47.

49. At the end of 1970, there were 11,261 "airport facilities" nationwide, including 10,040 airports, 790 heliports, and 431 seaplane bases. Of these, 3,805 were paved, 3,554 were lighted, and 6,828 were less than 3,000 feet long. "Table 3.2A—U.S. Civil and Joint-Use Airports, Heliports, and Seaplane Bases and Reported Abandonments on Record, by FAA Region and State: December 31, 1970," "Table 3.4A, U.S. Civil and Joint-Use Airports, Heliports, and Seaplane Bases on Record by Length of Longest Runway, by FAA Region and State: December 31, 1970," and "Table 38A, Airports on Record with FAA and Reported Based Aircraft: December 31, 1970," in FAA, *Statistical Handbook of Aviation*, 1972 ed., 49, 53, 58.

50. In 1980, there were 12,240 airports nationwide (excluding heliports, STOLports—that is, extremely short runways suitable only for specialized aircraft with short takeoff and landing [STOL] capability—and seaplane bases). "Table 3-2—U.S. Civil and Joint-Use Airports, Heliports, STOLports, and Seaplane Bases, and Reported Abandonments on Record, by FAA Region and State: December 31, 1980," and "Table 3-4—U.S. Civil and Joint-Use Airports, Heliports, STOLports, and Seaplane Bases on Record by Length of Longest Runway, by FAA Region and State: December 31, 1980," in FAA, *Statistical Handbook of Aviation*, 1980 ed. (Washington, DC: U.S. GPO, 1980), 32, 36.

51. Diane Ackerman, *On Extended Wings* (New York: Atheneum, 1985), 15.

52. Matt Thurber, "Happier Landings: How to Improve Airports," *Flying*, Feb. 1985, 22–24, quotes from 24.

53. Phillips 66 Aviation Performance Centers, "The Good News Is That You've Arrived Twenty Minutes Ahead of Schedule" [advertisement], *AOPA Pilot* 28, no. 9 (Sept. 1985): 16.

54. Kim Murphy, "Meadowlark: On Collision Course with Urban Sprawl?" *Los Angeles Times*, Orange County edition, Oct. 6, 1985, 1.

55. David T. Courtwright, *Sky as Frontier: Adventure, Aviation, and Empire* (College Station: Texas A&M University Press, 2005), 50. By 1929, the weekly journal *Aviation* featured a regular column titled "Our Hangar Flying Department."

56. "Selling Airplanes in Today's Market—a Panel," in *Proceedings of the 1949 Conference [on] Airport Management and Operations Held at the Extension Study Center, University of Oklahoma, October 13-14-15, 1949* (published ca. 1949 by *Southern Flight* magazine and the University of Oklahoma Extension Division, in cooperation with the CAA), 65, Smithsonian Libraries, NASM Library / ALM Special Collections.

57. Frank Trumbauer, assistant to the regional administrator, "Toot Your Own Horn," in *Proceedings of the 1950 Conference [on] Airport Management and Operations*, 50.

58. The Bookkeeper [pseud.], "Hangar Pests," VOX Pilot, *AOPA Pilot* (Feb. 1957): 96h.

59. Major George P. Haviland, USAF, "YOU Can Sell Flying," *AOPA Pilot* 1, no. 2 (Apr. 1958): 29.

60. Boyne, *Flying*, 8.

61. George H. Moloney Jr., "The Middle-Aged Student vs. 'The System,'" *AOPA Pilot* 15, no. 9 (Sept. 1972): 77.

62. Boyne, *Flying*, 8.

63. Embry Riddle Aeronautical Institute, "What's Linda Lacking?" [advertisement], *Flying*, Jan. 1958, 65; Ruth Downie, "The Pilot's Missus," *Flying*, Feb. 1947, 68; Thora Kron, "Ground School for Backseat Pilots," *AOPA Pilot* 4, no. 11 (Nov. 1961): 64; Don Downie, "The Headline That DIDN'T Happen," *AOPA Pilot* 9, no. 6 (June 1966): quote from 36.

64. Gill Robb Wilson, "Clubs Could Be Trumps," editorial, *Flying*, July 1956, 20.

65. Robert W. Helber Jr., "On Top and Nowhere to Go!" Never Again! *AOPA Pilot* 1, no. 2 (Apr. 1958): 36–38.

66. For the first of this long-running series (still published under the same title today), see Garland Lincoln, "I Learned about Flying from That! No. 1," *Popular Aviation*, May 1939, 18–19, 74. *Popular Aviation* became *Flying* in 1942, and Wilson became editor and publisher of *Flying* in 1952. "Gill Robb Wilson: 1893–1966," *Flying*, Dec. 1966, 55. From 1939 through 1958, *AOPA Pilot* was published first as a special insert to *Popular Aviation*, then *Flying* when the magazine changed names. *AOPA Pilot* was not included in newsstand copies or copies mailed to regular subscribers; only AOPA members received magazines with this special insert. *AOPA Pilot* was last published as an insert in the March 1958 issue of *Flying*. That same month *AOPA Pilot* 1, no. 1 (Mar. 1958) was also published as a full-length, stand-alone magazine for members only.

67. Examples of editorial defenses of these columns can be found in Wilson, "Clubs Could Be Trumps," 20; editor's addendum to Wilmer Carlson, "This Pilot Thought That Jet Flying Was Easy until He Suddenly Found Himself in Compressibility," I Learned about Flying from That! *Flying*, May 1958, 50, 77; editor's comment appended to W. G. Bradley, M.D., "To Thine Own Self Be True," Never Again! *AOPA Pilot* 21, no. 11 (Nov. 1978): 95.

68. Jon Hahn, "Food-Loving Pilots Help Web Site Take Off," *Seattle Post-Intelligencer*, Aug. 10, 1999, D2.

69. John Purner, *The $100 Hamburger: A Guide to Pilots' Favorite Fly-In Restaurants* (New York: McGraw-Hill, 1998).

70. Keith Connes, Winds Aloft, *Flying*, Aug. 1969, 120; Marion C. McDonald, "What Motivates a 'Pinch-Hitter'?" *AOPA Pilot* 16, no. 5 (May 1973): 54; Eric Weiner, "FAA Reviews Pilot Certification," Reporting Points, *Flying*, Nov. 1988, 12, 14.

71. Betty Denné Haesloop, "Five Four Delta, Where Are You?" Never Again! *AOPA Pilot* 8, no 7 (July 1965): 64–65.

72. *I Learned about Flying from That* (New York: Delacorte Press / Eleanor Friede, 1976); Nigel Moll, ed., *More I Learned about Flying from That* (New York: Macmillan, 1982); *I Learned about Flying from That*, vol. 3 (Blue Ridge Summit, PA: TAB

Books / McGraw-Hill 1993); "The Hundred Dollar Hamburger," http://www.100dollarhamburger.com/; Boyne, *Flying*, 8.

73. Robert T. Shaw, "A Voice in the Fog," Never Again! *AOPA Pilot* 7, no. 9 (Sept. 1964): 58–59.

74. Bradley, "To Thine Own Self Be True," 94–95.

75. Doris Purdy, "The Lady Inside . . . ," Never Again! *AOPA Pilot* 9, no. 4 (Apr. 1966): 73; V. Alan Mode, "Comedy—IFR?" VOX Pilot, *AOPA Pilot* 9, no. 7 (July 1966): 20–21.

76. Ackerman, *On Extended Wings*, 58.

77. Robert L. Caswell, "The Bang in the Night," Never Again! *AOPA Pilot* 3, no. 2 (Feb 1960): 36, 38.

78. Chester A. Adams, "Wild Dreams?" VOX Pilot, *AOPA Pilot* 3, no. 4 (Apr. 1960): 10.

79. K. O. Eckland, "The Old Aerodrome," *AOPA Pilot* 21, no. 5 (May 1978): 82.

80. Richard [L.] Collins, "Discover Flying," editorial, *Flying*, June 1969, 40.

81. "Saturated Sky," *Time*, Aug. 2, 1968, 24.

82. Quotes from ibid.; David H. Scott, Washington Advisory, *Flying*, Oct. 1968, 28; "The Airport Crisis," Staff Report, *Flying*, July 1969, 59. During and after the crisis, the airline industry's Air Transport Association consistently maintained that "private aircraft carrying a handful of persons or less" was a major reason for airline delays. Richard Witkin, "Airlines Suggest Tighter Controls: Call for Surveillance of All Craft in Congested Areas," *New York Times*, July 26, 1968, 2.

83. Donald Bain, *The Case against Private Aviation* (New York: Cowles Book, 1969); "Donald Bain: Biography," http://www.donaldbain.com/bio.htm.

84. Burton Bernstein, "The Piper Cub vs. the 747: Private Planes and the Big Airliners Are Getting in Each Other's Way, Literally as Well as Figuratively," *New York Times Magazine*, Mar. 1970, 34–35, 94, 98, 102.

85. Gordon Baxter, *How to Fly, For People Who Are Not Sure They Want To* (New York: Summit Books, 1981), 57–69, quotes from 57, 62–63, 68.

86. Mark Twombly, "Cessna 140," *AOPA Pilot* 28, no. 7 (July 1985): 50, 54.

87. Ibid., 50–51, 54.

88. Ibid., 51, 54.

89. Ibid., 51, 54, photograph on 54.

Chapter 6 · Gendered Communities

1. Milton W. Horowitz, Ph.D., "For Men Only?" *Flying*, Aug. 1965, 30–33, quotes from 31.

2. Bart E. May, Bruce M. Fonoroff, and Dolly Kearn, "Dear Dr. Horowitz" [three separate letters to the editor], Flying Mail, *Flying*, Oct. 1965, 14 (Kearn), 134 (May, Fonoroff). For all 17 letters to the editor on the subject, see Flying Mail in the October and November 1965 issues of *Flying*.

3. Robert B. Parke, "The Feminine Case," editorial, *Flying*, Aug. 1965, 28.

4. Early history of the organization from James Thurber and Eames, "Beyond Keewee and Modock," The Talk of the Town, *New Yorker*, Jan. 7, 1928, 9; Harry A. Bruno, "Quiet Birdmen," letters, *Time*, July 24, 1939, 4; full official name from "Card, Membership, Quiet Birdmen," object no. A20010304000, Smithsonian National Air and Space Museum, viewable online at http://airandspace.si.edu/collections/artifact.cfm?id=A20010304000; rule quoted from "Booklet, Quiet Birdmen," object no.

A19890646000, Smithsonian National Air and Space Museum, 13, viewable online at http://airandspace.si.edu/collections/artifact.cfm?id=A19890646000. Although undated, the booklet refers to several specific dated events, the last being September 3, 1939 (12), and elsewhere refers to "World War I" instead of the "Great War," indicating it was published during or after World War II (7).

5. Gordon Baxter, "The Airperson's World," Bax Seat, *Flying*, Mar. 1982, 74–75.

6. Tracey L. Potter, "Re: [DCPILOTS] The Ancient and Secret Order of Quiet Birdmen," comment posted on DCPilots.com, Jan. 19, 2001, accessible online at http://web.archive.org/web/20030928033422/listarchives.his.com/dcpilots-l/dcpilots-l.0101/msg00441.html.

7. "Booklet, Quiet Birdmen," 20. At some point the book's owner crossed out "250" and wrote "500" in the margin, suggesting that the minimum flight time had doubled after the booklet was published.

8. Glenn Jordan, "Federal Judge Candidate Leaves All-Male Club," *Hartford (CT) Courant*, Aug. 7, 1992, n.p., http://articles.courant.com/1992-08-07/news/0000114061_1_covello-confirmation-hearing-all-male.

9. Martha Lunken, "A Name I Won't Utter," Unusual Attitudes, *Flying*, Mar. 2010, 70–71; "Martha Lunken, Contributing Editor," Meet the Editors, *Flying*, Dec. 8, 2009, http://www.flyingmag.com/bio/martha-lunken-contributing-editor.

10. Lunken, "A Name I Won't Utter," 70–71.

11. Because of the secretive nature of the QBs, it is difficult for an outsider to determine when the Somerville Hangar first adopted this logo or whether it remains in use today. After 2004, the Somerville Hangar changed its publicly viewed homepage to a more innocuous site that required members to log in to access any content. To view an archived version of the homepage featuring this logo dated March 31, 2004, see Somerville Hangar of the Quiet Birdmen, "Ye Ancient and Secret Order of Quiet Birdmen," http://web.archive.org/web/20040331112736/http://www.somqb.org/. Readers who wish to delve more deeply into the symbolism of these images might consider that the fighter plane is modeled after a World War II–era P-47 Thunderbolt, nicknamed the "jug" by pilots because its stout profile resembled milk bottles of the era (or perhaps because some considered the heavily armed and armored plane a "juggernaut" in combat); the topless sunbather is baring her "jugs," slang since the first half of the twentieth century for female breasts; and the brown jugs marked "XXX" have long served as a visual representation of "extra strong" liquor or moonshine in popular culture.

12. Fred Weick and James R. Hansen, *From the Ground Up: The Autobiography of an Aeronautical Engineer* (Washington, DC: Smithsonian Institution Press, 1988), 188–91, 300, 339, 368, 400.

13. A collection of Wilson's poems, accompanied by large-format aviation photographs, were published as a coffee table book under the same title as Wilson's monthly column. Gill Robb Wilson, *The Airman's World* (New York: Random House, 1957). Biographical information on Wilson from "Gill Robb Wilson: 1893–1966," *Flying*, Dec. 1966, 54–55; Richard Bach, "The Legend of Gill Robb Wilson," *Flying*, Dec. 1966, 56–57.

14. According to Dr. Jeff Underwood and Doug Lantry of the National Museum of the U.S. Air Force, the term dates to the Vietnam War era. Jesse Sheidlower, "Balls in

the Air: Where Does the Expression 'Balls to the Wall' Come From?" *Slate Magazine*, posted Feb. 10, 2006, http://www.slate.com/articles/news_and_politics/explainer/2006/02/balls_in_the_air.html.

15. Baxter, "The Airperson's World," 74–75.

16. Tom Wolfe, *The Right Stuff* (New York: Farrar, Straus and Giroux, 1979), 50–51, 66.

17. Caroline L. Davey and Marilyn J. Davidson, "The Right of Passage? The Experiences of Female Pilots in Commercial Aviation," *Feminism & Psychology* 10, no. 2 (May 2000): 211–12, 218.

18. Sarah Byrn Rickman, *Nancy Batson Crews: Alabama's First Lady of Flight* (Tuscaloosa: University of Alabama Press, 2009), 17, 84.

19. Deborah G. Douglas, *American Women and Flight since 1940* (Lexington: University Press of Kentucky, 2004), 8, 20. 118–123; "Commemorating the 30th Anniversary of the Ninety-Nines, Inc., 1929–1959," originally published ca. 1959, reproduced online by The Ninety-Nines, Inc., http://www.ninety-nines.org/index.cfm/thirty_years.htm; "Sixty and Counting: 60th Anniversary Commemorative Collection, 1929–1989," originally published ca. 1989, reproduced online by The Ninety-Nines, Inc., http://www.ninety-nines.org/index.cfm/sixty_years.htm.

20. George Putnam, "The Story of the 49½ers," excerpted from "The Forgotten Husband," *Pictorial Review*, Dec. 1932, reproduced online by The Ninety-Nines, Inc., http://www.ninety-nines.org/index.cfm/49_5_story.htm.

21. Barbara A. Goodwin and Robert I. Goodwin, interviewed by the author at the Goodwin residence, Kalamazoo, MI, August 12, 2004, audio recording in author's collection; Douglas, *American Women and Flight*, 119–20, quote from 139–40.

22. Beatrice Edgerly, "Ladies' Day in the Air," *AOPA Pilot* 1, no. 4 (June 1958): 34–35.

23. Baxter, "The Airperson's World," 74.

24. Edgerly, "Ladies' Day in the Air," 34.

25. This long-distance, all-women air race, no longer sponsored by the Ninety-Nines, continued after 1977 as the "Air Race Classic." Glenn H. Buffington and Carolyn J. Van Newkirk, Ed.D., "About the ARC and Racing," Air Race Classic, http://www.airraceclassic.org/historyt.asp.

26. Betty Ryan Wolfe, "Women's Air Race," *Flying*, July 1952, 50.

27. Douglas, *American Women and Flight*, 120–21.

28. Richard Bach, "The Invisible 99s," *Flying*, Aug. 1965, 38–41, quotes from 39, 41.

29. "30th Anniversary of the Ninety-Nines, Inc."

30. Cobb joined the Ninety-Nines as a young pilot on August 8, 1952, and became a life member ca. 1965. Email correspondence between the author and Denise Neil-Binion, Executive Director, 99s Museum of Women Pilots, Oct. 14, 2014 (copy in author's files); email correspondence between the author and Ruth Lumis, Director of the Jerrie Cobb Foundation, Inc., Oct. 9, 2014 (copy in author's files). Other information about Cobb from Margaret A. Weitekamp, *Right Stuff, Wrong Sex: America's First Women in Space Program* (Baltimore: Johns Hopkins University Press, 2004).

31. Bach, "The Invisible 99s," 39; "Table 4.9: Active Airmen Certificates Held, Dec. 31, 1957–65," and "Table 4.13, Women Actively Engaged in Aviation, Dec. 31, 1960–65," both in U.S. Department of Transportation, FAA, *Statistical Handbook of Aviation*, 1976 ed. (Washington, DC: U.S. GPO, 1976), 77, 82, tables.

32. "Wanted—A Slogan," *AOPA Pilot* (June 1940): 34e; Lew Townsend, "AOPA Members Cast Large Shadow," *AOPA Pilot* 12, no. 5 (May 1969): 94–97; AOPA, "History of AOPA," http://www.aopa.org/info/history.html. Membership numbers and total number of active pilots from "AOPA Membership Tops 111,000," *AOPA Pilot* 8, no. 4 (Apr. 1965): 42; Kathryn Opulewsky, "A Remarkable Half Century," *AOPA Pilot* 51 no. 3 (Mar. 2008): 34–35; "Table 7.10: Estimated Active Pilot Certificates Held, by Category and Age Group of Holder: 1980, 1979, 1976," in FAA, *Statistical Handbook of Aviation*, 1980 ed. (Washington, DC: U.S. GPO, 1980), 128, table.

33. "Colorado Breakfast Flight," *AOPA Pilot* (Mar. 1946): 58c.

34. "Summer Round Up Flight," *AOPA Pilot* (Sept. 1947): 50g; William R. White, "Summer Round-up Flight," *AOPA Pilot* (Nov. 1947): 58g.

35. "Rehoboth Flight Breaks Record," *AOPA Pilot* (Nov. 1948): 50c.

36. "Private Flying Makes a Lot of Sense," editorial, *AOPA Pilot* (Nov. 1949): 50a.

37. "AOPA's Fifth Annual Roundup Draws 1,300 Flyers to Gala Beach Weekend," *AOPA Pilot* (Nov. 1952): 50b.

38. "Annual AOPA Roundup Flight Draws 525 Planes, 1,500 Pilots and Friends," *AOPA Pilot* (Dec. 1953): 50c.

39. "Nantucket Island Site Chosen for Roundup Flight," *AOPA Pilot* (Sept. 1954): 50a; "Seventh Annual Round-up Flight Best, So Say the Few Who Got to Nantucket," *AOPA Pilot* (Dec. 1954): 50d.

40. "Flight-Cruise to Bermuda Set for Annual Round-up," *AOPA Pilot* (May 1955): 50c; "Spaces for AOPA Flight-Cruise Are Going Very Fast," *AOPA Pilot* (June 1955): 50e.

41. "AOPA Round-up's Luxury Liner Has Brilliant Record," *AOPA Pilot* (July 1955): 50c.

42. "Flight-Cruise to Bermuda Set," 50c.

43. Quotes from "Spaces for AOPA Flight-Cruise Are Going Very Fast," 50e; "Flight-Cruise Date Nearing Fast, Get Reservations Now," *AOPA Pilot* (Aug. 1955): 50c; "Last Call for Bermuda Flight-Cruise Tells How to Figure Reservation Costs," *AOPA Pilot* (Sept. 1955): 50c.

44. "It Sure Was Fair Weather When AOPA Cruisers Got Together," *AOPA Pilot* (Nov. 1955): 50c.

45. "AOPA'ers Ready for Mississippi Flight, Oct. 12," *AOPA Pilot* (Oct. 1956): 82c; "Mississippi Flight," *AOPA Pilot* (Dec. 1956): 50d; C.P.M., "Hollywood Scores Again," *AOPA Pilot* 13, no 12 (Dec. 1970): 41.

46. "Luxury Liner Chartered for AOPA Cruise," *AOPA Pilot* (Dec. 1956): 50c; "Itinerary for Flight-Cruise Is Announced," *AOPA Pilot* (Feb. 1957): 96e; "'Pork Chops and Kidney Stew' Dancers Will Help AOPA Make It a Real Plantation Party," *AOPA Pilot* (Sept. 1957): 66d–66e; "More than 100 AOPA'ers and Guests Enjoy Cruise to Bermuda, Nassau," *AOPA Pilot* (Sept. 1957): 66e; "AOPA Party in Mississippi Is Big Success," *AOPA Pilot* (Dec. 1957): 82d–82e, 82h.

47. AOPA, "Announcing AOPA's Caribbean Flight Cruise" [advertisement], *AOPA Pilot* 3, no. 1 (Jan. 1960): 15; "Itinerary Set for AOPA Jamaican Cruise," *AOPA Pilot* 3, no. 2 (Feb. 1960): 70–71.

48. This shift from a purely social agenda to a combination of social and educational activities began with the 1961 Plantation Party. The first exhibits by aviation equipment manufacturers were added that year as well. Najeeb E. Halaby Jr., head of

the FAA, made his first appearance at the annual AOPA convention in 1962. AOPA renamed its annual convention in 1979, but continued to call the event's final banquet the "Plantation Party" for several more years. "AOPAers Will Be Hobnobbing at St. Petersburg Again," *AOPA Pilot* 4, no. 5 (May 1961): 68–69; Max Karant, "360° Rating Tested in Florida," *AOPA Pilot* 4, no. 12 (Dec. 1961): 29; Max Karant, "Two Thousand Attend AOPA Event," *AOPA Pilot* 5, no. 12 (Dec. 1962): 32; Phil Van Ostrand, "AOPA's 1979 Convention," *AOPA Pilot* 22, no. 12 (Dec. 1979): 26; Mary F. Silitch, "The San Diego Show: New Aircraft, Old Friends at the Twenty-Fifth AOPA Convention," *AOPA Pilot* 24, no. 2 (Feb. 1981): 26–27; Thomas A. Horne, "Something for Everyone," *AOPA Pilot* 25, no. 1 (Jan. 1982): 24.

49. "Impressive Program Arranged for AOPA Party," *AOPA Pilot* 13, no. 9 (Sept. 1970): 102; C.P.M., "Hollywood Scores Again," 41–47, quote from 44.

50. Las Vegas remained a perennial favorite, but later conventions also took place in San Diego, San Antonio, and Orlando. Calendar: 1975 Events, *Flying*, Jan. 1975, 121; Calendar: A Year's Worth, *Flying*, Jan. 1976, 76; Horne, "Something for Everyone," 23.

51. Don Downie, "AOPA Wins Again at Las Vegas," *AOPA Pilot* 14, no. 12 (Dec. 1971): 37.

52. Charles P. Miller, "Weather Couldn't Stop Them," *AOPA Pilot* 7, no. 12 (Dec. 1964): 52–53.

53. Charles P. Miller, "Fabulous Las Vegas," *AOPA Pilot* 8, no. 12 (Dec. 1965): 80.

54. "First Come, First Served," photograph by Murray Spitzer, in C.P.M., "Hollywood Scores Again," 40.

55. Charles P. Miller, "The Big Party," *AOPA Pilot* 9, no. 12 (Dec. 1966): 63.

56. Silitch, "The San Diego Show," 26–27.

57. In the first half of the twentieth century, the term *companionate marriage* described "a form of marriage which provides for divorce by mutual consent and in which neither partner has any legal responsibilities towards the other" (Oxford English Dictionary). In their seminal 1945 book, *The Family: From Institution to Companionship*, Ernest W. Burgess and Harvey J. Locke reiterated this prewar definition, then introduced the term *companionship family* to describe an emerging trend in domestic relationships that simultaneously emphasized long-term commitment, a romantic relationship, childrearing, and togetherness for couples and families. Later in the postwar era, sociologists relabeled Burgess and Locke's companionship family and started calling it companionate marriage, despite the fact that the pre- and postwar definitions of companionate marriage described completely different concepts. Ernest W. Burgess and Harvey J. Locke, *The Family: From Institution to Companionship* (New York: American Book, 1945), 416–19; Andrew J. Cherlin, "The Deinstitutionalization of American Marriage," *Journal of Marriage and Family* 66, no. 4 (Nov. 2004): 851; *Oxford English Dictionary Online*, "companionate marriage," http://www.oed.com.

58. Historian Joanne Meyerowitz observes in her introduction to a book of 15 essays on the lived experiences of American women during the immediate postwar era: "While some women fit the stereotype [personified by June Cleaver, Donna Reed, and Harriet Nelson], many others did not. To state the obvious, in the years following World War II, many women were not white, middle-class, married, and suburban; and many white, middle-class, married, suburban women were neither wholly domestic nor

quiescent." If one extends an equal degree of diversity for American men, then the situation becomes exponentially more complex. Joanne Meyerowitz, ed., *Not June Cleaver: Women and Gender in Postwar America, 1945–1960* (Philadelphia: Temple University Press, 2004), 1–2.

59. Holly K. Robinson, "Ideas," VOX Pilot, *AOPA Pilot* (June 1954): 50h; Mrs. Clint Allen, "Flying Wives," VOX Pilot, *AOPA Pilot* 11, no. 8 (Aug. 1968): 24.

60. Don Taylor, "On Vacation—with Mother Goose," Travel, *AOPA Pilot* 5, no. 12 (Dec. 1962): 82–83; Anne S. Moore, "Know When to Stop: Keeping a Young Family Happy When Traveling in a Lightplane," *AOPA Pilot* 21, no. 3 (Mar. 1978): 106–8.

61. Townsend, "AOPA Members Cast Large Shadow," 94.

62. "It Sure Was Fair Weather," 50c.

63. "Impressive Program Arranged for AOPA Party," 102; C.P.M., "Hollywood Scores Again," 44.

64. "Anti-flying wife" from Lois C. Philmus, "Women Banish Fear of Flying," *AOPA Pilot* 6, no. 5 (May 1963): 29. Other authors and editors of the era used this and similar labels.

65. Lois C. Philmus, "Beware the Skirt Barrier," *AOPA Pilot* 3, no. 6 (June 1960): 39, 67.

66. Don Downie, "A New Flying Concept: The 'Pinch-Hitter,' " *AOPA Pilot* 6, no. 12 (Dec. 1963): 23.

67. James R. Old, M.D., "Accolade for Pinch-Hitter Course," VOX Pilot, *AOPA Pilot* 8, no. 1 (Jan. 1965): 20; Hank Strickland, " 'Helpful Hint' to Operators," VOX Pilot, *AOPA Pilot* 8, no. 7 (July 1965): 18.

68. Emma Claypool Moore, "So Dad Took Up Flying," *AOPA Pilot* 5, no. 8 (Aug. 1962): 36.

69. Philmus, "Women Banish Fear of Flying," 29.

70. Selma Cronan, "Pinch-Hitter Enthusiasm," VOX Pilot, *AOPA Pilot* 8, no. 7 (July 1965): 14.

71. Katy Dance, "Feminine Funk," Flying Mail, *Flying*, Aug. 1969, 94.

72. "What to Do When the Pilot Dies," Modern Living, *Time*, Oct. 18, 1963, 54.

73. Downie, "A New Flying Concept," 24–25.

74. Fred F. Brown Jr., M.D., "Course Converts Wife," VOX Pilot, *AOPA Pilot* 12, no. 10 (Oct. 1969): 20.

75. Sara Cowart (wife of David B. Cowart), "Pinch-Hitter Course Saves Marriage," *AOPA Pilot* 14, no. 11 (Nov. 1971): 59.

76. Marian C. McDonald, "What Motivates a 'Pinch-Hitter'?" *AOPA Pilot* 16, no. 5 (May 1973): 53–54.

77. Max Karant, "Welcome Aboard, Honey!" editorial, *AOPA Pilot* 6, no. 12 (Dec. 1963): 20; AOPA Foundation, Inc., "Pinch-Hitter Course Sets Stage for a New Air Age . . . of Women!" [advertisement], *AOPA Pilot* 6, no. 12 (Dec. 1963): 44.

78. Thomas A. Horne, "ASF at 50: The AOPA Air Safety Foundation's Legacy of Safety, and a View to the Future," *AOPA Pilot* 43, no. 11 (Nov. 2000): 63.

79. Pilot statistics for 1963 and 1978 from "Table 4.9—Active Airman Certificates Held: Jan. 1, 1956–61" and "Table 4.11—Women Actively Engaged in Aviation: Jan. 1, 1960–64," both in FAA, *Statistical Handbook of Aviation*, 1964 ed. (Washington,

DC: U.S. GPO, 1964), 54, 56, tables; "Table 7.1—Active Airmen Certificates Held: December 31, 1969–1978" and "Table 7.2—Women Actively Engaged in Aviation: December 31, 1969–1978," both in FAA, *Statistical Handbook of Aviation*, 1978 ed. (Washington, DC: U.S. GPO, 1978), 88, 89, tables.

80. According to a 1998 AvWeb article, the AOPA Air Safety Foundation claimed that about 45 percent of Pinch-Hitter graduates went on to enroll in traditional flying lessons. The same article reported that "in the thirty five years since the AOPA Air Safety Foundation (ASF) first offered the Pinch-Hitter course, over 11,000 non-pilots have taken the full course, which until recently included flight training. An estimated ten times that number have attended the four hour ground school portion of the course." It is unclear which figure this 45 percent applies to—the 11,000 graduates of the full course including flight training, or the approximately 110,000 graduates of the four-hour ground-only course with no flight training. Doug Ritter, "The Pilot's Incapacitated—Now What?" Safety, AvWeb, Jan. 5, 1998, http://www.avweb.com/news/safety/183023-1.html. Other information from Horne, "ASF at 50," 63; FAA, *Statistical Handbook of Aviation*, published annually starting in 1959; U.S. Department of Transportation, FAA, *U.S. Civil Airmen Statistics* (annual tables for 1999–2013), http://www.faa.gov/data_research/aviation_data_statistics/civil_airmen_statistics/.

81. Cherlin, "The Deinstitutionalization of American Marriage," 852.

82. Horne, "Something for Everyone," 24.

83. Bill Nelson, "Paul Poberezny," Our Interesting Members, *AOPA Pilot* 13, no. 10 (Oct. 1970): 50–51; "Paul Poberezny: Founder, Experimental Aircraft Association," EAA, Aug. 2010, www.eaa.org/media/paul_poberezny.pdf; EAA, "Chronology: The 1980s," http://www.eaa.org/about/chronology.asp.

84. Nelson, "Paul Poberezny," 50; Robert V. Rozelle, "EAA's Guiding Force: Paul H. Poberezny," *Aviation Travel & Times* 1, no. 2 (date unknown, reprinted by the EAA and distributed as a handout at its 1981 convention), 22, Smithsonian National Air and Space Museum, NASM Tech. Files J1-1981-600-1.

85. D. A. Lande, *Oshkosh—Gateway to Aviation: 50 Years of EAA Fly-Ins* (Random Lake, WI: Times Printing, 2002), 4–6, 28, 31–32, 40–41, 51, 61.

86. William Garvey, "Oshkosh: EAA Mecca for the World's Most Pampered Planes," *AOPA Pilot* 17, no. 10 (Oct. 1974): 26; Edward G. Tripp, "Oshkosh Odyssey," *AOPA Pilot* 22, no. 11 (Oct. 1979): 71; "What's Up—and Down—at Oshkosh Air Show," *New York Times*, Aug. 3, 1981, A8.

87. Use of the term *warbird* to describe a former military aircraft (rather than one in current service) dates to at least the early 1960s, when *Flying* carried an article about an organization dedicated to "the salvation and flying of the WW II warbirds." Steve Randall, "Confederate Air Force," *Flying*, Oct. 1962, 88.

88. Howard Levy, "Homebuilts at Oshkosh," *AOPA Pilot* 13, no. 10 (Oct. 1970): 39.

89. Ibid., 38; "Air Shows at Oshkosh," *Sport Aviation* 19, no. 10 (Oct. 1970): 29; Don and Sherry Wood, "Oshkosh 1970," Convention Comments, *Sport Aviation* 19, no. 10 (Oct. 1970): 28.

90. "For The Ladies . . . ," in *Eleventh Annual Fly-In Convention, Experimental Aircraft Association* [convention program] ([Hales Corner, WI]: EAA, 1963), 13, NASM Tech. Files DE-960000-02 (EAA).

91. *14th Annual Experimental Aircraft Association International Fly-In Convention* [convention program] (Franklin, WI: EAA, 1966), 17, NASM Tech. Files DE 960000-02.

92. "EAA Women's Activities," in *26th Annual International Convention & Sport Aviation Exhibition* [convention program] (Oshkosh, WI: EAA, 1978), 78–81, NASM Tech. Files DE-960000-02 (EAA).

93. Beth Blagaich, "Women of EAA," *Sport Aviation* 28, no. 1 (Jan. 1979): 52.

94. "EAA Women's Activities," in *29th Annual International EAA Convention & Sport Aviation Exhibition* [convention program] (Oshkosh, WI: EAA, 1981), 9, 83–86, NASM Tech. Files J1-1981-600-1; Lande, *Oshkosh—Gateway to Aviation*, 59.

95. Philmus, "Women Banish Fear of Flying," 29.

96. "EAA Women's Activities," in *26th Annual International EAA Convention*, 78.

97. John Burton, "Preparing for Our Convention: Dedication . . . Camaraderie . . . Fulfillment," *Sport Aviation* 28, no. 9 (Sept. 1985): 40; Jayne Schiek, "Women of EAA: Activities at Oshkosh," *Sport Aviation* 27, no. 4 (Apr. 1978): 35.

98. Schiek, "Women of EAA" (1978), 35.

99. Jayne Schiek, "Women of EAA: Activities at Oshkosh," *Sport Aviation* 26, no. 6 (June 1977): 58–60; Schiek, "Women of EAA" (1978) 35; Blagaich, "Women of EAA," 53; Burton, "Preparing for Our Convention," 39; "Colorado Breakfast Flight," 58c.

100. As explained in the introduction to this book, proponents of the concept of *hegemonic masculinity* argue that only a minority of men had to act in a certain way in order to influence everyone, and that men "who received the benefits of [the resulting] patriarchy without enacting a strong version of masculine dominance could be regarded as showing a complicit masculinity." R. W. Connell and James W. Messerschmidt, "Hegemonic Masculinity: Rethinking the Concept," *Gender & Society* 19, no. 6 (Dec. 2005): quote from 832; R. W. Connell, *Masculinities*, 2nd ed. (Berkeley: University of California Press, 2005).

Conclusion

1. Claude Grahame-White, "No Place in the Air for Women," *Mobile (AL) Register*, Sept. 3, 1911.

2. Walter J. Boyne, *Flying: An Introduction to Flight, Airplanes, and Aviation Careers* (Englewood Cliffs, NJ: Prentice-Hall, 1980), 8.

3. "Table 10: Civil Aircraft Production in the United States," and "Table 19: Active Pilots by Type of Certificate," both in U.S. Department of Transportation, FAA, Office of Aviation Economics, Aviation Forecast Division, *Aviation Forecasts Fiscal Years, 1971–1982* (Washington, DC: U.S. GPO, Jan. 1971), 40, 49; General Aviation Manufacturers Association, *General Aviation Statistical Databook & Industry Outlook 2010* (Washington, DC: GAMA, 2010), 22.

4. Louis J. Budd, "Mark Twain as an American Icon," in *The Cambridge Companion to Mark Twain*, ed. Forrest G. Robinson (New York: Cambridge University Press, 1995), 7.

5. Thomas B. Haines, "The Effect of Tort Reform on General Aviation and Prospects for Additional Reform" (MBA Thesis, Mount St. Mary's University, 2006); John Frank, *Cessna 172 Skyhawk Buyers Guide* (Santa Maria, CA: Cessna Pilots Association, 2009), 7–9, 19, 26.

6. Any licensed pilot can fly an aircraft certified by the FAA as LSA, which is restricted in size, speed, and complexity, but holders of a Sport Pilot certificate may fly only LSA aircraft, which keeps them out of the cockpit of larger traditional personal planes (including the four-seat Cessna 172). EAA, "Light-Sport Aircraft Overview," EAA Government Advocacy, http://www.eaa.org/govt/lsa_overview.asp.

7. BRS Aviation, "BRS History," http://brsparachutes.com/brs_history.aspx.

8. Lane Wallace, "Van's Air Force," *Sport Aviation* 61, no. 6 (June 2012): 28–36, quote from 32 (all parenthetical additions to this quote appeared in the original article).

9. Thomas B. Haines, "Make a Difference," Waypoints, *AOPA Pilot* 49, no. 7 (July 2006): 32.

10. Average age of Mr. General Aviation in 1969 based on data from "Table 7.9—Airmen Certificates by Categories in Each Age Group of Holder: December 31, 1969," in U.S. Department of Transportation, *FAA, Statistical Handbook of Aviation*, 1970 ed. (Washington, DC: U.S. GPO, 1970), 169, table. Other information from "Average Age of Active Pilots by Category" and "Estimated Active Airmen Certificates Held," both in U.S. Department of Transportation, FAA, *U.S. Civil Airmen Statistics* (annual tables for 1999–2013; 1999 tables include data for 1990–99), http://www.faa.gov/data_research/aviation_data_statistics/civil_airmen_statistics/.

11. "An Important Notice to All AOPA Members: Your Assistance Is Urgently Required," Project Pilot, *AOPA Pilot* 49, no. 7 (July 2006): 75 (special three-page foldout insert).

12. Although Congress failed to enact a formal resolution to officially designate June as "National Learn to Fly Month" in 1989, the General Aviation Taskforce, a consortium of "representatives from all aspects of the general aviation industry," moved forward with a program that combined a national advertising campaign with a toll-free telephone number to connect prospective pilots to 1 of 500 participating flight schools where they could take a $25 introductory flight. Brian O'Lena, "Twenty Years of Young Eagles," *EAA Sport Aviation* 61, no. 9 (Sept. 2012): 114–15; AOPA, "Industry Be-A-Pilot Program Enters New Phase, Names President and Chief Executive Officer," June 3, 2008, http://www.aopa.org/News-and-Video/All-News/2008/June/3/Industry-Be-A-Pilot-program-enters-new-phase-names-president-and-chief-executive-officer; Thomas B. Haines, "Giving Back," AOPA Project Pilot, *AOPA Pilot* 49, no. 7 (July 2006): 60–61; Erik Lindbergh, "Sharing Your Passion for Flight," AOPA Project Pilot, *AOPA Pilot* 49, no. 7 (July 2006): 62–62; "AOPA Project Pilot: How It Works," AOPA Project Pilot, *AOPA Pilot* 49, no. 7 (July 2006): 69; Designating National Learn to Fly Month, H.R. Res. 212, 101st Cong. (1989–2000) (not enacted); William Garvey, "Game Plan Man," From the Tower, *Flying*, Oct. 1988, 10 (description of General Aviation Taskforce); William Garvey, "An Invitation to Fly," *Flying*, June 1989, 56–58 (description of National Learn to Fly Month).

13. Phil Boyer, "'You're Not a Real Pilot,' or How I Learned to Fly a Taildragger," *AOPA Pilot* 46, no. 1 (Jan. 2003): 64–71; Don Downie, "Taming the Taildragger: Toil, Tears, and Sweat—But Worth the Trouble," *AOPA Pilot* 17, no. 9 (Sept. 1975): 41–44.

14. Dave Hirschman, "Not a Fashion Statement," and Ian J. Twombly, "Don't Be 'That Guy,'" in Dogfight: Flight Suits, *AOPA Pilot* 56, no. 2 (Feb. 2013): 91.

15. Dave Hirschman, "Single Supremacy," and Peter A. Bedell, "Twinosaurus? Not Extinct Yet," in Dogfight: Twins vs. Single, *AOPA Pilot* 56, no. 6 (June 2013): 89–90.

16. Jim Mitchell et al., "How Pink Is the Sky? A Cross National Study of the Gendered Occupation of Pilot," *Employment Relations Record* 5, no. 2 (July 2005): 43–60.

17. Richard [L.] Collins, "The Great Debate: Boy v. Girl Pilots," Debate, *Air Facts Journal*, Jan. 9, 2012, http://www.airfactsjournal.com/2012/01/the-great-debate-boy-v-girl-pilots/.

18. Charles P. Miller, "Biggest Plantation Party," *AOPA Pilot* 6, no. 12 (Dec. 1963): 85.

I learned to fly in 1984 during the summer between high school graduation and my freshman year of college. As an 18-year-old who had scrimped and saved for several years while working at a local pizza joint, this accomplishment took on all the more importance since I earned my private pilot's license without financial assistance from my parents. The next summer a friend and I flew a borrowed 1947 Aeronca Champ—with a 65 hp engine we started manually by standing out front and yelling "Contact!" before swinging the propeller by hand—to my first-ever visit to the Experimental Aircraft Association's annual fly-in convention at Oshkosh, Wisconsin. Decades later I smile at our youthful audacity, flying into what for one week each year is the world's busiest airport in a tiny fabric-covered airplane with no electrical system, lights, or navigation equipment beyond the plane's magnetic compass and a map. We arrived near sunset and were directed into the "antique-classic" parking area. Only after we had secured the Champ for the night did we learn that at least one of us had to be a member of the EAA to park in this special section. I lost the coin toss, and thus it was I who shelled out enough money to pay for an hour of flying (a full eight hours' wages making pizzas) in order to join the organization. One of the benefits of belonging, aside from not having to wake up early the next morning to move the airplane to general parking on the far side of the field, was the EAA's colorful, information-packed magazine that started arriving in my mailbox every month. Instead of letting my membership expire, I renewed it again, and again, and again.

Fourteen years, several hundred hours in the air, and several detours in life later, my passion for flying remained unabated when I arrived at the University of Delaware to pursue a Ph.D. in history. And yet I had no intention of studying anything to do with aviation. After growing up reading entertaining but ultimately intellectually unstimulating coffee table books about airplanes, aviators, and aerial combat, it wasn't until I happened to pick up Joseph J. Corn's pathbreaking book *The Winged Gospel: America's Romance with Aviation, 1900–1950* (New York: Oxford University Press, 1983) over the winter break that first year in grad school that I realized that I could combine what had previously seemed—to me, at least—two completely unrelated interests: private aviation and the social and cultural history of technology.

That very next semester, as my first graduate-level research and writing seminar sent me in search of a project, I discovered that quite by chance I had chosen to attend graduate school just down the road from the archived records of the Aircraft Owners and Pilots

Association, housed at the Hagley Museum and Library in nearby Wilmington, Delaware. The resulting paper, in which I struggled mightily to make sense of the AOPA's Pinch-Hitter Program, would eventually form the cornerstone of my dissertation and feature prominently in this book, as well.

In his seminal review essay, "Aviation History in the Wider View," *Technology and Culture* 30, no. 3 (July 1989): 643–56, James R. Hansen cautions that even the most well-intentioned academically trained researchers can "go native" when their enthusiasm for aviation clouds their ability to approach the subject critically. "When that happens, and the historian becomes part of the culture he is studying, his ability to evaluate and represent the subject as anything other than a type of self-portrait is lost" (647). Although I was a pilot long before I became a historian, I consider myself first and foremost a social and cultural historian of technology who happens to specialize in aviation, but whose intellectual interests range far beyond the cockpit and airport to include a wide variety of topics related to gender, race, class, and technology. As a scholar, my real interest in aviation has everything to do with what this activity tells us about the wider world in which we live, not why one particular airplane is "better" than another, or how—and even whether—flight has made this world a better place.

Beyond this extended autobiographical introduction, I have divided this essay into two parts. I begin by describing the secondary literature that has most influenced my approach to researching and writing this book. I consider these to be "must-read" titles for anyone interested in similar lines of enquiry, regardless of whether they are exploring the history of aviation or the gendering of technology more broadly. I then summarize the kinds of primary sources I drew on to complete this project. Naturally, all primary and secondary sources used in this book are fully cited in the endnotes.

Gender and the History of Technology

There is a burgeoning literature on gender history that dates back more than a quarter century, tracing its roots to—yet different from—an earlier tradition of women's history. Although my own work is informed by numerous scholars in this field, I recommend that readers start with Joan Wallach Scott's classic yet still pertinent *Gender and the Politics of History*, rev. ed. (New York: Columbia University Press, 1999), first published in 1989. Scott's arguments that gender is a socially constructed category, and that gender (like other socially constructed categories including race and class) ultimately boils down to power relationships among different groups, are fundamental to understanding an entire generation of scholarly work that followed her prizewinning book, including the histories of gender and technology that shaped my own views on and approaches to understanding the past.

In 1997 the Society for the History of Technology devoted a special edition of its quarterly journal to gender. Although this issue's six articles serve as valuable case studies, the two introductory essays are indispensable reading for anyone interested in understanding the deeply embedded and intertwined historical relationship between gender and technology: Nina E. Lerman, Arwen Palmer Mohun, and Ruth Oldenziel, "Versatile Tools: Gender Analysis and the History of Technology," and "The Shoulders We Stand On and the View from Here: Historiography and Directions for Research," *Technology and Culture* 38, no. 1 (Jan. 1997): 1–8, 9–30. Several edited collections of essays, including Roger Horowitz and Arwen P. Mohun, eds., *His and Hers: Gender, Consumption, and Technology*

(Charlottesville: University of Virginia Press, 1998), and Roger Horowitz, ed., *Boys and Their Toys: Masculinity, Class, and Technology in America* (New York: Routledge, 2001), provide yet more case studies. These not only offer insight into the complex relationship among gender, technology, and society but also can serve as intellectual models for how one might approach similar topics.

While there are numerous scholarly essays and articles on the connections between technology and masculine identity (and countless recent popular pieces of varying length and quality by journalists and pundits), there are still surprisingly few book-length histories specifically devoted to the topic. Some scholars address this subject obliquely in the process of answering other questions. Even though Virginia Scharff's *Taking the Wheel: Women and the Coming of the Motor Age* (Albuquerque: University of New Mexico Press, 1991) focuses on how women found—and in some cases forced—their way into the male-dominated driver's seat, she does so in part by considering the early and lasting cultural connections between automobiles and masculinity in America. A handful of other scholars deal with the topic head-on. For instance, Arwen Palmer Mohun's *Steam Laundries: Gender, Technology, and Work in the United States and Great Britain, 1880–1940* (Baltimore: Johns Hopkins University Press, 1999), examines how the male owners and managers of steam-powered commercial laundry services actively attempted to re-gender laundry—traditionally viewed as unskilled "women's work"—into a respectable masculine endeavor involving considerable technical expertise and business savvy. Ruth Oldenziel's *Making Technology Masculine: Men, Women and Modern Machines in America, 1870–1945* (Amsterdam: Amsterdam University Press, 1999) explores how men explicitly set out to define engineering as a masculine pursuit during the professionalization of this field in the late nineteenth and early twentieth centuries. All three of these books helped me to better understand how supposedly natural assumptions regarding gender and technology were in fact socially constructed.

Defining Masculinity

I rely on both sociologists and gender historians not only for definitions of masculinity but also to help me understand how individuals, subcultures, and entire societies created, reinforced, and challenged masculine ideals and identity throughout recent history. Although she focuses on an earlier era than my own, Gail Bederman's *Manliness and Civilization: A Cultural History of Gender and Race in the United States, 1880–1917* (Chicago: University of Chicago Press, 1995) proved useful both as a model for understanding this topic in historical perspective and also for background information on what the American masculine ideal looked like during the two decades in which many of the subjects of my book—the so-called World War II generation of pilots who established postwar private flying—were growing up. Sociologist Michael S. Kimmel's *Manhood in America: A Cultural History*, 2nd ed. (New York: Oxford University Press, 2006), originally published in 1996, examines the evolution of concepts of masculinity from 1776 to the present and was therefore useful in helping me understand this concept in the context of postwar American society. A more theoretical and less historical approach to this topic, R. W. Connell's *Masculinities*, 2nd ed. (Berkeley: University of California Press, 2005), originally published in 1995, demonstrates how multiple versions of masculinity coexist within the same society, reveals that individuals routinely lay claim to multiple and seemingly contradictory masculine identities, and also explains the concept of *hegemonic masculinity* in which the actions of a few can shape the experiences of (and power relationships among) many, both men and women. A more recent study by historian Miriam G. Reumann, *American Sexual Character: Sex, Gender, and National Identity in the Kinsey Reports* (Berkeley: University of California Press, 2005), seeks to both contextualize and study the effects of two controversial works published in the decade immediately following World War II by Harvard-trained zoologist (and later human sex researcher) Alfred Kinsey. Portions of Reumann's work, read alongside Kimmel's *Manhood in America,* provide a useful description and analysis of the so-called masculinity crisis in America at a time when postwar private flying was in full swing. Not surprisingly, these and other scholars continue to expand and refine concepts described in earlier works. Useful article-length essays along these lines include R. W. Connell and James W. Messerschmidt, "Hegemonic Masculinity: Rethinking the Concept," *Gender & Society* 19, no. 6 (Dec. 2005): 829–59, and Gail Bederman, "Why Study 'Masculinity,' Anyway? Perspectives from the Old Days," *Culture, Society & Masculinities* 3, no. 1 (Spring 2011): 13–25.

In my introduction and occasionally throughout the rest of the book, I refer to the male norms described in sociologist Patricia Cayo Sexton's *The Feminized Male: Classrooms, White Collars, and the Decline of Manliness* (New York: Random House, 1969). Although Sexton's analysis is long since outdated, scholars with a serious interest in masculinity may still wish to consult her work for two reasons. First, her contemporary descriptions of normative masculinity in postwar American society are still useful in that they reflect, even if they fail to explain, mainstream (white, middle-class, heterosexual) expectations and ideals of the era. Second, many far more recent scholarly essays and books on masculinity, especially those written by nonhistorians, refer back to Sexton's list (in some cases critically, at other times unquestioningly) as the starting point for observations, discussion, and analysis.

The New Aviation History and the History of Women in Aviation

In his introductory essay to *Innovation and the Development of Flight* (College Station: Texas A&M University Press, 1999), editor Roger D. Launius uses the phrase "New Aviation History" to describe an emerging literature that employs intellectual tools and approaches pioneered in other historical fields to explore old topics in new ways, as well as to identify previously ignored or overlooked questions related to the history of human flight. To fully understand the extent to which these works depart from traditional aviation history, anyone interested in studying or writing on the subject should see Hansen's descriptions in "Aviation History in the Wider View," mentioned previously in this essay.

One of the books that helped start this trend, and remains a must-read to this day, is Joseph Corn's *Winged Gospel.* Corn's book is about neither airplanes nor pilots, but instead focuses on the meanings that everyday Americans attached to aviation, airplanes, and pilots during the first half of the twentieth century. I was lucky enough to have Corn serve as an outside reader for my dissertation, and his influence is everywhere in my own work, from the questions that I ask to the types of evidence that I use. Published some two decades after *Winged Gospel,* A. Bowdoin Van Riper's *Imagining Flight: Aviation and Popular Culture* (College Station: Texas A&M University Press, 2004) reexamines how flight was portrayed throughout the twentieth century in books, movies, and newspapers. Meanwhile, David T. Courtwright's *Sky as Frontier: Adventure, Aviation, and Empire* (College Station: Texas A&M University Press, 2005) draws direct parallels between the history of American aviation and the history of earlier earthbound geographic frontiers, particularly those found in the nineteenth-century American West. Both *Imagining Flight* and *Sky as Frontier* build upon rather than contradict Corn's basic arguments, and both also inform my understanding of the strong connections between aviation and masculinity in America.

There are many fine books about women in aviation, most of them focusing on the impressive accomplishments of individual female fliers in the face of often substantial and sometimes staggering odds. I relied on several such books for evidence as I sought to piece together trends in the gendering of aviation and also for personal anecdotes to illustrate and personalize these broader trends. However, Deborah G. Douglas's *American Women and Flight since 1940* (Lexington: University Press of Kentucky, 2004) stands apart from most literature on the subject. Rather than recovering and retelling an individual woman's history (or the history of a single group, as is the case of numerous excellent books about the Women Airforce Service Pilots, or WASP, of World War II fame), Douglas instead explores the experiences of women in all aspects of aviation, including not only civilian and military pilots but also air traffic controllers, mechanics, and members of the airline cabin crew, all within the broader context of American history.

Notes on Primary Sources

To gain insight into the varied roles that masculinity played in postwar private flying, I drew on hundreds of articles, editorials, monthly columns, news items, cartoons, advertisements, and letters to the editor published in various aviation periodicals, including *AOPA Pilot, Flying, Sport Aviation, Plane & Pilot,* and *Aviation Consumer,* plus more specialized periodicals such as the monthly newsletter (1967–94) and later magazine (starting in 1994) for members of the American Bonanza Society. Researchers should note that from 1939 through 1958, *AOPA Pilot* was published first as a special insert to *Popular Aviation*, then *Flying* when the magazine changed names during World War II. Only AOPA members received magazines with this special insert, and it was not included in otherwise identical magazines mailed to non-AOPA subscribers or sold in newsstands. *AOPA Pilot* last appeared in this format in the March 1958 issue of *Flying*. That same month, *AOPA Pilot* 1, no. 1, was published as a full-length members-only magazine. Several university libraries nationwide, as well as the National Air and Space Museum (NASM) branch of the Smithsonian Institution's Library, have copies of the full-fledged magazine version of *AOPA Pilot*. However, the only full run of pre-March 1958 *AOPA Pilot* inserts to *Popular Aviation* and *Flying* I know of is held at the Hagley Museum and Library in Wilmington, Delaware.

I used these periodicals to help explore masculinity in aviation from the perspective of the people who constructed and experienced it firsthand. Letters to the editor—published under headings like "VOX Pilot" in *AOPA Pilot* and "Flying Mail" in *Flying*—are especially useful because they probably represent the opinions, concerns, and experiences of everyday fliers more accurately than the prescriptive articles penned by professional editors and magazine staffers. The same is true of articles submitted by readers—most of whom are pilots—including brief confessionals regarding near-accidents published in monthly columns like *AOPA Pilot*'s "Never Again!" and *Flying*'s "I Learned about Flying from That!" But I also found other types of sources helpful as I researched each chapter-length case study and have listed key examples here.

For instance, in reconstructing the demographics of pilots for chapter 1, I found the following U.S. government publications especially useful: U.S. Department of Commerce, Civil Aeronautics Administration (CAA), *Statistical Handbook of Civil Aviation* (Washington, DC: U.S. GPO, published annually 1944–58); U.S. Department of Transportation, Federal Aviation Administration (FAA), *Statistical Handbook of Aviation* (Washington, DC: U.S. GPO, published annually starting in 1959); U.S. Department of Transportation, FAA, *U.S. Civil Airmen Statistics* (annual tables from 1999 available online). As described briefly in the text and more fully documented in the endnotes, I also found U.S. Census Bureau reports useful in teasing out statistics that the CAA (and later FAA) neglected to document, including what percentage of pilots actually flew for a living and, even more important, the racial demographics of postwar civil aviation.

As I switched focus from who pilots were to how people became pilots in chapter 2, I relied heavily on autobiographical descriptions of learning to fly. Although most came from numerous articles and letters published in the aviation periodicals just described, book-length memoirs by newly minted private pilots including Frank Kingston Smith's *Week-End Pilot* (New York: Random House, 1957) and Diane Ackerman's *On Extended Wings* (New York: Atheneum, 1985) provided a degree of detail simply not available in other sources. I also found prescriptive literature such as flight instructor handbooks, when used judiciously in conjunction with contemporary comments and recollections by real-life pilots, students, and instructors, useful in piecing together how the student-instructor relationship helped screen and acculturate new members of the aviation community throughout the postwar era.

A quarter century ago, James Hansen observed in his review essay that "a qualified disdain for the buff and for those who concentrate on the nuts and bolts of flying machines has been a persistent theme of those who wish for aviation history with the wider view." However, a few paragraphs later Hansen also noted that books produced by these buffs "contain mountains of raw information; without them, scholars would have a much harder time finding the precise data that flesh out subjects and put major trends and evolutionary developments into focus" (646). I cannot imagine completing chapters 3 and 4, both of which examine the controversies surrounding specific airplanes, without the help of what some scholars might dismiss as "buff literature." Examples include Mitch Mayborn and Bob Pickett, *Cessna Guidebook*, vol. 1, 2nd ed. (Dallas: Flying Enterprise Publications, 1976); Cleo M. Bickford and Buddy McGown, eds., *The 170 Book*, 4th ed. (Lebanon, MO: International Cessna 170 Association, 1999); John M. Frank, *150/152 Model History* (Santa Maria, CA: Cessna Pilots Association, July 17, 2006); John Frank, *Cessna 172 Skyhawk Buyers Guide* (Santa Maria, CA: Cessna Pilots Association, 2009); Edward H. Phillips, *Beechcraft: Pursuit of Perfection* (Eagan, MN: Flying Books, 1992). I also relied heavily on sources like Bill Clarke's *The Illustrated Buyer's Guide to Used Airplanes*, 5th ed. (New York: McGraw-Hill, 2000), plus numerous shorter articles on various used aircraft that were written by fliers, for fliers. These sources include not only technical information about every aircraft mentioned in my book but also anecdotal information collected from owners and former owners of these planes, essentially a written version of the advice sought out by prospective buyers and proffered by other pilots in informal hangar flying sessions down at the local airport throughout the postwar era.

For my discussion of airports in chapter 5—including descriptions of the physical environment and the types of activities that took place in these spaces—I rely equally on top-down efforts at reform as well as bottom-up observations and complaints from pilots, wives of pilots, and airport operators. These appeared in aviation periodicals, conference proceedings from meetings of airport managers, even books for prospective flyers such as Walter Boyne's *Flying: An Introduction to Flight, Airplanes, and Aviation Careers* (Englewood Cliffs, NJ: Prentice-Hall, 1980). Statistics about the numbers and types of airports nationwide come from some of the same sources used to describe pilot demographics in chapter 1, especially the CAA's *Statistical Handbook of Civil Aviation* and the FAA's *Statistical Handbook of Aviation*.

Chapter 6, on aviation organizations, relies heavily on information that appeared in the membership magazines published by the AOPA and EAA. Notable exceptions include case studies on the Quiet Birdmen and the Ninety-Nines. For the former, I pieced together fragments from the very limited number of primary sources I was able to find, plus a handful of first- and secondhand accounts about the nature of this highly secretive all-male organization. For the latter, which by no means seeks to shield its activities from public view, I relied on contemporary accounts by outsiders, combined with well-researched secondary sources such as Douglas's *American Women and Flight*.

In the process of working on this project, from its inception as a seminar paper in the spring of 1999 to final revisions in the spring of 2015 prior to publication as a book, innumerable individuals—pilots and scholars alike—served as helpful sources for information ranging from general observations to very specific details. In many cases their advice and anecdotes pointed me in new directions and led me to ask new questions of the printed sources that I used to write this book. They also corrected me when I was wrong and pointed out, sometimes bluntly but more often politely, errors or incorrect assumptions I made along the way. I mention this here as a humble reminder, to myself as well as to others, that some of our most important "sources" are the people we meet and talk to about our work, even though most of the them will never end up in either the formal acknowledgments section or the endnotes.

INDEX

Italicized page numbers refer to illustrations.